M000122176

HOLT CALIFORNIA

Earth Science

Interactive Reader and Study Guide

HOLT, RINEHART AND WINSTON
A Harcourt Education Company
Orlando • Austin • New York • San Diego • London

Copyright © by Holt, Rinehart and Winston

All rights reserved. No part of this publication may be reproduced or transmitted in any form or by any means, electronic or mechanical, including photocopy, recording, or any information storage and retrieval system, without permission in writing from the publisher.

Requests for permission to make copies of any part of the work should be mailed to the following address: Permissions Department, Holt, Rinehart and Winston, 10801 N. MoPac Expressway, Building 3, Austin, Texas 78759.

HOLT and the **"Owl Design"** are trademarks licensed to Holt, Rinehart and Winston, registered in the United States of America and/or other jurisdictions.

Printed in the United States of America

If you have received these materials as examination copies free of charge, Holt, Rinehart and Winston retains title to the materials and they may not be resold. Resale of examination copies is strictly prohibited and is illegal.

Possession of this publication in print format does not entitle users to convert this publication, or any portion of it, into electronic format.

ISBN-13: 978-0-030-92478-1
ISBN-10: 0-03-092478-2

17 18 19 0868 17 16 15 14 13 12
0360625

Contents

Copyright © by Holt, Rinehart and Winston. All rights reserved.

Copyright © by Holt, Rinehart and Winston. All rights reserved.

CHAPTER 11 Rivers and Groundwater

CHAPTER 12 Exploring the Oceans

CHAPTER 13 The Movement of Ocean Water

CHAPTER 14 The Atmosphere

CHAPTER 15 Weather and Climate

Copyright © by Holt, Rinehart and Winston. All rights reserved.

CHAPTER 16 Interactions of Living Things

CHAPTER 17 Biomes and Ecosystems

Copyright © by Holt, Rinehart and Winston. All rights reserved.

Name ______________________ Class ______________ Date ____________

CHAPTER 1 The Nature of Earth Science

SECTION 1

Thinking Like a Scientist

BEFORE YOU READ

After you read this section, you should be able to answer these questions:

- What qualities do all good scientists have?
- Why is it important for scientists to be ethical?
- What are ways you can think like a scientist every day?

What Makes a Good Scientist?

You may notice things that happen throughout the day. Why does the firetruck sound like it has a higher pitch just as it passes by? How do clouds form? When you observe things around you and ask questions, you are acting like a scientist.

There are several features that all good scientists have. Scientists are curious. They do not like to accept facts without proof. Scientists are also open-minded, creative, and ethical.

It is important for scientists not to give up. The inventor Thomas Edison once said that he never failed. Instead, he just found 10,000 ways that did not work. ☑

STUDY TIP

Summarize As you read, keep a list of the different features that good scientists have. When you are finished reading, write a paragraph or two telling why each feature is important to a good scientist.

READING CHECK

1. Identify Name three qualities that good scientists have.

____

____

____

CURIOSITY

Scientists are curious about the world around them. An example of a scientist whose curiosity led her to make important discoveries is Jane Goodall.

Jane Goodall wanted to know about where chimpanzees lived, what they ate, and how they behaved. She has spent years studying chimpanzees. Her studies have helped other scientists to know more about chimpanzees and other primates.

Jane Goodall has studied chimpanzees for more than 40 years. Her curiosity has helped her, and other scientists, to learn a lot about chimpanzees.

TAKE A LOOK

2. Describe How did Jane Goodall's curiosity help scientists learn about chimpanzees?

____

____

____

____

Copyright © by Holt, Rinehart and Winston. All rights reserved.

SKEPTICISM

Skepticism is the practice of questioning accepted ideas or claims. All good scientists are skeptical. They look for proof before they believe new information. ☑

Skepticism helped scientist Rachel Carson to discover a major threat to the environment. Rachel Carson was a biologist in the 1950s. At that time, many new chemicals were being used to kill insects.

The companies that made the chemicals said that the chemicals would hurt only insects. Carson was skeptical. She wanted proof that the chemicals wouldn't hurt other living things.

Carson did a lot of research on the chemicals. She found that some of the chemicals were harmful to birds. Because of her work, a chemical called DDT was banned from use. Many kinds of birds were saved from harm by her work.

READING CHECK

3. Define What does *skepticism* mean? Give an example of a time when you were skeptical.

____

____

____

Rachel Carson helped save bald eagles and other birds. She showed skepticism by questioning the claims of chemical makers.

TAKE A LOOK

4. Explain How did Rachel Carson show the quality of skepticism?

____

____

OPEN-MINDEDNESS

Keeping an open mind means being willing to consider new ideas. This can be hard to do. Most people have a very hard time believing in ideas that are different from what they believe. Scientists should be open to new ideas, even if the ideas are different from what the scientists are used to.

Sometimes, considering a different idea can help a scientist to make an important new discovery. A famous astronomer, Nicolaus Copernicus, lived in the 1400s. At that time, most scientists believed that the Earth was the center of the universe. But Copernicus studied the stars very closely. Based on his observations, he realized that Earth revolved around the sun.

Say It

Investigate Find out more about Copernicus or another scientist whose unpopular ideas changed science. Present your findings to your class or a small group.

Copyright © by Holt, Rinehart and Winston. All rights reserved.

CREATIVITY

Another feature scientists have is creativity. They use their imaginations to think about the world in new ways. For example, Andy Michael is a scientist who uses his imagination to connect earthquakes to music! ☑

Andy Michael is a scientist who studies earthquakes. He also plays the trombone. He realized that the stress building up inside the Earth is like the beat in a song. He wrote a song called "Earthquake Quartet #1." He used the music to show how stress builds up in the Earth before an earthquake.

Writing earthquake music helped Michael think about earthquakes in a new way. His ideas have helped scientists learn more about earthquakes.

Andy Michael used his creativity to write "Earthquake Quartet #1." Writing music helped him think about earthquakes in new ways.

ETHICAL RESPONSIBILITY AND HONESTY

Ethics are rules about right and wrong. Scientists have a duty to be ethical. Being ethical means they must not cause people or animals unnecessary harm. It also means that scientists must be honest about their work. They must not make up data or copy other people's work.

READING CHECK

5. Describe Think of some talents that you have. How can you use your talents to help you learn science?

____

____

____

____

Critical Thinking

6. Explain Why is it unethical for a person to take credit for someone else's idea?

____

____

____

____

Copyright © by Holt, Rinehart and Winston. All rights reserved.

Say It

Discuss Think of a scientist you know or have read about. What has he or she studied? What qualities do you especially admire? Discuss your ideas in a group.

Who Can Be a Scientist?

What do you think of when someone talks about a scientist? Do you picture a man who has crazy white hair and who wears thick glasses? Is he also wearing a white lab coat? This is a common image of a scientist, but it is not what most scientists are like.

There are scientists all over the world. They have very different backgrounds. Some are a lot like you. The pictures below show some famous scientists from around the world.

A Mae Jemison was a scientist on the space shuttle. Now, she is using space technology to help people in West Africa.

B David Ho is a researcher who has developed new treatments for the virus that causes AIDS.

C Stephen Hawking is a physicist who has taught us about black holes in space.

Copyright © by Holt, Rinehart and Winston. All rights reserved.

How Can You Be Like a Scientist?

Why is it important for you to learn science? Learning science increases your **scientific literacy**. When you have scientific literacy, you understand how science works. You also understand the role of science in society. ☑

Even if you do not become a scientist, you can still think like a scientist. Thinking like a scientist can help you in your daily life. Science teaches you how to ask questions and find answers.

Science also helps you make careful observations. It teaches you to think carefully about information. Science can help you decide if the information is true. Even if you are not a scientist, being skeptical and asking questions can help you know when to believe someone else's claims.

skin-klear
Kills pimple germs dead!
You'll never wash your face again!
SAFE FOR ALL SKIN TYPES
skin-klear
ASTRINGENT
Skin-klear acne wash gets rid of pimples instantly!

READING CHECK

7. Define What is scientific literacy?

____

____

____

TAKE A LOOK

8. Analyze Claims The advertisement makes many claims about the product. Pick one claim and explain how you can figure out if it is true.

____

____

____

____

Copyright © by Holt, Rinehart and Winston. All rights reserved.

CRITICAL THINKING

Scientists have good critical-thinking skills. When you think critically about something, you think in a clear and logical way. You can do this by gathering information, asking questions, and making educated guesses. You should also try not to let your feelings or beliefs get in the way of your thinking. ☑

READING CHECK

9. List What are two ways that you can help yourself to think critically?

____

____

The key to critical thinking is studying the information. You should think about the following things when you decide whether the information makes sense:

1. Is the person who is giving you the information trying to sell you something or make you believe something?
2. How was the information gathered?
3. Was the research done scientifically?
4. Are there any other sources that back up the information?
5. How may your opinion or beliefs affect how you interpret information? ☑

READING CHECK

10. Describe Give an example of how a person's opinion or beliefs may affect how they interpret information.

____

____

____

____

You should think critically about information from all sources. This includes information from the TV, the Internet, and newspapers and magazines.

How Does Science Affect Your Life?

Science is not limited to the laboratory or the classroom. Science is a process. It is a way of thinking about the world.

Scientists can affect your daily life by their discoveries. Many scientists work to improve the lives of others. Mario Molina is an example of such a scientist. He has worked hard to protect Earth's ozone layer.

When Molina was a student in the 1970s, he studied chemicals called CFCs. Molina discovered that CFCs can damage Earth's ozone layer. The ozone layer protects living things on Earth from the sun's harmful rays.

Molina warned leaders and other scientists about his discovery. It took a long time for people to believe him. He worked for many years teaching others about the connection between CFCs and damage to the ozone. Finally, in the 1990s, the use of CFCs was banned in most of the world.

Critical Thinking

11. Brainstorm Name a science topic that interests you. What are two ways you can learn more about the topic?

____

____

____

Copyright © by Holt, Rinehart and Winston. All rights reserved.

How Can You Learn More About Science?

Learning science does not stop when you leave the classroom. You can read about science topics that interest you. Try going to a museum or a zoo or on a nature hike.

You can enter a science fair. You can also join a science club. Your library and teachers are great resources to find out about other opportunities. Some more ways to learn more about science are described below.

SPECIAL SCIENCE PROGRAMS

You can learn about science by being part of a special science program. These programs can offer exciting, hands-on activities. Students interested in the ocean can explore tide pools or go whale watching. They can even explore the ocean by using an ROV, or remotely operated vehicle.

These students are learning to think like scientists by taking part in a special science program.

TAKE A LOOK

12. Predict What is one thing that these students may be learning about?

____

____

COLLABORATING WITH OTHERS

Your class can participate in science projects that connect classrooms around the world. One such project is the JASON Project.

Each year, the scientists working on the project lead students, teachers, and other scientists on a virtual two-week research trip. Students talk online with researchers and do online experiments. They also keep notes in their online journals and share their observations with others.

Copyright © by Holt, Rinehart and Winston. All rights reserved.

Name ______________________ Class ______________ Date ______________

Section 1 Review

SECTION VOCABULARY

scientific literacy the understanding of the methods of scientific inquiry, the scope of scientific knowledge, and the role of science in society	**skepticism** a habit of mind in which a person questions the validity of accepted ideas

1. Explain Why is it important for scientists to be skeptical?

____

____

2. List What are five qualities that all good scientists have? For each quality, include an example from a scientist you know or have read about.

____

____

____

____

____

3. Analyze What are three questions you can ask yourself if you want to think critically about new information?

____

____

____

4. Infer Why is it important for scientists not to give up easily?

____

____

____

5. Explain Why is it hard for most people to be open-minded?

____

____

____

Copyright © by Holt, Rinehart and Winston. All rights reserved.

SECTION 2
Scientific Methods in Earth Science

BEFORE YOU READ

After you read this section, you should be able to answer these questions:

- What are the steps used in scientific methods?
- How is a hypothesis tested?
- Why do scientists share their findings with others?

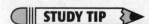

California Science Standards

6.7.a, 6.7.d

How Do Scientists Learn About the World?

Imagine you are standing in a thick forest. Suddenly, you hear a booming noise, and you feel the ground begin to shake. You notice a creature's head looming over the treetops.

The creature's head is so high that its neck must be 20 m long! Then, the whole animal comes into view. Now you know why the ground is shaking. The giant animal is *Seismosaurus hallorum*, the "earthquake lizard."

This description of *Seismosaurus hallorum* is not just from imagination. Since the 1800s, scientists have gathered information about dinosaurs and their environment. Using this knowledge, scientists can recreate what dinosaurs may have been like hundreds of millions of years ago.

How do scientists piece all the information together? How do they know if they have discovered a new species of dinosaur? Asking these questions is the first step in using scientific methods to learn more about the world.

STUDY TIP

Outline As you read this section, make a chart showing the ways that David Gillette used the steps in scientific methods to learn more about the dinosaur bones he studies.

Seismosaurus hallorum was one of the largest dinosaurs that ever lived.

Math Focus

1. Make Comparisons When a *Seismosaurus* held its head up as high as it could, it could have been 25 m tall. How many times taller than you was *Seismosaurus*?

____

Copyright © by Holt, Rinehart and Winston. All rights reserved.

What Are Scientific Methods?

Scientific methods are a series of steps that scientists use to answer questions and to solve problems. Although each question is different, scientists can use the same methods to find answers. ☑

Scientific methods have several steps. Scientists may use all of the steps or just some of them. They may even repeat some of the steps.

The goal of scientific methods is to come up with reliable answers and solutions. These answers and solutions must be able to stand up to the skepticism of other scientists.

READING CHECK

2. Define What are scientific methods?

____

____

____

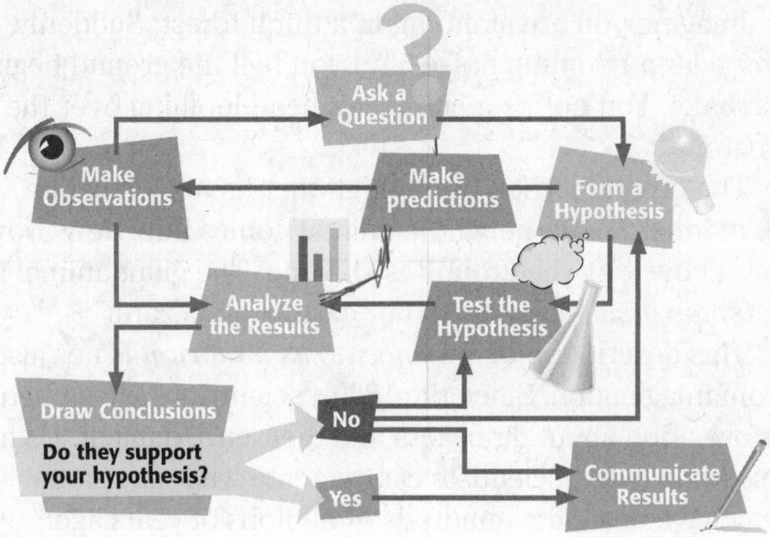
Ask a Question
Make Observations
Make predictions
Form a Hypothesis
Analyze the Results
Test the Hypothesis
Draw Conclusions
Do they support your hypothesis?
No
Yes
Communicate Results

TAKE A LOOK

3. Use Models Starting with "Ask a question," trace two different paths through the figure to "Communicate results." Use a colored pen or marker to trace your paths.

Why Is It Important to Ask a Question?

Asking a question helps scientists focus on the most important things they want to learn. The question helps to guide the research that the scientist does.

David D. Gillette is a scientist who studies fossils. In 1979, he studied some bones from New Mexico. He knew they came from a dinosaur, but he did not know which kind.

Gillette began his study by asking, "What kind of dinosaur did these bones come from?" We will use a table to follow Gillette as he tried to answer his question by using scientific methods.

Step in scientific methods	How did David Gillette apply this step?
Asking questions	He wondered what kind of dinosaur the bones came from.

Copyright © by Holt, Rinehart and Winston. All rights reserved.

How Do Scientists Form a Hypothesis?

When scientists want to investigate a question, they form a hypothesis. A **hypothesis** (plural, *hypotheses*) is a possible answer to a question. It is sometimes called an educated guess.

The hypothesis is a scientist's best answer to the question. But a hypothesis can't be just any answer. The hypothesis must be tested to see if it is true.

From his observations and knowledge about dinosaurs, Gillette formed a hypothesis about the bones. He said that the bones came from a new kind of dinosaur that had not been discovered yet. To test his hypothesis, Gillette had to do a lot of research.

Step in scientific methods	How did David Gillette apply this step?
Forming a hypothesis	He thought that the bones came from a new kind of dinosaur.

CALIFORNIA STANDARDS CHECK

6.7.a Develop a hypothesis.

4. Hypothesize Imagine you are given a small cardboard box. It is heavy. When you shake it, there is a metallic clinking sound. Form a hypothesis about what is in the box. What are some ways you can test your hypothesis without opening the box?

Why Do Scientists Make Predictions?

Before scientists test a hypothesis, they make predictions. A prediction is a guess based on a hypothesis.

Most predictions are stated in an if-then form. For example, Gillette could make this prediction based on his hypothesis: "If the bones are from a dinosaur that has not been discovered, then some of the bones will not match any dinosaur bones that have been studied before."

Scientists may make many predictions based on one hypothesis. After predictions are made, scientists do experiments to see which, if any, of the predictions are true. If a prediction turns out to be true, it suggests that the hypothesis may also be true.

Step in scientific methods	How did David Gillette apply this step?
Making predictions	

TAKE A LOOK

5. Identify In the table, fill in the prediction that David Gillette made based on his hypothesis.

Copyright © by Holt, Rinehart and Winston. All rights reserved.

How Do Scientists Test a Hypothesis?

To see if an idea can be proven scientifically, scientists must test the hypothesis. They do this by gathering data. **Data** (singular, *datum*) are any pieces of information gathered through observation or experimentation. Scientists then use the data to see if the hypothesis is correct. ☑

READING CHECK

6. Define What are data?

____

____

____

____

TESTING WITH EXPERIMENTS

To test a hypothesis, a scientist may perform a controlled experiment. A **controlled experiment** tests only one factor, or variable, at a time. All other variables remain constant. By changing only one variable, scientists can see the results of changing just one thing.

During experiments, scientists must keep accurate records of everything that they do and observe. This includes failed attempts, too. Accurate record keeping is needed for overcoming the skepticism of other scientists.

Critical Thinking

7. Compare and Contrast What is the difference between a controlled experiment and observation?

____

____

____

____

TESTING WITHOUT EXPERIMENTS

Sometimes, it is not possible to do a controlled experiment. In such cases, scientists depend on observation to test their hypotheses. By observing nature, scientists can collect large amounts of data. If the data support a hypothesis, the hypothesis is probably correct.

To test his hypothesis, Gillette took hundreds of measurements of bones. He compared his measurements with the measurements of bones from known dinosaurs. He also visited museums and talked with other scientists.

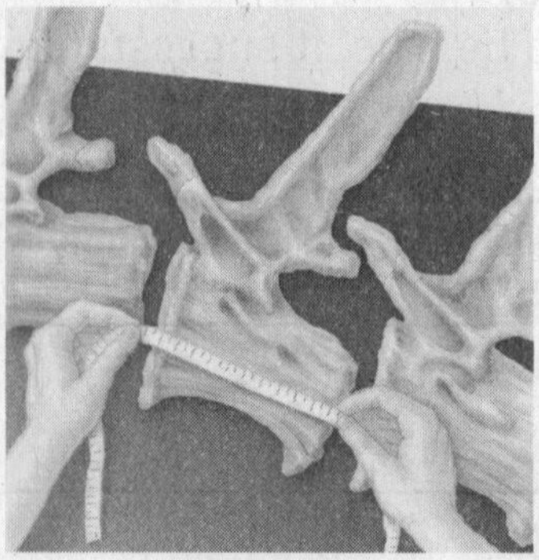

To test his hypothesis, Gillette took hundreds of measurements of the sizes and shapes of the bones.

TAKE A LOOK

8. Identify In the table, fill in the way that David Gillette tested his hypothesis.

Step in scientific methods	How did David Gillette apply this step?
Testing hypotheses	

Copyright © by Holt, Rinehart and Winston. All rights reserved.

How Do Scientists Analyze Results?

When scientists finish collecting data, they must analyze the results. Analyzing results helps scientists explain their observations. Their explanations are based on the evidence they collected. ☑

To arrange their data, scientists often make tables and graphs. Gillette organized his data in a table that showed the sizes and shapes of his dinosaur bones. When he analyzed his results, he found that the mystery dinosaur's bones did not match the bones of any known dinosaur.

	Hip bone of *Diplodocus*	**Hip bone of *Apatosaurus***	**Hip bone of unknown dinosaur**
Top view		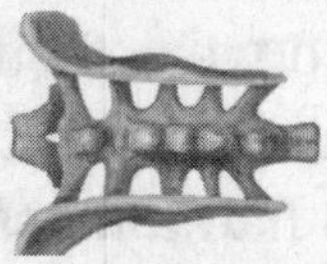	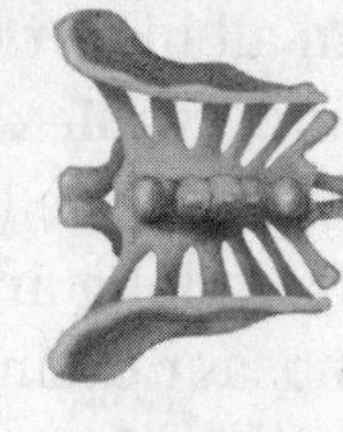
Side view	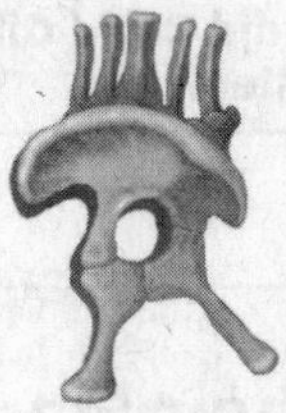	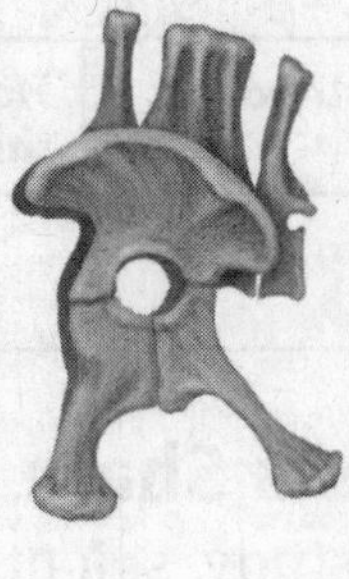	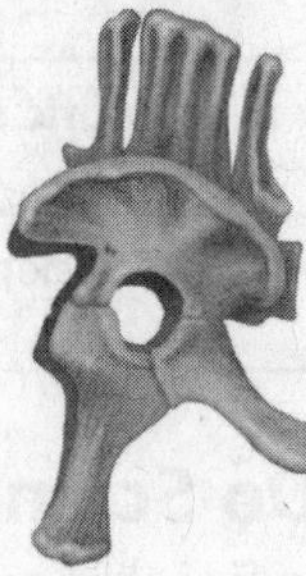

Step in scientific methods	**How did David Gillette apply this step?**
Analyzing results	

✓ **READING CHECK**

9. Explain Why do scientists analyze their data?

____

____

____

TAKE A LOOK

10. Explain Why did David Gillette include bones from *Diplodocus* and *Apatosaurus* when he analyzed his results?

____

____

____

TAKE A LOOK

11. Identify In the table, fill in the way that David Gillette analyzed his results.

Copyright © by Holt, Rinehart and Winston. All rights reserved.

What Are Conclusions?

After analyzing results from experiments, scientists must decide if the results agree with, or support, the hypotheses. This is called *drawing conclusions*. Finding out that a hypothesis is not true can be as valuable as finding out that a hypothesis is true.

What happens if the results do not support the hypothesis? Scientists may repeat the investigation to check for mistakes. Sometimes scientists repeat experiments hundreds of times.

Another option is to look at the original question in a different way. A scientist can then ask another question and make a new hypothesis.

From all his work, Gillette concluded that the bones found in New Mexico were from an unknown dinosaur. From his data, he learned that the new dinosaur was about 35 m long and had a mass of 30 to 70 metric tons. The dinosaur definitely fits the name Gillette gave it—*Seismosaurs hallorum*, or the "earthquake lizard."

Critical Thinking

12. Infer How can finding out that a hypothesis is not true be useful for a scientist?

TAKE A LOOK

13. Identify In the table, fill in David Gillette's conclusions about his dinosaur bones.

Step in scientific methods	How did David Gillette apply this step?
Drawing conclusions	

Why Do Scientists Share Their Findings?

After finishing a study, scientists share their results with others. They write reports and give presentations. They can also put their results on the Internet.

Sharing information allows others the chance to repeat the experiments for themselves. If other scientists get different results, more studies must be done to find out if the differences are significant. ☑

READING CHECK

14. Describe Why is it important for scientists to share their results?

In many cases, the results of an investigation are reviewed year after year as new evidence is found. In the case of *Seismosaurus*, the debate still continues. Some wonder if *Seismosaurus* is a new genus. Other scientists believe it belongs to the genus *Diplodocus*. The best way to resolve this issue is to find another set of bones from *Seismosaurus* to study.

Copyright © by Holt, Rinehart and Winston. All rights reserved.

Name ______________________ Class ____________ Date ____________

Section 2 Review

6.7.a, 6.7.d

SECTION VOCABULARY

controlled experiment an experiment that tests only one factor at a time by using a comparison of a control group with an experimental group **data** any pieces of information acquired through observation or experimentation	**hypothesis** a testable idea or explanation that leads to scientific investigation **scientific methods** a series of steps followed to solve problems

1. Describe How can a scientist test a hypothesis if she or he cannot do a controlled experiment?

____

____

____

2. Explain Why is it important for scientists to ask questions?

____

____

3. Apply Procedures You observe that your tongue sticks to a very cold ice pop when you lick it. You ask yourself, "Why does my tongue stick to the ice pop?" Make a hypothesis about why your tongue sticks to the ice pop.

____

____

____

4. Identify How could you share the results of an experiment with the rest of your class? Give two ways.

____

____

5. Infer Why might a scientist need to repeat a step in scientific methods?

____

____

____

Copyright © by Holt, Rinehart and Winston. All rights reserved.

SECTION 3

Safety in Science

BEFORE YOU READ

After you read this section, you should be able to answer these questions:

- Why should you follow safety rules when learning science?
- What are the six elements of safety?
- What should you do if there is an accident?

STUDY TIP

Summarize As you read this section, underline the important ideas. When you have finished reading, use the ideas you underlined to write a summary of the section.

Why Are Safety Rules Important?

All safety rules have two purposes. Safety rules help prevent accidents. They also help prevent injuries if an accident does happen.

PREVENTING ACCIDENTS

To be safe while doing science activities, it is a good idea to learn some safety rules. The most important safety rule is to follow directions. The directions of a science activity are made to help you prevent accidents. Following directions will also make your work easier, which will help you get better results. ☑

READING CHECK

1. Describe Why should you follow directions in science?

____

____

____

____

PREVENTING INJURIES

If an accident takes place, you or someone nearby can get hurt. Following safety rules after an accident can help prevent injuries. For example, you should never touch or try to clean up a spilled chemical unless you know how to do it safely. If you are using chemicals in a lab, you should learn how to use them safely.

Safety rule	How it prevents accidents or injuries
Wear goggles when working in the lab.	
Throw away broken glass in a specially labeled container.	
Tell your teacher if a chemical spills in the lab.	

TAKE A LOOK

2. Identify In the table, fill in the way that each safety rule can help prevent accidents or injuries.

Copyright © by Holt, Rinehart and Winston. All rights reserved.

What Are the Elements of Safety?

Safety has many parts. To be safe, you need to recognize safety symbols. You also need to follow directions, be neat, use equipment correctly, and clean up after experiments.

SAFETY SYMBOLS

Signs and symbols have special meanings when they are used in science. Some of these symbols are safety symbols. They tell you what to do to prevent injuries or accidents.

Look at the safety symbols in the figure below. Each symbol tells you something important. For example, the "animal safety" symbol tells you to be careful when using animals in experiments. You should follow your teacher's directions about how to handle the animals. You should always wash your hands carefully after touching animals.

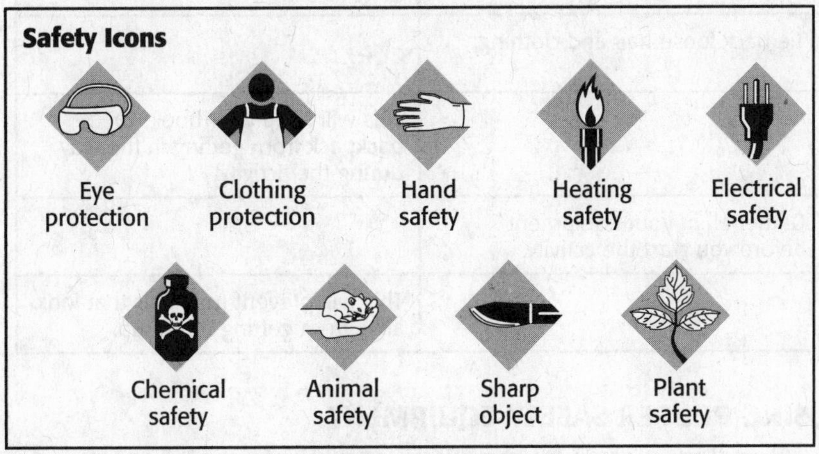
Safety Icons
Eye protection
Clothing protection
Hand safety
Heating safety
Electrical safety
Chemical safety
Animal safety
Sharp object
Plant safety

TAKE A LOOK

3. Investigate Look around your classroom for safety symbols like the ones in the figure. Give two examples of places where safety symbols are found in your classroom.

____

____

READING AND FOLLOWING DIRECTIONS

If you want to bake cookies, you use a recipe. The recipe provides all the directions on how to make cookies. When scientists work in a lab, they also follow directions.

Before starting a science activity, read the directions very carefully. If you do not understand them, ask your teacher to explain them. If you can't finish some of the directions, you should stop your experiment and ask your teacher for help.

When you read, understand, and follow directions, you will get better results. You will also reduce the chance of having an accident.

Copyright © by Holt, Rinehart and Winston. All rights reserved.

NEATNESS

Before starting any experiment, you should clear your work area of anything you don't need for the lab. Some objects can get in the way and can cause an accident. Long hair and loose clothing can get in the way, too. They should be tied back.

During an experiment, keep your table or desk clean. Gather all the equipment you need for the activity before you start. Arrange your equipment and materials so that they are easy to find and pick up.

Label your materials clearly. Some lab materials look alike and can get mixed up if they are not labeled.

As you collect data, you should record your findings carefully in your data table or notebook. Neatly recorded data are easier to read and analyze.

TAKE A LOOK

4. Explain In the table, fill in the blank spaces with things you should do in a lab experiment, and the reasons for doing them.

What to do	Why to do it
Tie back loose hair and clothing.	
	This will keep your books or backpack from getting in the way during the activity.
Gather all of your equipment before you start the activity.	
	This will prevent materials that look alike from getting mixed up.

USING PROPER SAFETY EQUIPMENT

Goggles, gloves, and aprons are all pieces of safety equipment that you may use. Some of the safety symbols tell you what safety equipment you need.

For example, when you see the symbol for eye protection, you must put on safety goggles. Your goggles should be clean and fit properly. Your teacher can help you adjust them for a proper fit. ☑

✓ READING CHECK

5. Identify Why is it important to wear goggles during lab activities?

____

____

____

The chemicals that you use may not always be dangerous. However, you should always wear aprons, goggles, and protective gloves whenever you use chemicals.

You should wear protective gloves when handling animals, too. Different gloves are available for different uses. For example, if you are handling warm or hot objects, you should wear heat-resistant gloves.

Copyright © by Holt, Rinehart and Winston. All rights reserved.

PROPER CLEANUP PROCEDURES

At the end of a science activity, you must clean up your work area. Put caps back on bottles or jars. Return everything to its proper place. Spills and accidents are less likely to happen when everything is put away correctly. ☑

Wash your glassware and check for chips and cracks. If you find any damaged glassware, notify your teacher. It should be carefully thrown away in a special container.

If you have extra chemicals, follow your teacher's directions for throwing them away. Once your work area is clear, you should wipe it with a wet paper towel. Finally, wash your hands carefully with soap and water.

READING CHECK

6. Explain Why is it important to clean up correctly after a lab activity?

____

____

____

What Should You Do If There Is an Accident?

Even when all the safety rules are followed, accidents may still happen. If an accident happens, try to remain calm. Panicking may make the situation worse. You may be scared, but staying in control will help keep you and others safe.

Being prepared and having a plan of action will help you do the right things if an emergency happens.

THINGS TO KNOW BEFORE AN ACCIDENT

Take a look around your classroom or work area. Find out where these things are:

- the fire extinguisher
- the emergency shower and eyewash station
- the first-aid kit
- all the exits from the room
- a phone that you can use to call for help

Find out when and how to use the fire extinguisher, the emergency shower, and the eyewash station. Find out what phone number you can call in case there is an emergency. Make sure that the number is clearly written on or near the phone.

Critical Thinking

7. Identify Why is it important to plan ahead for an accident?

____

____

____

____

Copyright © by Holt, Rinehart and Winston. All rights reserved.

STEPS TO FOLLOW AFTER AN ACCIDENT

If an accident happens, follow these steps:

Step 1: Remain calm.

Try to stay calm and figure out what happened. Look around you for clues, but do not touch anything.

STOP

Step 2: Secure the area.

Make sure that no one (including you!) is in danger. If other students are coming over to see what happened, tell them to stay away so that they don't get hurt.

Step 3. Tell your teacher.

If an accident happens, you must tell your teacher. Tell your teacher, even if the accident is small. Explain exactly what happened. Give your teacher details, such as where the accident happened, what chemicals were spilled, or if glass was broken. If you can't find your teacher, call for help.

Step 4. Help your teacher.

Ask your teacher if there is anything you can do to help. If not, stay out of your teacher's way.

CARING FOR INJURIES

If an accident happens, someone may need first aid. **First aid** is emergency care for someone who has been hurt. Find the first-aid kit in your classroom. Become familiar with the items in it, such as bandages and protective gloves.

If an accident happens, you can help your teacher give first aid to someone who is hurt. However, you should not try to give first aid to someone unless you have first-aid training.

Critical Thinking

8. Design a Plan You are heating water on a hot plate. You suddenly notice that some papers near the hot plate have caught fire. What should you do? What could you have done to prevent this accident?

____

____

____

____

____

____

Copyright © by Holt, Rinehart and Winston. All rights reserved.

Name ______________________ Class ______________ Date ______________

Section 3 Review

SECTION VOCABULARY

first aid emergency medical care for someone who has been hurt or who is sick

1. Define Write your own definition of first aid.

____

____

2. Apply Concepts Fill in the table below to show that you understand safety symbols.

Safety symbol	What does it mean?	Where might you see it?
	eye protection	
		near a chemical storage closet

	electrical safety	
		in a lab activity using chemicals that could burn you

3. Identify Name four pieces of safety equipment in your classroom. Write down where each is located.

____

____

____

____

4. List What should you do if an accident happens? List four steps.

____

____

____

____

Copyright © by Holt, Rinehart and Winston. All rights reserved.

SECTION 1

Tools and Measurement

California Science Standards

6.7.b

BEFORE YOU READ

After you read this section, you should be able to answer these questions:

- How do tools help scientists?
- How do scientists measure length, area, mass, volume, and temperature?

What Tools Do Scientists Use?

STUDY TIP

Compare As you read this section, make a table that compares how scientists measure length, area, mass, volume, and temperature. Include the tools and units of measurement that scientists use.

Scientists collect and record a lot of information, or data. To gather and study the data, scientists use many tools. A *tool* is anything that helps a person do a job.

Different tools help scientists gather different kinds of data. Some tools help scientists see things. Some tools help scientists analyze, or examine, information. Some tools help scientists take measurements.

TOOLS FOR SEEING OBJECTS

Scientists can use tools to help them see objects clearly. Different tools help scientists study objects of different sizes. If the objects are very small, scientists may use a microscope or a magnifying glass to study them. If the objects are very far away, scientists may use a telescope or binoculars to study them. ☑

READING CHECK

1. Identify What are two tools that scientists can use to see things that are very small?

____

____

A telescope is a tool that can help you see things, such as the moon, that are very far away.

TOOLS FOR EXAMINING DATA

The measurements that scientists collect from experiments are called *data* (singular, *datum*). Scientists analyze, or examine, data using many different tools. Calculators can help scientists do calculations quickly. Graphs and charts are tools that can help scientists see patterns in data. Computers are very important tools for collecting, storing, and studying data. ☑

READING CHECK

2. List Give three tools that scientists use to examine data.

____

____

____

Copyright © by Holt, Rinehart and Winston. All rights reserved.

TOOLS FOR MEASURING OBJECTS

During an experiment, scientists record what is happening. They measure the changes that happen to the objects in the experiment. To make these measurements, scientists use many different tools.

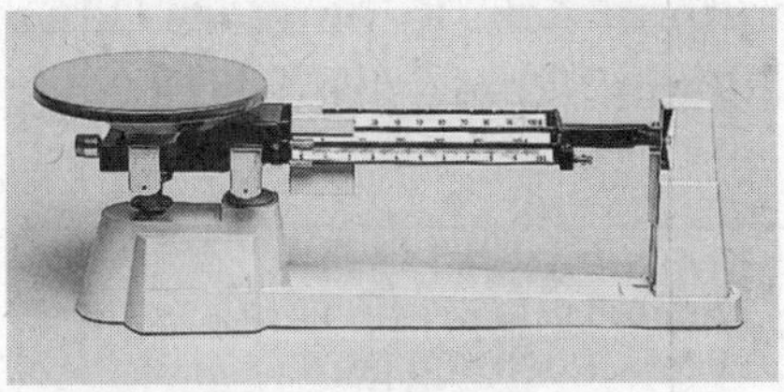

A **balance** is a tool that is used to measure mass.

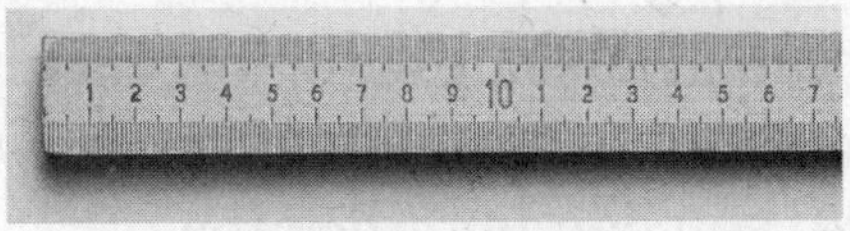
1 2 3 4 5 6 7 8 9 10 1 2 3 4 5 6 7

A **meterstick** is a tool that is used to measure length.

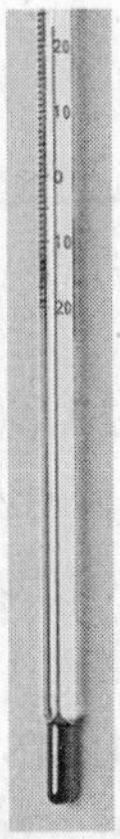

A **thermometer** is a tool that is used to measure temperature.

A **graduated cylinder** is a tool that is used to measure volume.

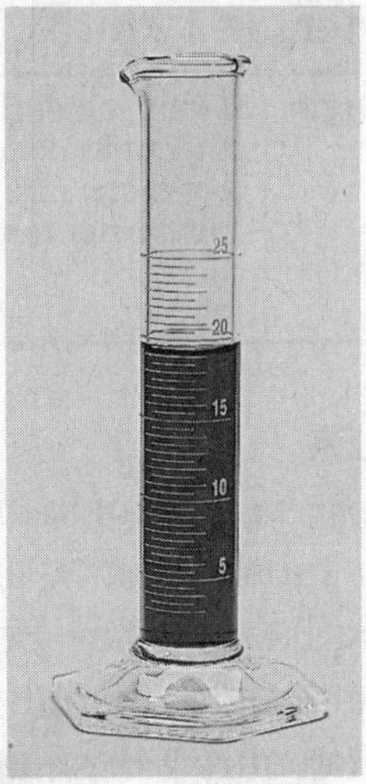
25
20
15
10
5

How Do Scientists Measure Objects?

Scientists make many measurements as they collect data. It is important for scientists to be able to share their data with other scientists. Therefore, scientists use units of measurement that are known to all other scientists. One system of measurement that most scientists use is called the International System of Units.

THE INTERNATIONAL SYSTEM OF UNITS

The *International System of Units*, or *SI*, is a system of measurement that scientists use when they collect data. This system of measurement has two benefits. First, scientists around the world can easily share and compare their data because all measurements are made in the same units. In addition, SI units are based on the number 10. This makes it easy to change from one unit to another.

It is important to learn the SI units that are used for different types of measurements. You will use SI units when you make measurements in the science lab.

CALIFORNIA STANDARDS CHECK

6.7.b Select and use <u>appropriate</u> tools and <u>technology</u> (including calculators, <u>computers</u>, balances, spring scales, microscopes, and binoculars) to perform tests, collect <u>data</u>, and display <u>data</u>.

Word Help: appropriate
correct for the use; proper

Word Help: technology
tools, including electronic products

Word Help: computer
an electronic device that stores, retrieves, and calculates data

Word Help: data
facts or figures; information

3. Identify List the tools that you should use to make the following measurements:

mass: ______________________

temperature: ______________________

length: ______________________

Critical Thinking

4. Predict Consequences What could happen if all scientists used different systems of measurement to record their data?

____

____

____

____

Copyright © by Holt, Rinehart and Winston. All rights reserved.

LENGTH

Length is a measure of how long an object is. The SI unit for length is the **meter** (m). Centimeters (cm) are used to measure small distances. There are 100 cm in 1 m. Micrometers (µm) are used to measure very small distances. There are 1 million µm in 1 m. Kilometers (km) are used to measure large distances. There are 1,000 meters in 1 km. ☑

Length	SI unit: meter (m) other units: kilometer (km) centimeter (cm) micrometer (µm)	1 cm = 0.01 m 1 km = 1,000 m 1 m = 1,000,000 µm 1 m = 100 cm 1 m = 0.001 km

READING CHECK

5. Identify What is the SI unit for length?

______

AREA

Area is a measure of how much surface an object has. For most objects, area is calculated by multiplying two lengths together. For example, you can find the area of a rectangle by multiplying its length by its width. Area is measured in square units, like square meters (m^2) or square centimeters (cm^2). ☑

Area	square meter (m^2) square centimeter (cm^2)	1 cm^2 = 0.0001 m^2 1 m^2 = 10,000 cm^2

READING CHECK

6. Explain How can you find the area of a rectangle?

______

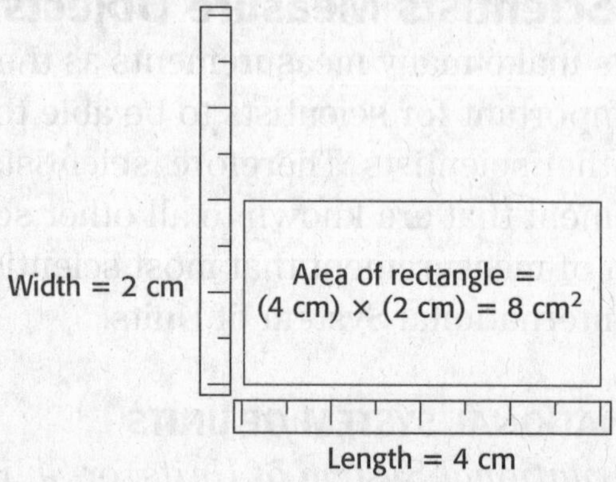
Width = 2 cm
Area of rectangle = (4 cm) × (2 cm) = 8 cm^2
Length = 4 cm

Math Focus

7. Calculate What is the area of a rectangle that has a width of 7 cm and a length of 4 cm?

______

MASS

Mass is the amount of matter in an object. The SI unit for mass is the kilogram (kg). The masses of large objects, such as people, are measured using kg. The masses of smaller objects, such as an apple or a thumbtack, are measured in grams (g) or milligrams (mg). There are 1,000 g in 1 kg. There are 1 million mg in 1 kg.

Mass	SI unit: kilogram (kg) other units: gram (g) milligram (mg)	1 g = 0.001 kg 1 mg = 0.000001 kg 1 kg = 1,000 g 1 kg = 1,000,000 mg

Math Focus

8. Convert How many mg are there in 1 g?

______

Copyright © by Holt, Rinehart and Winston. All rights reserved.

VOLUME

Volume is the amount of space an object takes up. You can find the volume of a box-shaped object by multiplying its length, width, and height together. You can find the volume of objects with many sides by measuring how much liquid they can push out of a container. ☑

Volume is most often measured in cubic units. For example, very large objects can be measured in cubic meters (m^3). Smaller objects can be measured in cubic centimeters (cm^3). There are 1 million cm^3 in 1 m^3.

The volume of a liquid is sometimes given in units of liters (L) or milliliters (mL). One mL has the same volume as one cm^3. There are 1,000 mL in 1 L. There are 1,000 L in one m^3.

Volume	cubic meter (m^3) cubic centimeter (cm^3) liter (L) milliliter (mL)	1 cm^3 = 0.000001 m^3 1 L = 0.001 m^3 1 mL = 1 cm^3 1 mL = 0.001 L 1 m^3 = 1,000 L

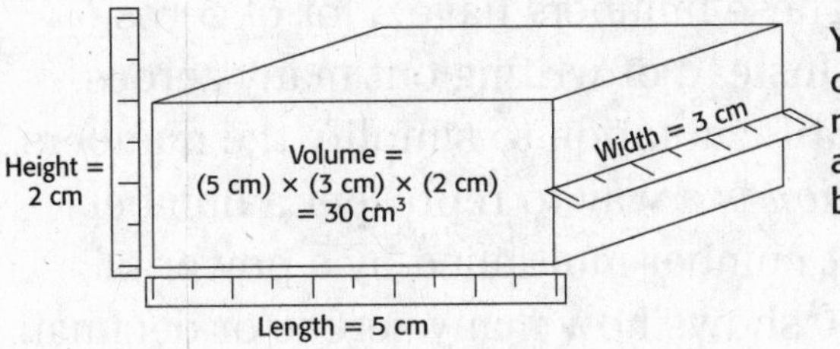
Height = 2 cm
Volume = (5 cm) × (3 cm) × (2 cm) = 30 cm³
Width = 3 cm
Length = 5 cm

You can find the volume of a box-shaped object by multiplying its length, width, and height together. This box has a volume of 30 cm^3.

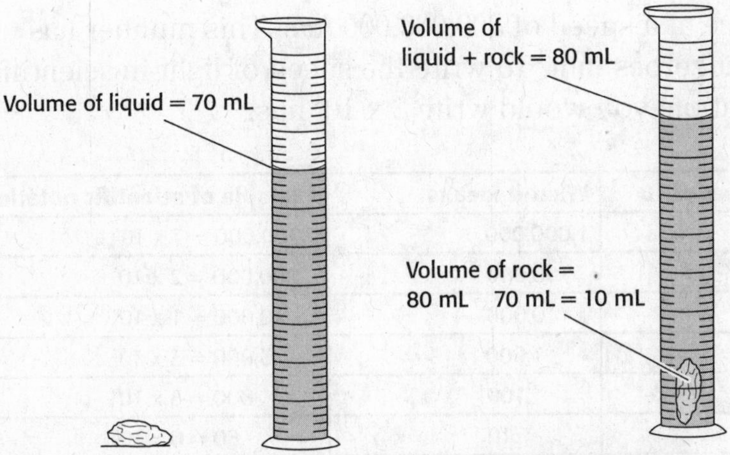
Volume of liquid = 70 mL
Volume of liquid + rock = 80 mL
Volume of rock = 80 mL 70 mL = 10 mL

You can find the volume of a more complicated object, such as this rock, by measuring how much liquid it pushes out of the way. The graduated cylinder has 70 mL of liquid in it before the rock is added.

The rock made the volume of material in the cylinder go up to 80 mL. The rock pushed 10 mL of liquid out of the way. The volume of the rock is 10 mL. Because 1 mL = 1 cm^3, the volume of the rock can also be written as 10 cm^3.

READING CHECK

9. Define What is volume?

____

____

Math Focus

10. Calculate What is the volume of a cube with 5 m-long sides?

____

TAKE A LOOK

11. Explain How do you know that the rock in the figure has a volume of 10 mL?

____

____

____

____

Copyright © by Holt, Rinehart and Winston. All rights reserved.

Name ______________________ Class ______________ Date ______________

Say It

Discuss Where have you seen volume, mass, or temperature measurements outside of science class? In a small group, talk about different places that these measurements are found.

TEMPERATURE

Temperature is a measure of how hot or cold an object is. The SI unit for temperature is the Kelvin (K). However, most people are more familiar with other units of temperature. For example, most people in the United States measure temperatures using degrees Fahrenheit (°F). Scientists often measure temperatures using degrees Celsius (°C).

It is easy to change measurements in °C to K. To change a temperature measurement from °C to K, you simply add 273 to the measurement. For example, 200 °C = 200 + 273 = 473 K. It is more complicated to change measurements in K or °C into °F. That is why scientists do not measure temperatures in °F. ☑

READING CHECK

12. Explain Why do scientists measure temperature in K or °C instead of °F?

____

____

____

Temperature	SI unit: Kelvin (K) other units: degrees Celsius (°C)	0°C = 273 K 100°C = 373 K

SCIENTIFIC NOTATION

Scientists often work with numbers that are very big or very small. These numbers have a lot of zeros or decimal places. Instead of writing out many zeroes, scientists use scientific notation to simplify the numbers. *Scientific notation* is a way to represent a number by using a different number multiplied by a power of 10. The power of 10 shows how many zeroes or decimal places the number has. For example, light moves through space at a speed of 300,000,000 m/s. This number has eight zeroes in it. To write the speed of light in scientific notation, you would write 3×10^8 m/s.

Power of 10	What it means	Example of scientific notation
10^6	1,000,000	7,000,000 = 7×10^6
10^5	100,000	200,000 = 2×10^5
10^4	10,000	10,000 = 1×10^4
10^3	1,000	3,000 = 3×10^3
10^2	100	800 = 8×10^2
10^1	10	60 = 6×10^1
10^{-1}	0.1	0.5 = 5×10^{-1}
10^{-2}	0.01	0.04 = 4×10^{-2}
10^{-3}	0.001	0.009 = 9×10^{-3}

Math Focus

13. Convert Write the number 5,000,000 in scientific notation.

____

Copyright © by Holt, Rinehart and Winston. All rights reserved.

Name ______________________ Class ______________ Date ______________

Section 1 Review

6.7.b

SECTION VOCABULARY

area a measure of the size of a surface or a region **mass** a measure of the amount of matter in an object **meter** the basic unit of length in the SI (symbol, m)	**temperature** a measure of how hot (or cold) something is; specifically, a measure of the average kinetic energy of the particles in an object **volume** a measure of the size of a body or region in three-dimensional space

1. Describe You can find the volume of a box-shaped object by multiplying its length, width, and height together. How can you measure the volume of an object if it is not shaped like a box?

____

____

____

2. Identify What are two units that scientists use to measure temperature?

____

____

3. Explain Why do scientists use scientific notation?

____

____

4. List Give two tools that you can use to see something that is far away.

____

____

5. Calculate What is the area of a room that is 10 m long and 12 m wide?

____

6. Identify Give one example of a tool that scientists use to find patterns in their data.

____

7. List Give three units that can be used to describe volume.

____

____

____

Copyright © by Holt, Rinehart and Winston. All rights reserved.

SECTION 2

Models in Science

California Science Standards

6.7.b, 6.7.c, 6.7.e

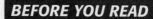

BEFORE YOU READ

After you read this section, you should be able to answer these questions:

- How do scientists use models?
- What are scientific theories and laws?

What Are Models?

Why do scientists use crash-test dummies to learn how safe cars are? By using crash-test dummies, scientists can learn how to make cars safer without putting real people in danger. A crash-test dummy is a model of a person. A **model** is something scientists use to represent an object or event in order to make it easier to study.

Scientists use models to study things that are very small, like atoms, or things that are very large, like Earth. Some scientists use models to predict things that haven't happened yet, or to study events that happened long ago. Some models, like crash-test dummies, allow scientists to study events without affecting or harming the things they are studying. ☑

STUDY TIP

Compare As you read, make a table to show the features of physical models and mathematical models.

READING CHECK

1. Identify Give two reasons scientists use models.

____

____

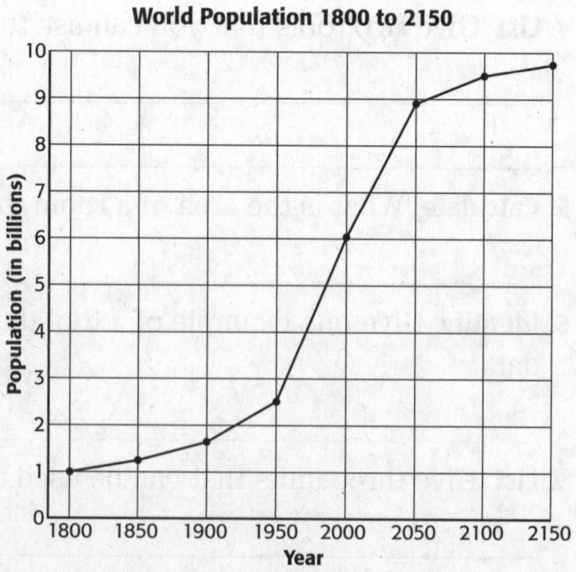
World Population 1800 to 2150
Population (in billions)
Year
0
1
2
3
4
5
6
7
8
9
10
1800
1850
1900
1950
2000
2050
2100
2150

This graph shows how the population of the world is changing with time. Notice that it has predictions for the population of the world in the future. Scientists used a model to make these predictions.

Math Focus

2. Read a Graph In about what year did the world's population reach 6 billion people?

____

Models can come in many forms. Many models are objects that you can see and touch. Other models are made up of data and mathematical equations. Some models are made up of both objects and data.

Copyright © by Holt, Rinehart and Winston. All rights reserved.

PHYSICAL MODELS

Physical models are models that you can see or touch. Many physical models look like the things they represent. For example, a globe is a physical model of Earth. Other physical models may look different from the things they represent. For example, a map is a physical model of Earth. A map can show more details about Earth's surface than a globe can. However, a flat map looks very different from the round Earth! ☑

MATHEMATICAL MODELS

A *mathematical model* is made up of mathematical equations and data. Some mathematical models are simple. These models help you to calculate things such as how far a car will travel in an hour. Other models are more complicated. These models can have many different parts related by complicated equations. ☑

Many mathematical models contain so much data that a person would need hundreds of years to sort through it all. Scientists use computers to help them build and use these models. Because computers can deal with large amounts of data, they can solve many mathematical equations at once. Some computers can solve 30 trillion (3×10^{13}) equations in only one second!

The model in this picture was produced by a computer. The computer combined huge amounts of data into a model that can predict changes in Earth's climate. Without computers, scientists would not be able to use complicated models like this.

READING CHECK

3. Define What is a physical model?

____

____

READING CHECK

4. Define What is a mathematical model?

____

____

TAKE A LOOK

5. Explain How do computers help scientists make models?

____

____

____

____

____

Copyright © by Holt, Rinehart and Winston. All rights reserved.

Critical Thinking

6. Analyze A crash-test dummy is a model of a person. Give two ways that a crash-test dummy is like a person, and two ways that they are different.

____

____

____

____

____

____

____

____

LIMITS OF MODELS

Models are very useful for scientists. However, models are not exactly like the things they represent. For example, a globe is a model of Earth. It shows the shape of Earth and the locations of oceans and land. However, it does not show the living things on Earth, or what Earth looks like inside. When you use a model, make sure that you know how it is different from the thing it represents.

Models are not perfect, but all models can change. Models can change if a scientist finds new data or thinks about ideas in a new way. Scientists are always working to make models better. For example, scientists can use technology, such as computers, to make their models more accurate.

HOW SCIENTISTS USE MODELS

Scientists often use models to study repeating patterns in nature. For example, the model in the figure on the previous page can help scientists predict repeating weather events, such as hurricanes.

How Do Scientists Explain Natural Events?

Many of the events that happen in the natural world form patterns. For example, the moon looks different every night, but it looks the same again about every month. Scientists study these patterns to learn more about how the world works. Scientists study how the moon changes to learn about its movements through space.

Scientists use their observations about the natural world to develop scientific laws and theories. A scientific **law** is a statement or equation that can predict what will happen in certain situations. A **theory** is a scientific explanation that connects and explains many observations.

TAKE A LOOK

7. Identify Fill in the blank boxes in the table with the terms *scientific theory* and *scientific law.*

Name	What it is
	an explanation that connects and explains evidence and observations
	a statement or equation that predicts what will happen in a certain situation

Copyright © by Holt, Rinehart and Winston. All rights reserved.

THEORIES AND OBSERVATIONS

Scientific theories are based on observations. They explain all of the observations about a topic that scientists have at a certain time. However, scientists are always discovering new information. This new information may show that a theory is incorrect. When this happens, the theory must be changed so that it explains the new information. Sometimes, scientists have to develop a totally new theory to explain the new and old information. ☑

READING CHECK

8. Explain What must scientists do if new information shows that a theory is incorrect?

____

____

____

For thousands of years, people believed that Earth was at the center of the solar system. They observed stars moving in a circle in the sky. This evidence seemed to support the theory that Earth was at the center of the solar system.

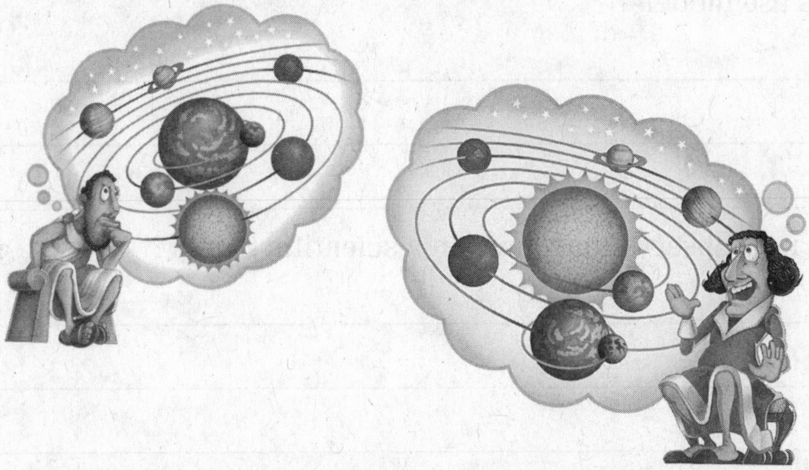

When the telescope was invented, scientists were able to make more detailed observations about the solar system. They saw that planets moved in ways that did not fit the theory of an Earth-centered solar system. Therefore, they developed a new theory about the shape of the solar system. This theory states that the sun is at the center of the solar system. All of the evidence that scientists have been able to collect supports this theory. Therefore, scientists now accept that this theory is correct.

TAKE A LOOK

9. Identify Why did scientists change their theory about the shape of the solar system?

____

____

____

THEORIES AND LAWS

Many people think that scientific theories become scientific laws, but this is not true. Actually, many scientific laws provide evidence to support scientific theories.

In 1665, Sir Isaac Newton discovered the *law of universal gravitation*. This law states that all objects in the universe attract each other with a force called *gravity*. Gravity holds the planets in their orbits as they travel around the sun. Scientists used this law to support the theory of a sun-centered solar system.

Copyright © by Holt, Rinehart and Winston. All rights reserved.

Name ______________________ Class ______________ Date ______________

Section 2 Review

6.7.b, 6.7.c, 6.7.e

SECTION VOCABULARY

law a descriptive statement or equation that reliably predicts events under certain conditions **model** a pattern, plan, representation, or description designed to show the structure or workings of an object, system, or concept	**theory** a system of ideas that explains many related observations and is supported by a large body of evidence acquired through scientific investigation

1. Identify How are scientific theories related to observations and evidence?

____

____

____

2. Explain Why do scientists use models?

____

____

____

3. Describe What effect can new observations have on a scientific theory?

____

____

____

____

4. List Give one example of a physical model and one example of a mathematical model.

____

____

____

5. Summarize How is a scientific theory different from a scientific law?

____

____

____

____

Copyright © by Holt, Rinehart and Winston. All rights reserved.

Name ____________________ Class ____________ Date ____________

CHAPTER 2 Tools of Earth Science

SECTION 3

Mapping Earth's Surface

BEFORE YOU READ

After you read this section, you should be able to answer these questions:

- What is a map?
- How are maps made?

California Science Standards

6.7.b, 6.7.f

What Is a Map?

A **map** is a model that shows the features of an object. Most of the maps that people use show the features of Earth's surface. Some maps show all of Earth's surface. Other maps show only part of Earth's surface. Maps can show natural features, such as rivers. They can also show features made by people, such as roads. Although maps can show different things, all maps of Earth's surface have one thing in common: they all show directions.

STUDY TIP

Discuss Read this section quietly to yourself. In a small group, try to figure out anything you didn't understand.

FINDING DIRECTIONS ON EARTH

Earth's shape is similar to a sphere. However, a true sphere has no top, bottom, or sides that can be used as reference points. *Reference points* are certain locations that never change. They can be used to define directions. ☑

Unlike a true sphere, Earth has two reference points. They are located where Earth's axis of rotation passes through Earth's surface. The reference points are called the North Pole and the South Pole. The North and South Poles are known as *geographic poles.* Since these poles never move, they are used as references points to define directions on Earth.

READING CHECK

1. Define What are reference points?

____

____

____

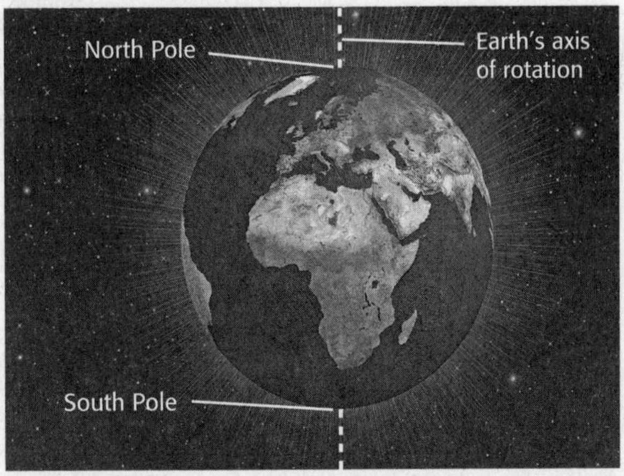

North Pole
Earth's axis of rotation
South Pole

The North Pole and the South Pole can be used as reference points to define directions on Earth.

TAKE A LOOK

2. Identify What are two reference points on Earth?

____

____

Copyright © by Holt, Rinehart and Winston. All rights reserved.

READING CHECK

3. Identify What does the needle on a compass point to?

____

____

READING CHECK

4. Define What are parallels?

____

____

____

TAKE A LOOK

5. Describe Which parallel is farther from the equator: 10°N latitude or 10°S latitude?

____

____

USING A COMPASS TO FIND DIRECTIONS

Earth's core produces a field that makes Earth act as a giant magnet. Earth has two *magnetic poles* located near the geographic poles. A *compass* is a tool that uses the Earth's natural magnetism to show direction. The needle on a compass points to the magnetic pole that is near the North Pole. Therefore, a compass can show you which direction is north. ☑

How Do People Find Specific Locations on Earth?

All of the houses and buildings in your neighborhood have addresses that give their locations. These addresses may include a street name and a number. You can tell someone exactly where you live by giving them your address. In a similar way, you can use latitude and longitude to give an "address" for any place on Earth.

LATITUDE

The **equator** is a circle halfway between the North and South Poles. It divides Earth into two *hemispheres*, or halves—the Northern Hemisphere and the Southern Hemisphere. *Lines of latitude*, or *parallels*, are imaginary lines on Earth's surface that are parallel to the equator. ☑

Latitude is the distance north or south from the equator. Latitude is measured in degrees. The equator represents 0° latitude. The North Pole is 90° north latitude and the South Pole is 90° south latitude. North latitudes are in the Northern Hemisphere and south latitudes are in the Southern Hemisphere.

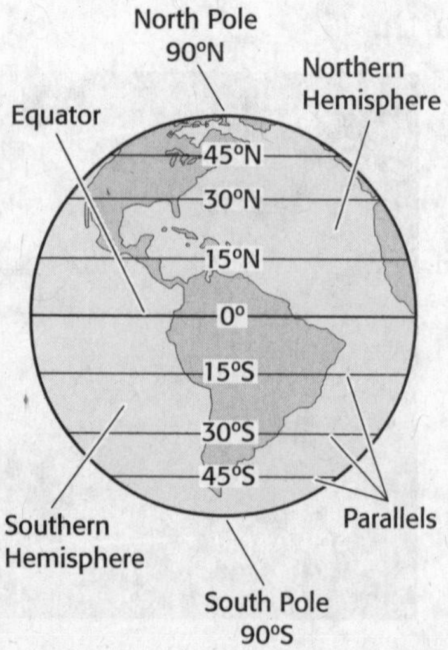
North Pole
90°N
Northern
Hemisphere
Equator
45°N
30°N
15°N
0°
15°S
30°S
45°S
Southern
Hemisphere
Parallels
South Pole
90°S

Copyright © by Holt, Rinehart and Winston. All rights reserved.

LONGITUDE

Lines of longitude, or *meridians*, are imaginary lines that link the geographic poles. Lines of longitude are similar to the lines on a basketball. The lines all touch at the poles. They are farthest apart at the equator. The **prime meridian** is the line that represents 0° longitude. **Longitude** is the distance east or west of the prime meridian. Like latitude, longitude is measured in degrees.

The prime meridian does not circle the whole globe. It runs from the North Pole, through Greenwich, England, to the South Pole. On the other side of the globe, the 180° meridian runs from the North to the South Pole. Together, the prime meridian and the 180° meridian split Earth into Western and Eastern Hemispheres.

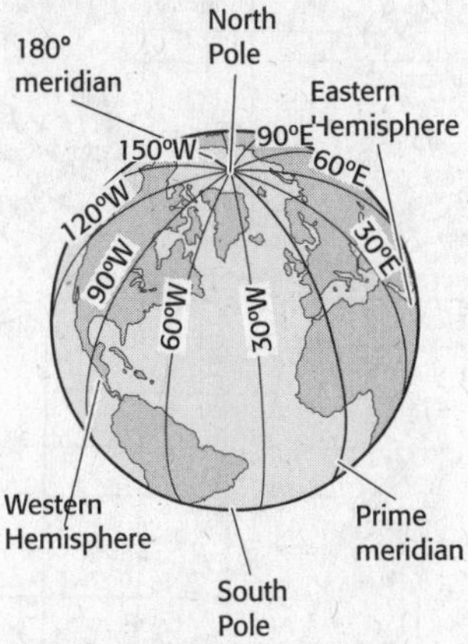
North Pole
180° meridian
Eastern Hemisphere
150°W
90°E
60°E
120°W
30°E
90°W
60°W
30°W
Western Hemisphere
Prime meridian
South Pole

Critical Thinking

6. Apply Concepts A friend asks you what the distance is between 80°E longitude and 90°E longitude. What else do you need to know in order to answer your friend's question? Explain why you need this piece of information.

____

____

____

____

____

____

TAKE A LOOK

7. Identify Where do all lines of longitude meet?

____

USING LATITUDE AND LONGITUDE

Lines of latitude and longitude cross to form a grid. This grid is shown on maps and globes. You can use the lines of latitude and longitude to tell someone the location of any point on Earth's surface. First, find the point on a map like the one below. Then, estimate the latitude and longitude of the point, using the lines closest to it.

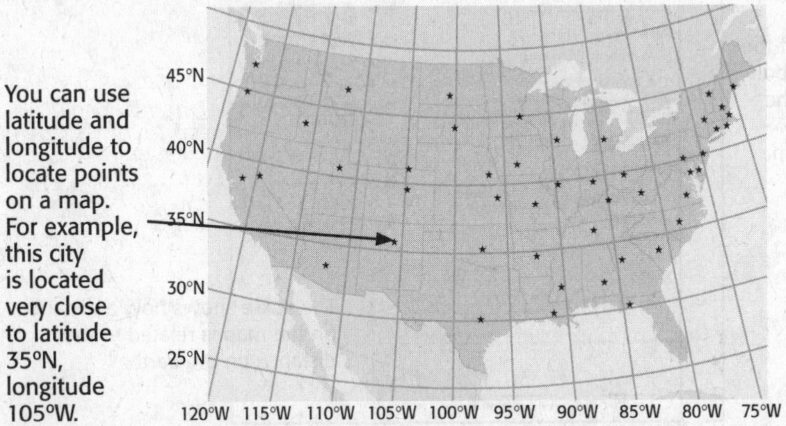
You can use latitude and longitude to locate points on a map. For example, this city is located very close to latitude 35°N, longitude 105°W.
45°N
40°N
35°N
30°N
25°N
120°W 115°W 110°W 105°W 100°W 95°W 90°W 85°W 80°W 75°W

TAKE A LOOK

8. Read a Map Circle the city on the map that is closest to latitude 45°N, longitude 100°W.

Copyright © by Holt, Rinehart and Winston. All rights reserved.

How Is Information Shown on Maps?

The information on a map is in the form of symbols. To read a map, you must understand the symbols on the map. Maps can show many different kinds of information. Almost all maps contain five pieces of information: a title, an indicator of direction, a scale, a legend, and a date. ☑

READING CHECK

9. Describe How is information shown on a map?

____

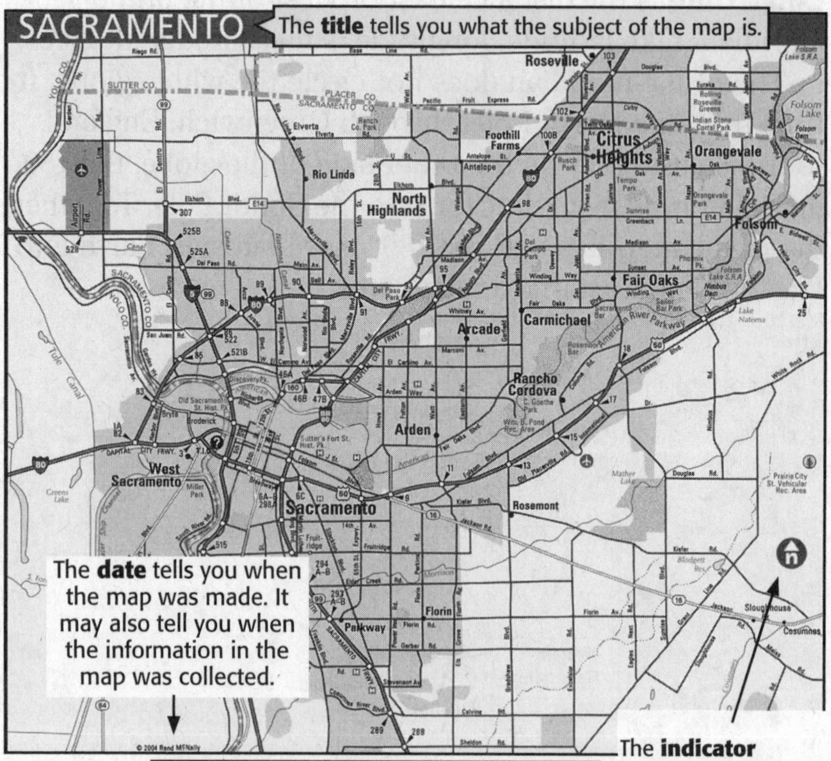
SACRAMENTO
Roseville
Citrus Heights
Orangevale
Folsom
Foothill Farms
Rio Linda
North Highlands
Fair Oaks
Carmichael
Arcade
Rancho Cordova
Arden
West Sacramento
Sacramento
Rosemont
Florin
Parkway
© 2004 Rand McNally

The **title** tells you what the subject of the map is.

The **date** tells you when the map was made. It may also tell you when the information in the map was collected.

The **indicator of direction** can show which way is north or give other information about the location of the map. It can be an arrow pointing north, like this one, a compass rose, or a latitude and longitude grid.

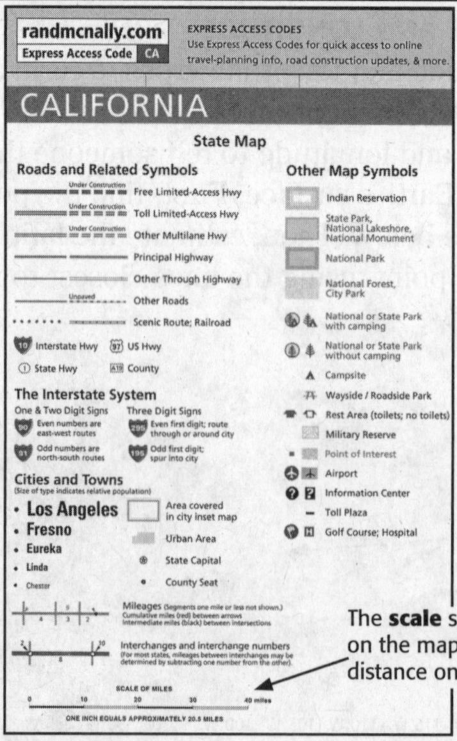
randmcnally.com
Express Access Code CA
EXPRESS ACCESS CODES
Use Express Access Codes for quick access to online travel-planning info, road construction updates, & more.
CALIFORNIA
State Map
Roads and Related Symbols
Free Limited-Access Hwy
Toll Limited-Access Hwy
Other Multilane Hwy
Principal Highway
Other Through Highway
Other Roads
Scenic Route; Railroad
Interstate Hwy
US Hwy
State Hwy
County
The Interstate System
One & Two Digit Signs
Three Digit Signs
Even numbers are east-west routes
Odd numbers are north-south routes
Even first digit; route through or around city
Odd first digit; spur into city
Cities and Towns
Los Angeles
Fresno
Eureka
Linda
Chester
Area covered in city inset map
Urban Area
State Capital
County Seat
Mileages
Interchanges and interchange numbers
Other Map Symbols
Indian Reservation
State Park, National Lakeshore, National Monument
National Park
National Forest, City Park
National or State Park with camping
National or State Park without camping
Campsite
Wayside / Roadside Park
Rest Area (toilets; no toilets)
Military Reserve
Point of Interest
Airport
Information Center
Toll Plaza
Golf Course; Hospital
SCALE OF MILES
0 10 20 30 40 miles
ONE INCH EQUALS APPROXIMATELY 20.5 MILES

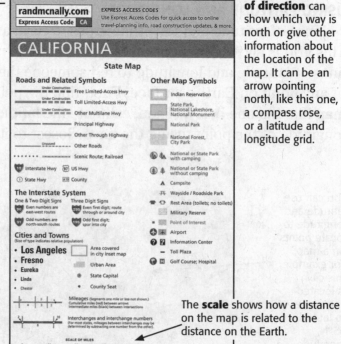

The **legend** tells what the symbols on the map mean.

The **scale** shows how a distance on the map is related to the distance on the Earth.

TAKE A LOOK

10. Identify What is the subject of the map in the figure?

____

Math Focus

11. Use Models How far is Arcade from Rio Linda? Use a ruler and the map's scale to find out. Give your answer in km.

____

Copyright © by Holt, Rinehart and Winston. All rights reserved.

How Are Maps Made?

The information used to make maps today comes from remote sensing. **Remote sensing** is a way to gather information about an object without actually touching or seeing the object. ☑

Today, most maps are made from photographs. Cameras on low-flying airplanes take the photographs. However, mapmakers are beginning to use new equipment that can be carried on both airplanes and satellites.

READING CHECK

12. Define What is remote sensing?

PASSIVE REMOTE SENSING

All things on Earth's surface give off radiation, such as heat or light. *Passive* remote-sensing equipment records the radiation given off by objects. A satellite collects the data and changes it to numbers. These numbers are then sent to a computer. The computer changes the numbers into a picture of the Earth's surface.

ACTIVE REMOTE SENSING

Active remote-sensing equipment makes its own radiation, mostly in the form of microwaves. It sends out the radiation toward Earth's surface. The radiation bounces off Earth's surface and travels back to the sensing equipment. The strength of the signal that bounces back can be used to produce an image of Earth's surface. ☑

Microwaves can travel through clouds and water. Therefore, active remote-sensing equipment can be used to map areas that are difficult to see.

READING CHECK

13. Compare How is active remote sensing different from passive remote sensing?

This picture of downtown San Francisco, California, was made using remote-sensing equipment.

Copyright © by Holt, Rinehart and Winston. All rights reserved.

GLOBAL POSITIONING SYSTEM

Satellites can keep you from getting lost. The *global positioning system* (GPS) can help you find where you are on Earth. GPS is a system of satellites that travel around the Earth. The satellites send radio waves to receivers on Earth. The receivers calculate the latitude, longitude, and elevation of a certain place. ☑

READING CHECK

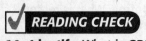

14. Identify What is GPS?

______

______

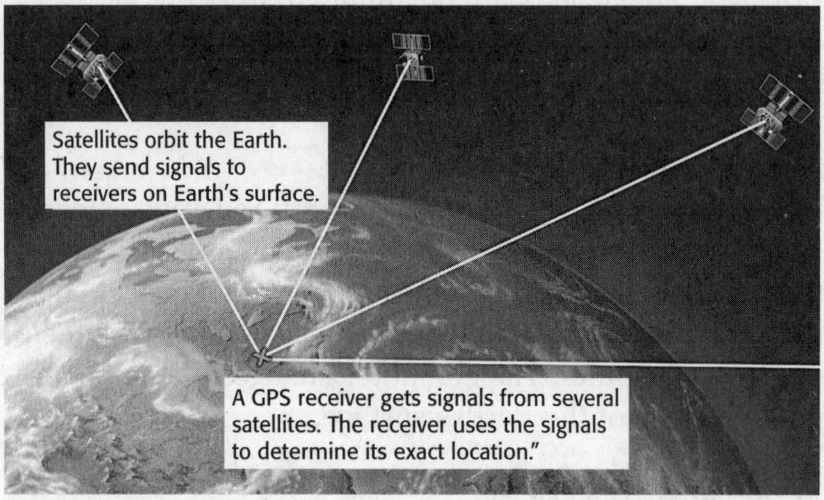
Satellites orbit the Earth. They send signals to receivers on Earth's surface.
A GPS receiver gets signals from several satellites. The receiver uses the signals to determine its exact location."

TAKE A LOOK

15. Describe What do GPS satellites do?

______

______

GPS is very common in people's lives today. Mapmakers use GPS to check the location of boundary lines between countries. Airplane and boat pilots use GPS for navigating. GPS receivers are even put into cars and watches!

GEOGRAPHIC INFORMATION SYSTEMS

A geographic information system (GIS) is a computerized system that shows information about an area. A GIS allows scientists to combine a lot of information into one model. A GIS organizes information in layers. Scientists can compare the layers in order to answer questions. The figure on the next page shows how scientists use GIS to combine information and find safe places for bears to live. ☑

READING CHECK

16. Define What is a GIS?

______

______

______

Copyright © by Holt, Rinehart and Winston. All rights reserved.

This image shows where bears were killed on the roads near Ocala National Forest in Florida. Each dot represents one bear.

This image shows the roads and places where people live near Ocala National Forest.

This image shows where parks are found near Ocala National Forest.

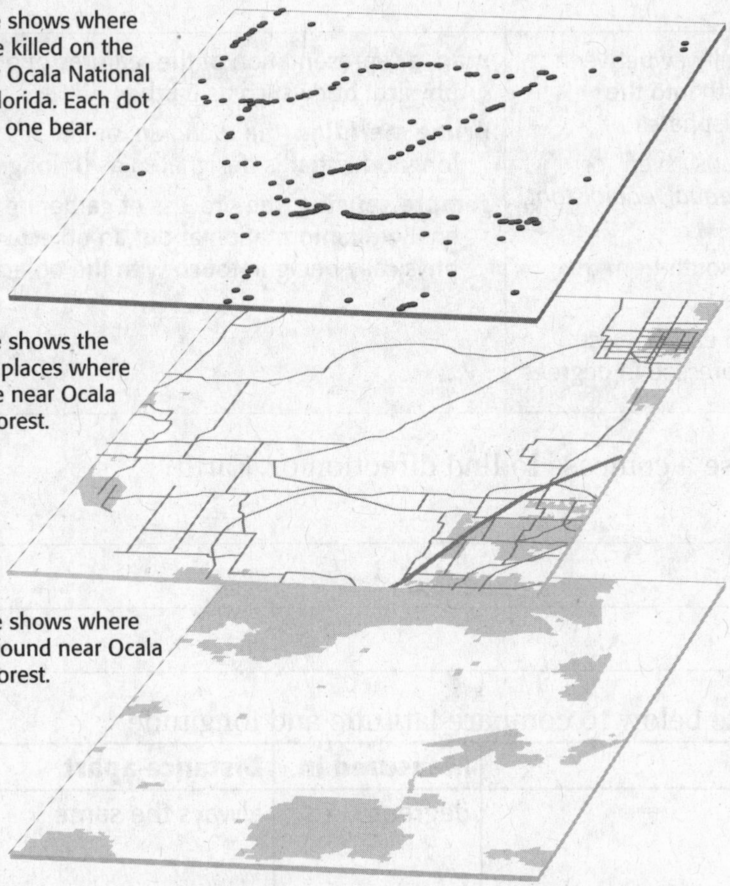

Scientists used GIS to combine the information in the images above. They used the information to learn where bears could safely live

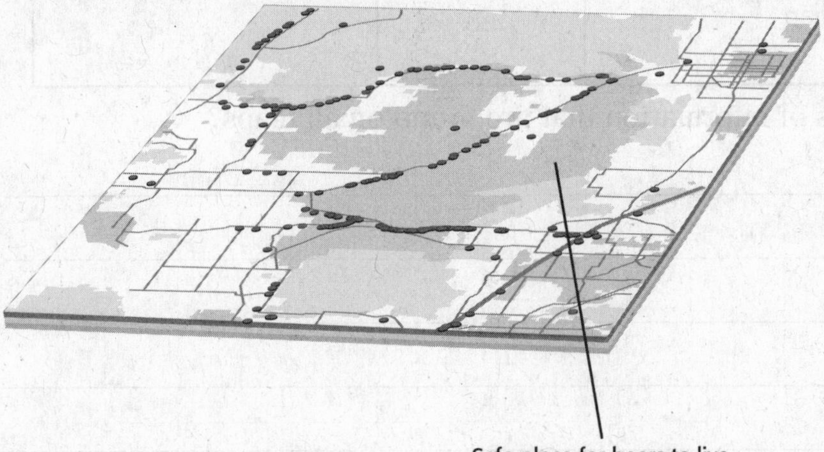

Safe place for bears to live.

Say It

Discuss In a small group, talk about some different problems that scientists can use GIS to solve.

TAKE A LOOK

17. Color On the maps, color the parks using green. Color the places where bears can live using purple.

Copyright © by Holt, Rinehart and Winston. All rights reserved.

Name ______________________ Class ____________ Date ____________

Section 3 Review

6.7.b, 6.7.f

SECTION VOCABULARY

equator the imaginary circle halfway between the poles that divides the Earth into the Northern and Southern Hemispheres

Wordwise The root *equ* means "even" or "equal." Other examples are *equal, equidistant,* and *equality.*

latitude the distance north or south from the equator; expressed in degrees

longitude the angular distance east or west from the prime meridian; expressed in degrees

map a representation of the features of a physical body such as Earth

prime meridian the meridian, or line of longitude, that is designated as 0° longitude

remote sensing the process of gathering and analyzing information about an object without physically being in touch with the object

1. Explain How can you use a compass to find direction on Earth?

____

____

____

2. Compare Fill in the table below to compare latitude and longitude.

Name	What it is	Measured in	Distance apart
latitude		degrees	always the same
longitude	distance east or west of the prime meridian		

3. List What are five pieces of information that are found on all maps?

____

____

____

____

____

4. Apply Concepts Why would a mapmaker use active remote sensing instead of passive remote sensing?

____

____

____

Copyright © by Holt, Rinehart and Winston. All rights reserved.

Name ____________________ Class ____________ Date ____________

CHAPTER 2 Tools of Earth Science

SECTION 4

Maps in Earth Science

BEFORE YOU READ

After you read this section, you should be able to answer these questions:

- What is a topographic map?
- What is a geologic map?

California Science Standards

6.7.f

What Is a Topographic Map?

If you were going hiking in the wilderness, you would want to take a compass and a map. Because there are no roads in the wilderness, you would not take a road map. Instead, you would take a topographic map.

A **topographic map** is a map that shows the surface features, or *topography*, of an area. Topographic maps show natural features, such as rivers and lakes. They show some features made by people, such as bridges. Topographic maps also show elevation. **Elevation** is the height of an object above the surface of the sea. The elevation at sea level is 0 m. ☑

STUDY TIP

Compare After you read this section, make a chart comparing the features of topographic maps and geologic maps.

READING CHECK

1. Define What is elevation?

____

____

CONTOUR LINES

How can a flat map show elevations? Contour lines are used to show elevation on a topographic map. *Contour lines* are lines that connect points on a map that are the same elevation. Each contour line on a map shows a different elevation. Here are some rules for using contour lines:

- Contour lines never cross. All points on a contour line are at the same elevation.
- The space between contour lines depends on the slope of the ground. Contour lines that are close together show a steep slope. Contour lines that are far apart show a gentle slope.
- Contour lines that cross a valley or stream are V-shaped. The V points toward the area of higher elevation, or upstream.
- The tops of hills, mountains, and depressions (dips) are shown by closed circles. Depressions are marked with short, straight lines inside the circle. The lines point toward the center of the depression.

Critical Thinking

2. Explain Why can two contour lines never cross?

____

____

____

____

____

____

____

Copyright © by Holt, Rinehart and Winston. All rights reserved.

CONTOUR INTERVALS AND RELIEF

Each contour line represents a certain elevation. The difference in elevation between any two contour lines is called the *contour interval.* For example, a map with a contour interval of 20 m has contour lines drawn at 0 m, 20 m, 40 m, and so on. The contour interval of a map is usually given in or near the map's legend. The contour interval on a map is based on the relief in the area. ☑

READING CHECK

3. Define What is a contour interval?

____

____

Relief is the difference in elevation between the highest and lowest points in the area on the map. Mountains have high relief. They are usually mapped with large contour intervals. Plains have low relief. They are usually mapped with small contour intervals.

INDEX CONTOURS

The many contour lines on a map can make it hard to read. An index contour is used to make reading the map easier. An *index contour* is a darker, heavier contour line that is labeled with an elevation. In most maps, every fifth contour line is an index contour. So, for example, a map with a contour interval of 20 m may have index contours at 0 m, 100 m, 200 m, and so on.

Math Focus

4. Calculate A map has index contours at 250 m, 500 m, and 750 m. What is the contour interval?

____

COLORS

Topographic maps use colors and symbols to show different features of Earth's surface. Buildings, bridges, and railroads are shown by special symbols drawn in black. Contour lines are brown. Major roads are red. Bodies of water are blue. Wooded areas are shaded in green. Cities are shaded in gray or red. ☑

READING CHECK

5. Identify How do topographic maps show information?

____

____

Topographic maps contain a lot of information. This information can be confusing at first. However, if you practice, you will be able to read topographic maps more easily. When you look at a topographic map, ask yourself these questions to help you read the map:

- What area does the map show?
- What is the contour interval of the map?
- What is the relief of the area in the map?
- What kinds of features are shown on the map?

The map on the next page is an example of a topographic map. You can use it to become more familiar with reading topographic maps.

Copyright © by Holt, Rinehart and Winston. All rights reserved.

Topographic Map of El Capitan

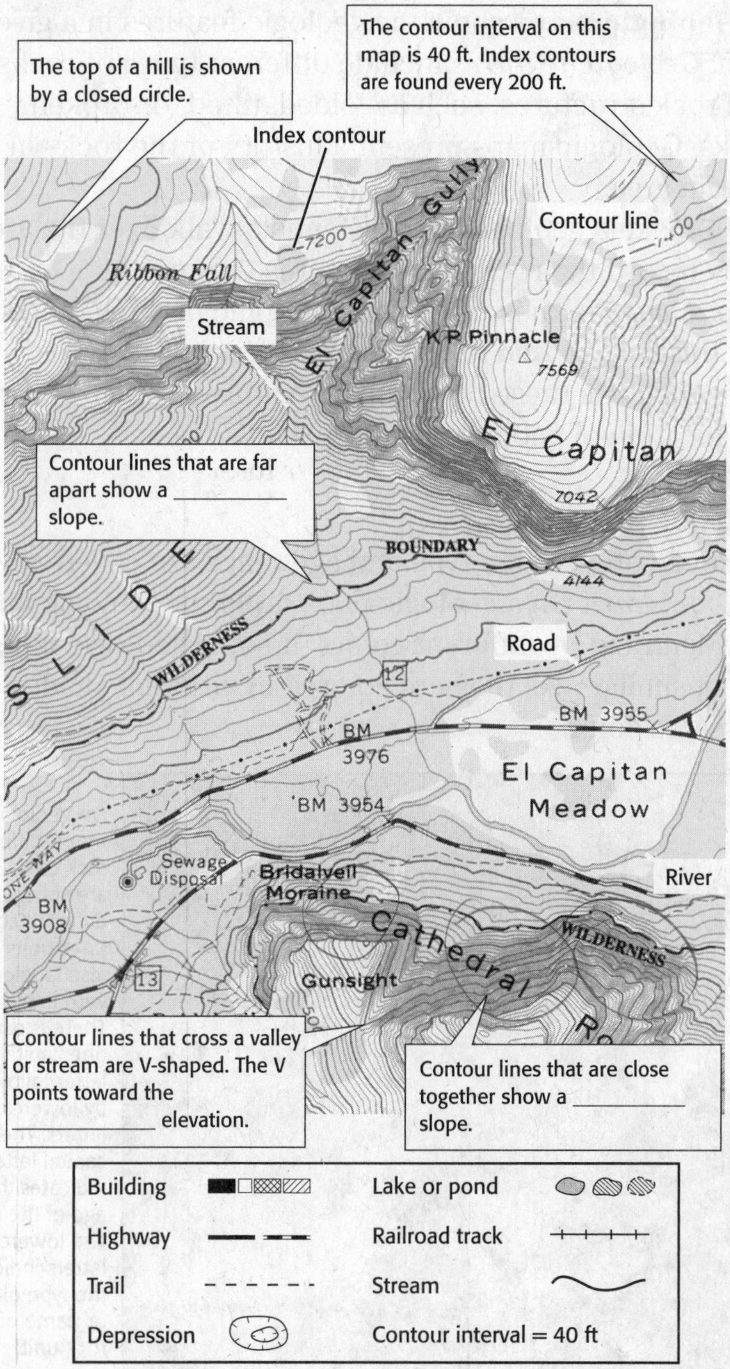
The top of a hill is shown by a closed circle.
The contour interval on this map is 40 ft. Index contours are found every 200 ft.
Index contour
Contour line
7200
Ribbon Fall
El Capitan Gully
Stream
K P Pinnacle
7569
El Capitan
7042
Contour lines that are far apart show a __________ slope.
BOUNDARY
4144
WILDERNESS
SLIDE
Road
12
BM 3955
BM 3976
El Capitan Meadow
BM 3954
ONE WAY
Sewage Disposal
Bridalveil Moraine
River
BM 3908
Cathedral
WILDERNESS
13
Gunsight
Contour lines that cross a valley or stream are V-shaped. The V points toward the __________ elevation.
Contour lines that are close together show a __________ slope.
Building
Lake or pond
Highway
Railroad track
Trail
Stream
Depression
Contour interval = 40 ft

CALIFORNIA STANDARDS CHECK

6.7.f Read a topographic map and a geologic map for evidence provided on the maps and construct and interpret a simple scale map.

Word Help: evidence
information showing whether an idea or belief is true or valid

Word Help: construct
to build; to make from parts

Word Help: interpret
to tell or explain the meaning of

6. Read a Map On the figure, circle the areas that have the steepest slope. Make a box around the areas that have the gentlest slope.

TAKE A LOOK

7. Identify Fill in the blanks in the figure to explain how to use contour lines.

Copyright © by Holt, Rinehart and Winston. All rights reserved.

What Is a Geologic Map?

Geologic maps show the geologic features in a given area. *Geologic features* include different types of rocks and rock structures, such as folded, tilted, or broken rocks. Geologic maps present a history of the rocks in a certain area.

To make a geologic map, scientists walk over the area. They record the rock and structures they see on a base map. A *base map* is often a topographic map. Geologists use this map to find hills, valleys, and other features. They then use these topographic features to help record the location of geologic features. ☑

READING CHECK

8. Explain How do scientists make a geologic map?

____

____

____

ROCK UNITS

Rocks of a certain type and age range are called a *geologic unit*. On a geologic map, different geologic units are identified by different colors. If some geologic units are of similar age, they are colored in different shades of the same color.

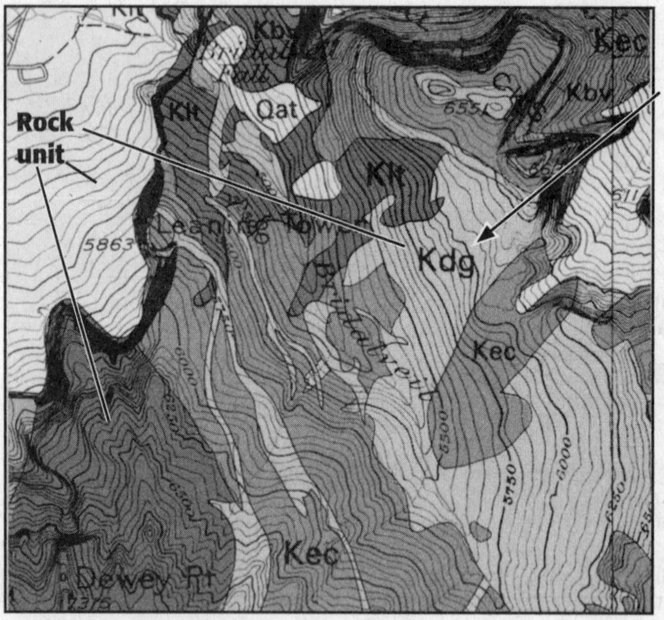
Rock unit
Kb
Kec
Kbv
Klt
Qat
Klt
Kdg
Kec
Kec
Leaning Tower
Dewey Pt
5863
6551

On geologic maps, different rock units are shown by different colors. Rock units are also labeled with letters. There is usually one capital letter followed by lowercase letters. The capital letter indicates the age of the rock. The lowercase letters indicate the type of rock or name of the rock unit.

TAKE A LOOK

9. Color On the map, shade the different rock units with different colors. Make sure to make a legend explaining your colors.

OTHER STRUCTURES

Besides rock units, other features can be shown on geologic maps. *Contact lines* show places where two geologic units meet. The shape of the contact line can show where rock has been folded. Other symbols show if rocks have been tilted. *Faults*, or breaks in rock, can also be shown on geologic maps.

Copyright © by Holt, Rinehart and Winston. All rights reserved.

Name ______________________ Class ____________ Date ____________

Section 4 Review

6.7.f

SECTION VOCABULARY

elevation the height of an object above sea level **geologic map** a map that records geologic information, such as rock units, structural features, mineral deposits, and fossil localities	**relief** the difference between the highest and lowest elevations in a given area; the variations in elevation of a land surface **topographic map** a map that shows the surface features of Earth

1. Define What are contour lines?

____

____

2. Explain What is the relationship between the relief of an area and the contour interval on a map of the area?

____

____

3. Describe What do the colors on a geologic map represent?

____

____

4. Identify Give three features that are shown on topographic maps.

____

____

____

5. Identify Give three features that are shown on geologic maps.

____

____

____

6. Calculate The highest point on a topographic map is marked as 345 m. The lowest contour interval is at 200 m. What is the relief of the area in the map?

____

7. Describe How is the top of a mountain shown on a topographic map?

____

____

Copyright © by Holt, Rinehart and Winston. All rights reserved.

Name ______________________ Class ______________ Date ______________

CHAPTER 3 Earth's Systems and Cycles

SECTION 1

The Earth System

California Science Standards

6.1.b, 6.4.a, 6.4.b, 6.4.d, 6.5.a, 6.5.b

BEFORE YOU READ

After you read this section, you should be able to answer these questions:

- What are the parts of the Earth system?
- How does energy move through the atmosphere, hydrosphere, and biosphere?

STUDY TIP

Compare In your notebook, make a chart showing the features of the atmosphere, biosphere, geosphere, and hydrosphere.

What Are the Parts of the Earth System?

What do the sky and the ground have in common? Like all other parts of Earth, they are always changing. Some parts of Earth change slowly. Other parts change very quickly. All of these changes are important in shaping the Earth system.

Scientists divide Earth into four parts: the geosphere, the hydrosphere, the atmosphere, and the biosphere. The *geosphere* is the mostly solid, rocky part of Earth. The *hydrosphere* is the part of Earth that is water. The *atmosphere* is the mixture of gases that surrounds Earth. The *biosphere* is the part of Earth where life is found. ☑

READING CHECK

1. Identify What are the four parts of the Earth system?

____

____

____

____

Energy and matter move through and between the four parts of the Earth system. This movement of energy and matter is what makes life on Earth possible.

Part of the Earth system	**Definition**
	the part of Earth that is water
Atmosphere	
	the part of Earth where life is found
Geosphere	

TAKE A LOOK

2. Define Fill in the blank spaces in the table.

Copyright © by Holt, Rinehart and Winston. All rights reserved.

What Are the Parts of the Geosphere?

The geosphere is the rocky part of Earth. It includes all of the rock on Earth's surface and all of the rock inside Earth. The geosphere is made up of layers.

LAYERS IN THE GEOSPHERE

Different parts of the geosphere are made of different materials. The geosphere can be divided into three layers based on composition: the crust, the mantle, and the core. The **crust** is the thin, outermost layer of Earth. The **mantle** is the layer of hot, solid rock beneath the crust. The **core** is the central part of Earth. ☑

Another way to think about the parts of the geosphere is to divide it into five layers based on physical features: the lithosphere, the asthenosphere, the mesosphere, the outer core, and the inner core.

READING CHECK

3. List What are the three layers of the Earth, based on composition?

____

____

____

Critical Thinking

4. Infer Where do you think the thickest crust on Earth is found?

____

____

Crust The crust is the thin, outermost layer of Earth. It is between 5 km and 70 km thick. Even though it is very thin compared with the rest of Earth, it is still too thick for people to drill or dig through. Even the deepest holes that people have drilled reach only a small part of the way through the crust.

Mantle The mantle is about 2,900 km thick. Even though it is very hot, the rock in the mantle is solid. Most of the rock in the mantle can flow like soft chewing gum.

Core The core has a radius of 3,400 km. It is made of iron and nickel.

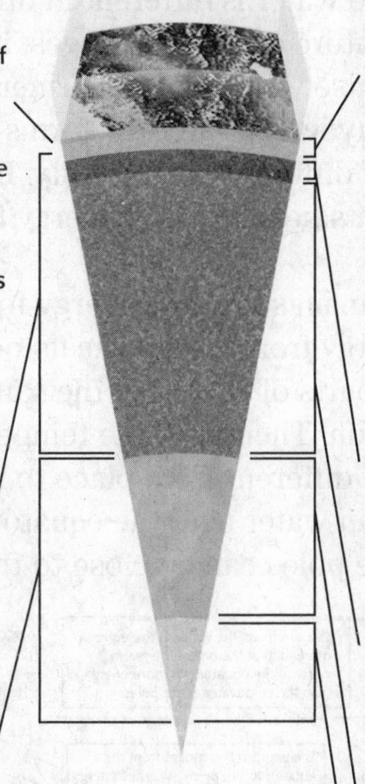

Lithosphere The lithosphere is the outer, rigid layer of Earth. It is between 15 km and 300 km thick. It contains all of the crust and a little bit of the mantle. The lithosphere is broken into several large pieces called tectonic plates. The tectonic plates move very slowly over Earth's surface.

Asthenosphere The asthenosphere is the solid layer of mantle that lies beneath the lithosphere. It can flow and move like putty.

Mesosphere The mesosphere is the solid layer of the mantle between the asthenosphere and the core.

Outer Core The outer core is made of liquid iron and nickel.

Inner Core The inner core is made of solid iron and nickel.

TAKE A LOOK

5. Identify What is the main difference between the inner core and the outer core?

____

____

Copyright © by Holt, Rinehart and Winston. All rights reserved.

CALIFORNIA STANDARDS CHECK

6.4.a Students know the sun is the major source of energy for phenomena on Earth's surface; it powers winds, ocean currents, and the water cycle.

Word Help: major
of great importance or large scale

Word Help: source
the thing from which something else comes

Word Help: energy
what makes things happen

Word Help: phenomenon
any facts or events that can be sensed or described scientifically

Word Help: cycle
a repeating series of changes

6. Explain How does the sun cause convection currents in the ocean?

____
____
____
____
____
____
____

What Are the Features of the Hydrosphere?

Earth's hydrosphere is made up of the water in, on, and around Earth. The hydrosphere includes all of the oceans, lakes, rivers, glaciers, and polar icecaps on Earth. Clouds, rain, and snow are also part of the hydrosphere. Even water that is found in rocks deep underground is part of the hydrosphere.

GLOBAL OCEAN

Most of the water on Earth is found in the oceans. This water is sometimes called the *global ocean*. The global ocean holds more than 97% of all of the water on Earth. People cannot use ocean water for drinking, because it is too salty.

ENERGY FLOW IN THE OCEAN

The water in the oceans moves because the temperature of the water is different in different places. When the temperature of water changes, its density also changes. Cold, dense water sinks. Warmer, less-dense water rises. A **convection current** forms when matter moves because of differences in density. In the ocean, convection currents help to move energy from one place to another.

The sun is the main source of energy for convection in the ocean. Energy from the sun heats ocean water. However, not all parts of Earth get the same amount of energy from the sun. Therefore, the temperature of ocean water can be very different from place to place. For example, the ocean water near the equator is warm, but the water near the poles can be close to freezing.

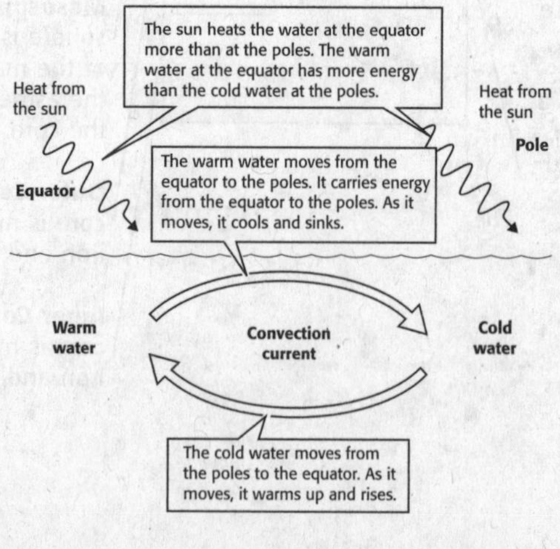
The sun heats the water at the equator more than at the poles. The warm water at the equator has more energy than the cold water at the poles.
Heat from the sun
Heat from the sun
Pole
Equator
The warm water moves from the equator to the poles. It carries energy from the equator to the poles. As it moves, it cools and sinks.
Warm water
Convection current
Cold water
The cold water moves from the poles to the equator. As it moves, it warms up and rises.

Copyright © by Holt, Rinehart and Winston. All rights reserved.

What Is the Atmosphere?

The atmosphere is the mixture of gases that surrounds Earth. The atmosphere reaches out from the surface of Earth for about 500 km. However, most of the gases in the atmosphere are found within about 8 km to 12 km of Earth's surface.

LAYERS IN THE ATMOSPHERE

Earth's atmosphere is made up of four layers. The *troposphere* is the layer that is closest to Earth's surface. It is the layer in which people live. Most weather happens in the troposphere. The *stratosphere* is the layer that is right above the troposphere. The *mesosphere* is the coldest layer of the atmosphere. The *thermosphere* is the uppermost layer of the atmosphere. ☑

READING CHECK

7. Draw In the space above, draw a picture of the four layers of the atmosphere. Label each layer.

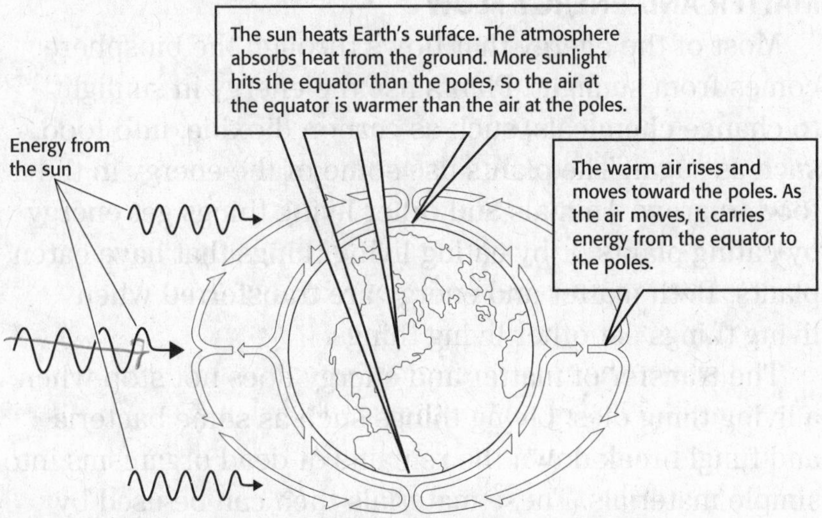
The sun heats Earth's surface. The atmosphere absorbs heat from the ground. More sunlight hits the equator than the poles, so the air at the equator is warmer than the air at the poles.
Energy from the sun
The warm air rises and moves toward the poles. As the air moves, it carries energy from the equator to the poles.

TAKE A LOOK

8. Identify On the figure, draw an arrow showing the main direction that energy moves in the atmosphere.

ENERGY FLOW IN THE ATMOSPHERE

The sun is the main source of energy for Earth's surface. The energy from the sun heats Earth's surface unevenly. The atmosphere absorbs energy from Earth's surface. Since the surface is heated unevenly, the atmosphere is, too. The uneven heating makes the air in the atmosphere move.

The movement of the air helps move heat energy through the atmosphere. The term **convection** is used to describe the transfer of energy by the movement of matter. Convection is an important way that energy flows through the atmosphere, just as it is in the hydrosphere.

Copyright © by Holt, Rinehart and Winston. All rights reserved.

Critical Thinking

9. Apply Concepts Why are the upper parts of the atmosphere not part of the biosphere?

____

____

What Is the Biosphere?

The biosphere is made up of *organisms*, or living things, and the parts of Earth where life is found. The biosphere includes Earth's surface, the lower part of the atmosphere, and most of the hydrosphere. Scientists continue to find life in new places. Therefore, the boundaries of the known biosphere are always changing.

FACTORS NEEDED FOR LIFE

The biosphere has many factors that living things need in order to survive. For example, most living things need liquid water and a certain range of temperatures in order to survive. Therefore, most living things are found near Earth's surface. ☑

READING CHECK

10. Explain Why are most living things found near Earth's surface?

____

____

____

____

MATTER AND ENERGY FLOW

Most of the energy that flows through the biosphere comes from sunlight. Plants use the energy in sunlight to change chemicals, such as carbon dioxide, into food, such as sugar. The plants use some of the energy in the food to grow. Animals and other living things get energy by eating plants or by eating living things that have eaten plants. Both matter and energy are transferred when living things eat other living things.

The transfer of matter and energy does not stop when a living thing dies. Living things such as some bacteria and fungi break down the remains of dead organisms into simple materials. These materials then can be used by other organisms. The bacteria and fungi get their energy from the remains of the dead organisms they break down.

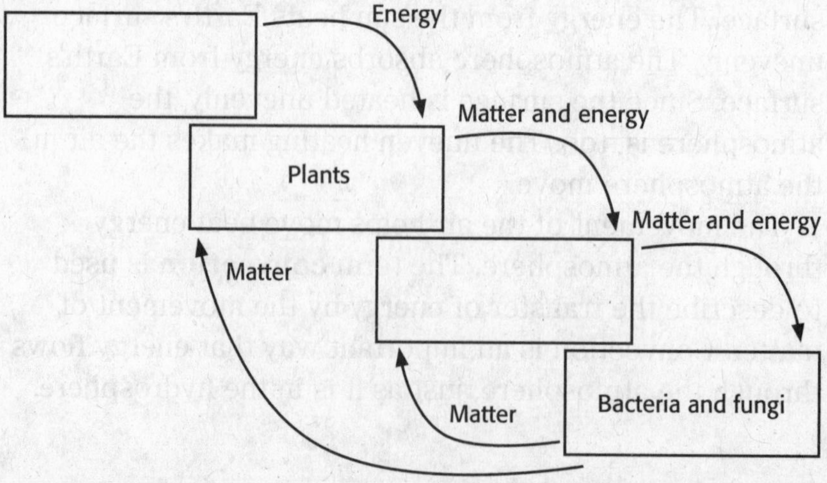
Energy
Matter and energy
Plants
Matter and energy
Matter
Matter
Bacteria and fungi

TAKE A LOOK

11. Identify Fill in the blank boxes in the figure to show how matter and energy flow in the biosphere.

Copyright © by Holt, Rinehart and Winston. All rights reserved.

Name ______________________ Class ______________ Date ______________

Section 1 Review

6.1.b, 6.4.a, 6.4.b, 6.4.d, 6.5.a, 6.5.b

SECTION VOCABULARY

convection the movement of matter due to differences in density; the transfer of energy due to the movement of matter

Wordwise The prefix *con-* means "with" or "together." The root *vect* means "to carry."

convection current any movement of matter that results from differences in density; may be vertical, circular, or cyclical

core the central part of the Earth below the mantle

crust the thin and solid outermost layer of the Earth above the mantle

mantle the layer of rock between the Earth's crust and core

1. Identify How does heat flow through the atmosphere and the hydrosphere?

____

2. Compare Give two differences between the mantle and the core.

____

____

3. Explain How does energy flow in the biosphere?

____

4. Describe How does matter flow in the biosphere?

____

____

5. Identify What is the main source of energy for convection in the atmosphere and the oceans?

____

6. Compare How is the lithosphere different from the asthenosphere?

____

____

____

7. Explain Why are water and air at the equator warmer than water and air at the poles?

____

____

Copyright © by Holt, Rinehart and Winston. All rights reserved.

SECTION 2
Heat and Energy

California Science Standards

6.3.a, 6.3.c, 6.3.d

BEFORE YOU READ

After you read this section, you should be able to answer these questions:

- What is temperature?
- What is heat?
- How does heat move from place to place?

What Is Temperature?

What makes an object have a certain temperature? The answer to that question has to do with the tiny particles that make up the object.

All matter is made up of tiny particles. These particles are always moving, even though we can't see them move. When particles are moving, they have a kind of energy called *kinetic energy*. The faster the particles move, the more kinetic energy they have. ☑

The particles in an object are all moving at different speeds and in different directions. We can't measure the kinetic energy of each particle by itself. However, we can measure the average kinetic energy of all of the particles in an object.

The average kinetic energy of these tiny particles is what gives an object its temperature. The **temperature** of an object is a measure of the average kinetic energy of all of the particles in the object. The temperature of an object does not depend on the size of the object. If the particles in an object have a lot of kinetic energy, the object will have a high temperature.

STUDY TIP

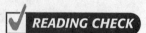

Discuss Read this section quietly to yourself. When you finish reading, talk about the section with a partner. Work together to figure out the parts that you don't understand.

READING CHECK

1. Explain How does the speed of a particle affect its kinetic energy?

____

____

____

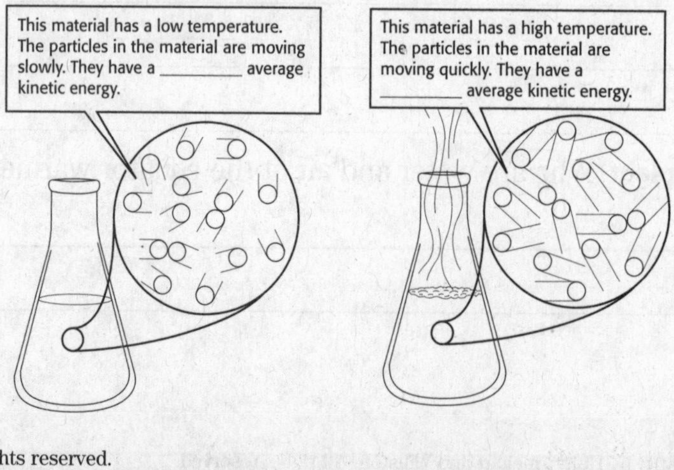
This material has a low temperature. The particles in the material are moving slowly. They have a _______ average kinetic energy.
This material has a high temperature. The particles in the material are moving quickly. They have a _______ average kinetic energy.

TAKE A LOOK

2. Identify Fill in the blank lines in the figure to describe the amount of kinetic energy in the particles.

Copyright © by Holt, Rinehart and Winston. All rights reserved.

What Is Heat?

What do you think of when you think about heat? Most people think of something that has a high temperature. However, heat also has to do with objects that have low temperatures. In science, **heat** is the energy that is transferred between objects that are at different temperatures. ☑

Why do some objects feel hot or cold? Heat moves from one object to another when their temperatures are different. Heat always moves from the object with a higher temperature to the object with a lower temperature. When you touch something that feels cold, heat moves from your body into the object. When you touch something that feels hot, heat moves from the object into your body.

READING CHECK

3. Define What is heat?

____

____

____

HEAT AND THERMAL ENERGY

If heat is transferred energy, what kind of energy is being transferred? The answer is thermal energy. **Thermal energy** is the total kinetic energy of the particles that make up a substance. Thermal energy is measured in joules (J).

The amount of thermal energy in a substance depends partly on its temperature. Something at a high temperature usually has more thermal energy than something at a lower temperature.

Thermal energy also depends on the number of particles in a substance. At the same temperature, objects made of many particles have more thermal energy than objects made of few particles.

Critical Thinking

4. Compare How is thermal energy different from temperature?

____

____

____

____

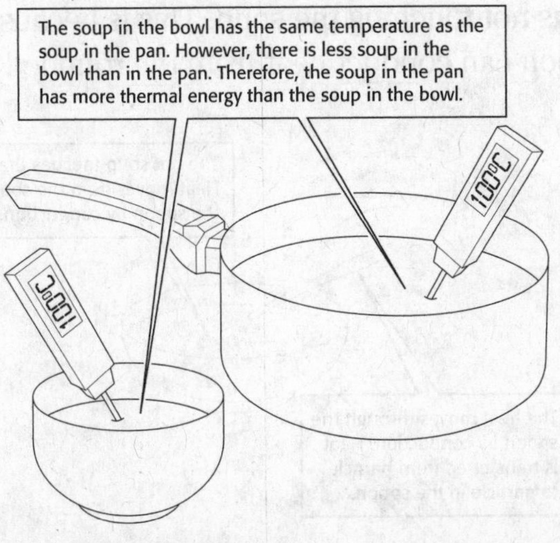
The soup in the bowl has the same temperature as the soup in the pan. However, there is less soup in the bowl than in the pan. Therefore, the soup in the pan has more thermal energy than the soup in the bowl.
100°C
100°C

TAKE A LOOK

5. Explain The soup in the bowl and the soup in the pan have the same temperature. Why does the soup in the pan have more thermal energy than the soup in the bowl?

____

____

Copyright © by Holt, Rinehart and Winston. All rights reserved.

REACHING THE SAME TEMPERATURE

When two objects with different temperatures touch, heat flows between them. The heat flows from the warmer object to the cooler object. Heat flows until both objects have the same temperature.

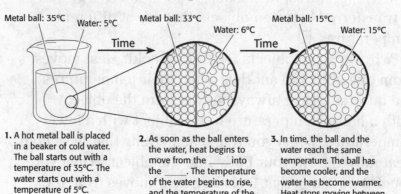

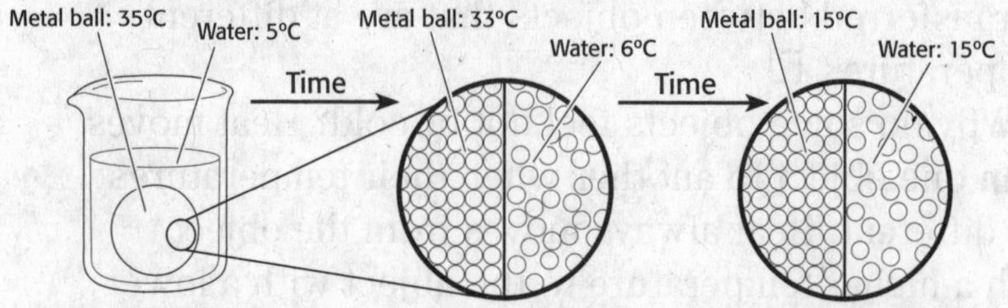
Metal ball: 35°C
Water: 5°C
Time
Metal ball: 33°C
Water: 6°C
Time
Metal ball: 15°C
Water: 15°C

1. A hot metal ball is placed in a beaker of cold water. The ball starts out with a temperature of 35°C. The water starts out with a temperature of 5°C.

2. As soon as the ball enters the water, heat begins to move from the ______ into the ______. The temperature of the water begins to rise, and the temperature of the ball begins to drop.

3. In time, the ball and the water reach the same temperature. The ball has become cooler, and the water has become warmer. Heat stops moving between the ball and water.

TAKE A LOOK

6. Identify Fill in the blank spaces to show the direction that thermal energy moves between the ball and the water.

Math Focus

7. Calculate By how many degrees did the temperature of the metal ball decrease when it transferred its thermal energy to the water?

____

How Does Heat Move?

Heat can move between objects in three main ways: conduction, convection, and radiation. During conduction, heat moves through matter, but the matter does not move. During convection, matter moves and carries heat with it. During radiation, no matter is needed to move the heat from place to place.

CONDUCTION

Conduction happens when heat moves from one object to another through direct contact. Imagine that you place a metal spoon into a bowl of hot soup. The particles in the rounded end of the spoon touch the particles in the soup. Heat moves from the soup into the spoon by conduction.

After a while, the handle of the spoon will get hot, even though it is not touching the soup. This is because the metal in the spoon can conduct heat from the rounded end to the handle.

The hot soup touches the spoon. Heat moves from the soup into the spoon by conduction.
The heat moves through the spoon by conduction. Heat is transferred from particle to particle in the spoon.

Critical Thinking

8. Apply Concepts What will happen to the temperature of the soup when the spoon is placed into the bowl? Explain your answer.

____

____

____

____

Copyright © by Holt, Rinehart and Winston. All rights reserved.

CONVECTION

Convection happens when matter carries heat from one place to another. For example, when you boil water in a pot, the temperature of the water near the hot burner increases. The particles in the hot water move farther apart. The hot water rises. As it rises, the hot water carries heat from the bottom of the pot to the top. ☑

In many cases, convection occurs because of thermal expansion. *Thermal expansion* happens when the volume of a substance increases because its temperature increases. Thermal expansion is what causes hot air to rise above cold air in the atmosphere. It is also what causes warm water to rise above colder water in the oceans. Thermal expansion can even cause rocks in the geosphere to move!

READING CHECK

9. Define What is convection?

____

____

____

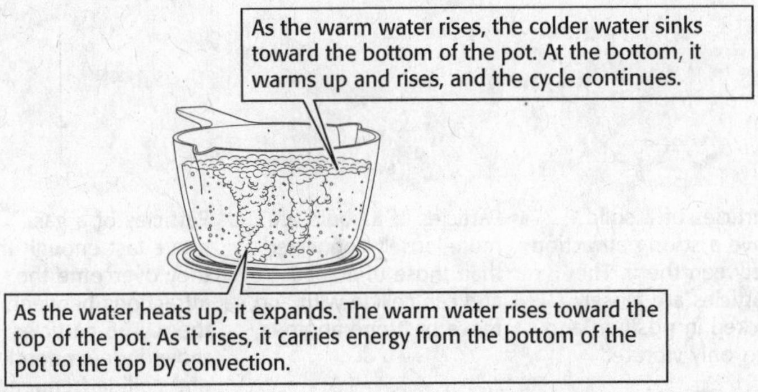
As the warm water rises, the colder water sinks toward the bottom of the pot. At the bottom, it warms up and rises, and the cycle continues.
As the water heats up, it expands. The warm water rises toward the top of the pot. As it rises, it carries energy from the bottom of the pot to the top by convection.

TAKE A LOOK

10. Identify On the figure, draw arrows showing how the warm and cold water moves in the pot.

RADIATION

Radiation happens when energy, such as heat, moves in waves between one object and another. Energy can move between objects by radiation even if there is no matter to carry the energy. For example, the sun produces a lot of energy. This energy travels to Earth by radiation through outer space. Even though there is almost no matter in outer space, the energy can travel to Earth by radiation.

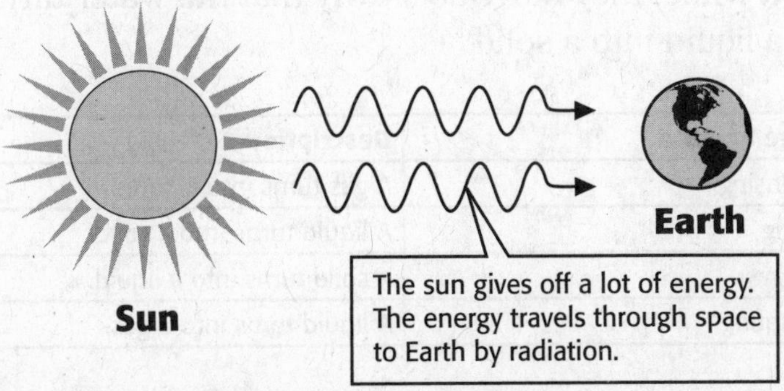
Sun
Earth
The sun gives off a lot of energy. The energy travels through space to Earth by radiation.

TAKE A LOOK

11. Compare How is radiation different from both convection and conduction?

____

____

____

Copyright © by Holt, Rinehart and Winston. All rights reserved.

How Can Heat Affect Matter?

What is the difference between ice and water? You probably answered that ice is a solid and water is a liquid. The chemicals that make up the water are the same as the chemicals that make up the ice. The difference between the ice and the water is the *state*, or physical form, that the matter is in. In general, substances can be in three different states: solid, liquid, or gas. ☑

READING CHECK

12. Identify What are the three states of matter?

____

____

Remember that matter is made of tiny particles. The state of a substance depends on the speed of its particles and on the attractions between them. The figure below shows how the particles in solids, liquids, and gases are different.

Particles of a solid have a strong attraction between them. The particles are closely locked in position and can only vibrate.

Particles of a liquid are more loosely connected than those of a solid and can collide with and move past one another.

Particles of a gas move fast enough that they overcome the attractions between them. The particles move independently and collide frequently.

One way to change matter from one state to another is to add heat. When heat moves into an object, the object's temperature goes up. The particles in the substance move faster. This can make a solid turn into a liquid or a liquid turn into a gas.

You can also change a substance's state by causing it to lose heat. When you put a glass of water in the freezer, heat moves from the water into the much colder freezer. The particles in the water move more and more slowly. After a while, they move so slowly that the water turns from a liquid into a solid.

CALIFORNIA STANDARDS CHECK

6.3.a Students know energy can be carried from one place to another by heat flow or by waves, including water, light and sound waves, or by moving objects.

Word Help: energy
what makes things happen

13. Apply Concepts Why does the ice in a glass of soda melt?

____

____

____

____

Change of state	Description
Condensing	A gas turns into a liquid.
Freezing	A liquid turns into a solid.
Melting	A solid turns into a liquid.
Evaporating	A liquid turns into a gas.

Copyright © by Holt, Rinehart and Winston. All rights reserved.

Name ________________________ Class ______________ Date ______________

Section 2 Review

6.3.a, 6.3.c, 6.3.d

SECTION VOCABULARY

conduction the transfer of energy as heat through a material

convection the movement of matter due to differences in density; the transfer of energy due to the movement of matter

heat the energy transferred between objects that are at different temperatures

radiation the transfer of energy as electromagnetic waves

temperature a measure of how hot (or cold) something is; specifically, a measure of the average kinetic energy of the particles in an object

thermal energy the kinetic energy of a substance's atoms

1. Explain How is temperature related to kinetic energy?

____

2. Explain Why does an ice cube feel cold when you touch it?

____

____

3. Describe You place a hot glass rod in a cup filled with a cold liquid. Describe what will happen to the rod and the liquid with time.

____

____

____

4. Apply Concepts Why does the energy from the sun travel to Earth by radiation instead of by convection or conduction?

____

____

____

5. Compare How is convection different from conduction?

____

____

____

____

Copyright © by Holt, Rinehart and Winston. All rights reserved.

Name ______________________ Class ______________ Date ______________

CHAPTER 3 Earth's Systems and Cycles

SECTION 3

The Cycling of Energy

California Science Standards

6.3.a, 6.3.c, 6.3.d, 6.4.a, 6.4.b, 6.4.c, 6.4.d

BEFORE YOU READ

After you read this section, you should be able to answer these questions:

- Where does the heat on Earth come from?
- How does heat move through the atmosphere, hydrosphere, and geosphere?

STUDY TIP

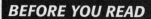

Compare As you read this section, make a chart comparing the ways that heat moves through Earth's systems.

How Does Energy Move from Place to Place?

Imagine that you are playing outside on a warm, sunny day. You see some clouds start to form in the sky. All of a sudden, a strong wind starts to blow, and the air feels much cooler. Soon, you hear thunder. It starts to rain. Why did the weather change so quickly? The answer has to do with how heat moves through Earth's atmosphere.

Remember that heat can be carried from one place to another in three ways: by radiation, by convection, or by conduction. During radiation, heat moves in the form of waves. During convection, matter moves from place to place and carries heat with it. During conduction, heat moves by heat flow. **Heat flow** means the transfer of heat from a warmer object to a cooler object. ☑

READING CHECK

1. Identify What are three ways heat can move?

____

____

____

Radiation, conduction, and convection are all important in moving heat through Earth's systems. Where does this heat come from? The sun is the main source of heat for Earth's systems. However, the inside of Earth also gives a little bit of heat to Earth's systems. The heat from the sun and from the inside of Earth moves through the geosphere, hydrosphere, biosphere, and atmosphere.

TAKE A LOOK

2. Recall Fill in the table to compare the three different ways that heat can move from place to place.

Type of heat transfer	How heat moves	Example
Radiation		
Conduction		
Convection		

Copyright © by Holt, Rinehart and Winston. All rights reserved.

How Does the Sun's Energy Get to Earth?

Earth is almost 150 million kilometers from the sun. However, the sun provides 99% of the energy on Earth. How does energy from the sun travel to Earth? The answer is radiation. Remember that energy can travel by radiation, even when there is no matter around to transmit the energy.

CALIFORNIA STANDARDS CHECK

6.4.b Students know solar energy reaches Earth through radiation, mostly in the form of visible light.

Word Help: energy
what makes things happen

Word Help: visible
that can be seen

3. Identify In what form is most of the energy that Earth gets from the sun?

____

ELECTROMAGNETIC SPECTRUM

Most of the energy that the sun gives off is light that you can see. However, the sun also gives off energy that you can't see. All of the energy from the sun travels in waves. These waves are called *electromagnetic radiation.* Electromagnetic radiation includes many different *wavelengths*, or kinds, of energy. Together, all of these wavelengths are called the **electromagnetic spectrum**.

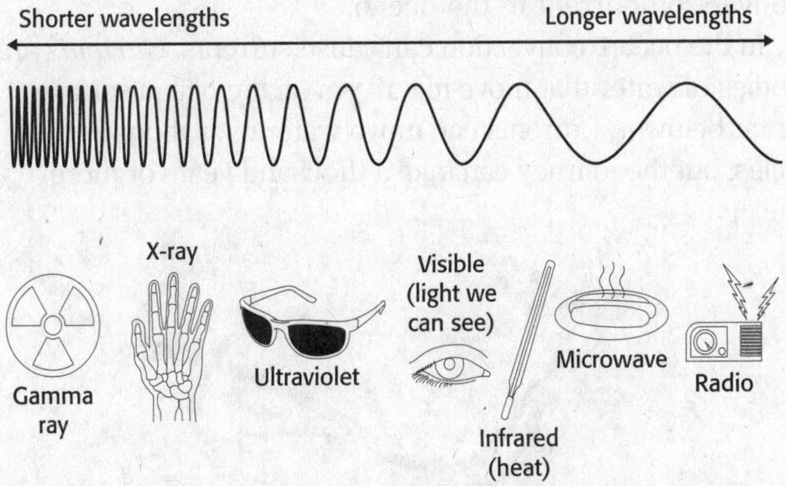
Shorter wavelengths
Longer wavelengths
Gamma ray
X-ray
Ultraviolet
Visible (light we can see)
Infrared (heat)
Microwave
Radio

The electromagnetic spectrum is made up of many different wavelengths of electromagnetic radiation. Most of the energy that the sun gives off is visible light that we can see. Other kinds of electromagnetic radiation are X-rays, microwaves, and radio waves.

Say It

Discuss You may have heard or seen the terms "X-ray," "infrared," and "ultraviolet" in other places. In a small group, talk about the ways that these words are used in everyday speech.

EARTH AND ENERGY FROM THE SUN

The energy from the sun travels through space by radiation. Some of the energy can pass through Earth's atmosphere and reach Earth's surface. The energy from the sun is the main force that drives the water cycle. The energy from the sun makes life on Earth possible.

Earth's geosphere, hydrosphere, atmosphere, and biosphere absorb a lot of the energy that Earth gets from the sun. As Earth's systems absorb the light energy, some of it turns into heat. Heat can move through Earth's systems by convection and conduction.

Copyright © by Holt, Rinehart and Winston. All rights reserved.

Name ______________________ Class ______________ Date ______________

How Does Convection Move Energy Through Earth's Systems?

Remember that convection happens when moving matter carries heat from one place to another. Convection happens when matter is heated unevenly. Most of the heat that moves through Earth's systems moves by convection. Convection moves energy in the hydrosphere, atmosphere, and geosphere. ☑

READING CHECK

4. Identify What is the main way that energy moves through the atmosphere, hydrosphere, and geosphere?

____

CONVECTION IN THE HYDROSPHERE

Most of the convection in the hydrosphere happens in the oceans. Remember that convection in the oceans happens when ocean water is colder in some places than in other places. The colder water is denser than the warmer water. The difference in density produces a convection current in the ocean.

In the ocean, convection can cause currents. *Currents* are bodies of water that move like rivers on the ocean surface or far below it. Currents can move water over thousands of miles, but the journey can take a thousand years or more!

Critical Thinking

5. Predict Consequences Would there still be convection in the oceans if the Earth stopped getting energy from the sun? Explain your answer.

____

____

____

____

____

____

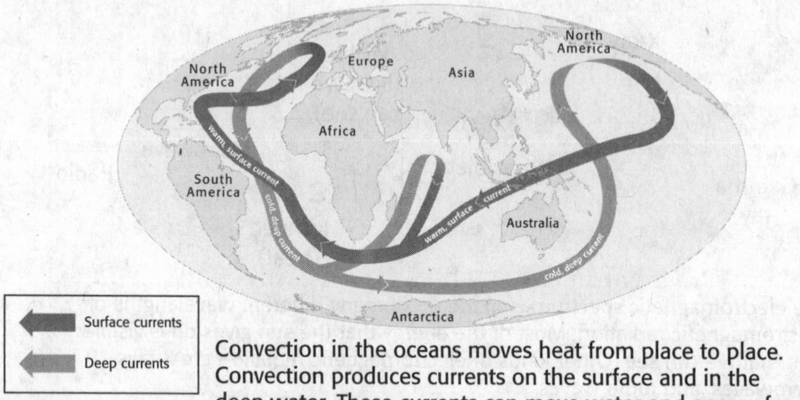
North America
Europe
Asia
North America
Africa
South America
Australia
Antarctica
warm surface current
cold deep current
warm surface current
cold deep current
Surface currents
Deep currents

Convection in the oceans moves heat from place to place. Convection produces currents on the surface and in the deep water. These currents can move water and energy for thousands of miles.

CONVECTION IN THE ATMOSPHERE

Convection in the atmosphere, just like convection in the oceans, happens because of uneven heating. The sun heats Earth's surface. The ground warms up the air by conduction. Then, the warm air rises and moves across Earth's surface. As it rises and moves, it loses some of its heat and gets cooler. The cooler air sinks toward Earth. ☑

The movement of air by convection can cause winds. It also helps transfer heat from one place to another on Earth's surface.

READING CHECK

6. Explain Why does convection happen in the atmosphere?

____

Copyright © by Holt, Rinehart and Winston. All rights reserved.

CONVECTION IN THE GEOSPHERE

Earth may seem very solid and rigid to you. However, inside Earth, solid rock is flowing slowly from place to place. Energy from inside Earth heats up the rock in Earth's mantle. The rock in the mantle is under a lot of pressure, so it does not melt. Instead, the heated rock acts like soft taffy or chewing gum. It is still solid, but it moves slowly from place to place.

As the rock heats up, it rises toward Earth's surface. At the same time, cooler, denser rock sinks deeper into the mantle. This movement produces convection currents like those in the ocean. In this way, heat from inside Earth moves to Earth's surface. Scientists think that convection currents in the mantle make tectonic plates move.

CALIFORNIA STANDARDS CHECK

6.4.c Students know heat from Earth's interior reaches the surface primarily through convection.

Word Help: primarily
mainly

7. Explain How does heat move from inside Earth to Earth's surface?

____

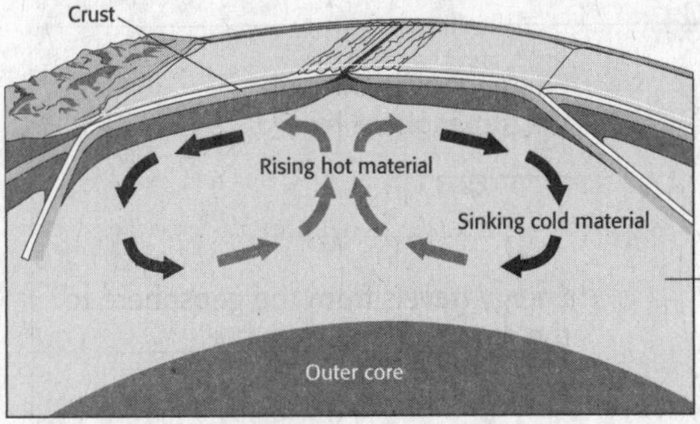
Crust
Rising hot material
Sinking cold material
Mantle
Outer core

Convection in the geosphere moves ______________ and matter from the inside of Earth to the surface.

TAKE A LOOK

8. Identify Fill in the blank line in the figure.

How Does Conduction Move Energy Through Earth's Systems?

Conduction happens when heat moves between objects that are touching. Heat moves from the warmer object to the cooler object. Heat can move by conduction between the geosphere and the atmosphere where the air touches the ground.

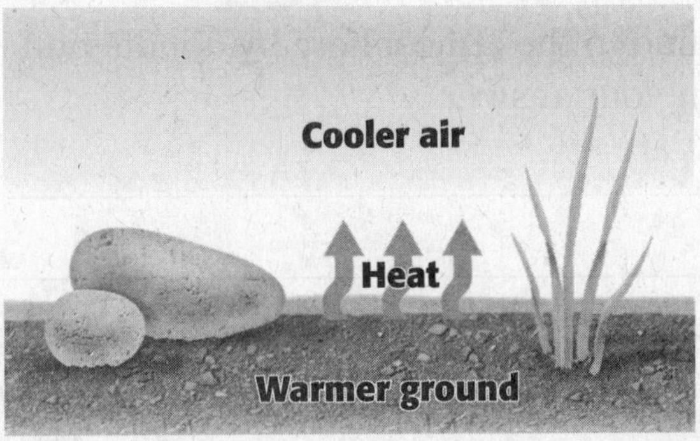
Cooler air
Heat
Warmer ground

Radiation from the sun warms the ground. If the ground is warmer than the air, heat can move from the ground into the air by conduction. Conduction happens only within a few centimeters of Earth's surface, where air particles are touching the ground.

TAKE A LOOK

9. Explain How does energy move from the geosphere to the atmosphere?

____

Copyright © by Holt, Rinehart and Winston. All rights reserved.

Name ______________________ Class ______________ Date ______________

Section 3 Review

6.3.a, 6.3.c, 6.3.d, 6.4.a, 6.4.b, 6.4.c, 6.4.d

SECTION VOCABULARY

electromagnetic spectrum all of the frequencies or wavelengths of electromagnetic radiation	**heat flow** another term for heat transfer, the transfer of energy from a warmer object to a cooler object

1. List What are three ways that heat can move from one place to another?

____

____

____

2. Identify On the figure below, fill in the blanks with the terms *radiation*, *conduction*, and *convection* to show how energy moves in Earth's systems. Hint: You may need to use some of the terms more than once.

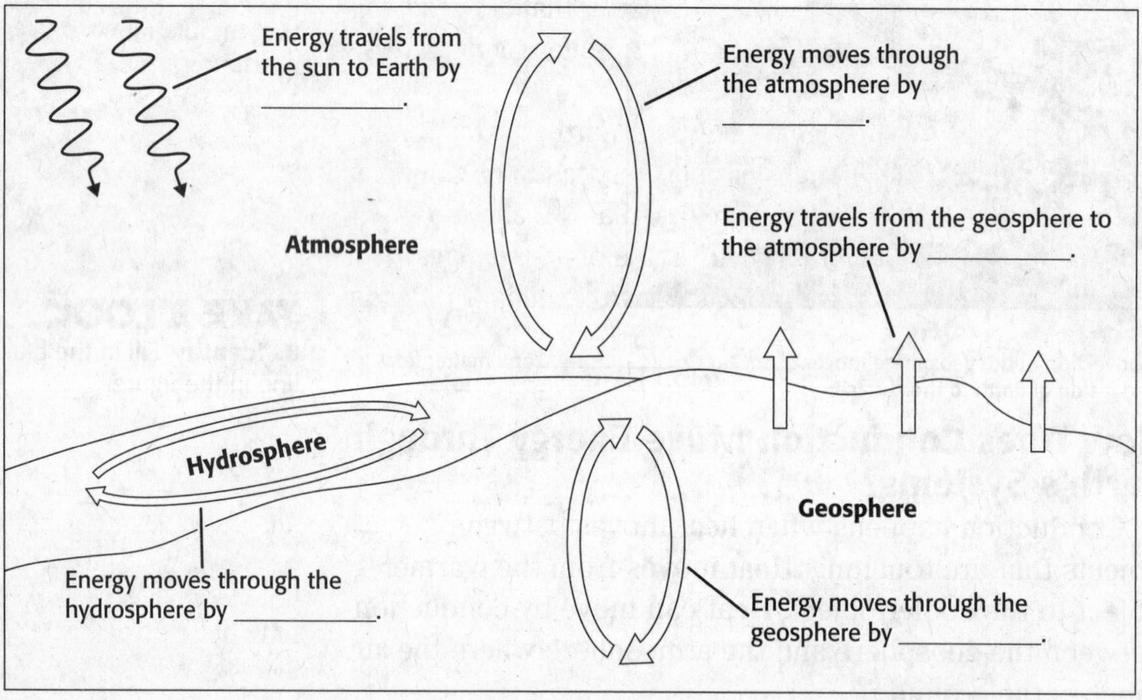

Energy travels from the sun to Earth by __________.
Energy moves through the atmosphere by __________.
Atmosphere
Energy travels from the geosphere to the atmosphere by __________.
Hydrosphere
Geosphere
Energy moves through the hydrosphere by __________
Energy moves through the geosphere by __________.

3. Apply Concepts Will heat move from the ground to the atmosphere by conduction if the air is warmer than the ground? Explain your answer.

____

____

____

Copyright © by Holt, Rinehart and Winston. All rights reserved.

Name ______________________ Class ______________ Date ______________

CHAPTER 3 Earth's Systems and Cycles

SECTION 4

The Cycling of Matter

BEFORE YOU READ

After you read this section, you should be able to answer these questions:

- What is the rock cycle?
- What is the water cycle?
- How do nitrogen, carbon, and phosphorus move through Earth's systems?

California Science Standards

6.4.a, 6.5.a, 6.5.b

How Fast Does Earth Change?

You probably know people who recycle. Did you know that Earth also recycles? Not very much new matter comes into the Earth system. Therefore, the matter that is already here has to be recycled. Water, carbon, nitrogen, phosphorus, and even rocks move through cycles on Earth.

As matter moves through the parts of Earth, the matter changes. Some of the changes happen very fast. For example, floods can move a lot of rock from one place to another in only a few days. On the other hand, some of the changes happen very slowly. For example, it can take millions of years for a tall mountain to form. Because Earth is so old, even these very slow changes have affected Earth a great deal.

STUDY TIP

Compare As you read, make a chart comparing the ways that carbon, nitrogen, and phosphorus move through Earth's systems.

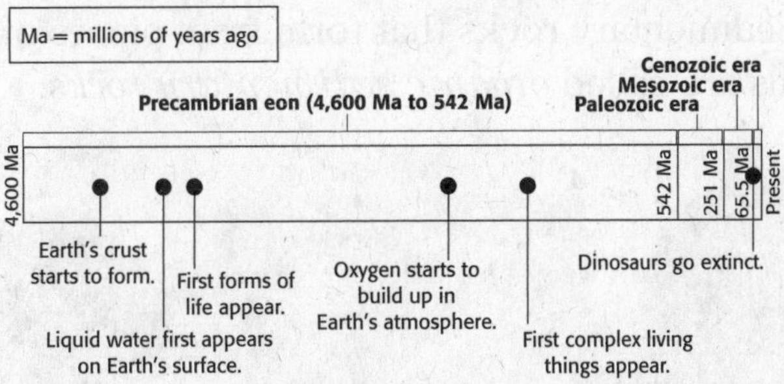

Ma = millions of years ago
Precambrian eon (4,600 Ma to 542 Ma)
Cenozoic era
Mesozoic era
Paleozoic era
4,600 Ma
542 Ma
251 Ma
65.5 Ma
Present
Earth's crust starts to form.
First forms of life appear.
Liquid water first appears on Earth's surface.
Oxygen starts to build up in Earth's atmosphere.
First complex living things appear.
Dinosaurs go extinct.

Earth is 4.6 billion years old. This timeline shows when some important events happened on Earth. Human-like animals did not appear on Earth until about 4 million years ago.

Math Focus

1. Calculate For what fraction of Earth's history have human-like animals existed? Show your work.

____

____

____

What Are the Different Kinds of Rocks?

Rocks can look very different from each other. In many cases, they look different because they form in different ways. Scientists put rocks into three groups based on how they form. *Sedimentary rocks* form when pieces of rock stick together to form a hard rock. *Igneous rocks* form when melted rock cools and hardens. *Metamorphic rocks* form when rock changes because of heat or pressure.

Copyright © by Holt, Rinehart and Winston. All rights reserved.

Name ______________________ Class ______________ Date ____________

SEDIMENTARY ROCKS

What do sand, clay, and pebbles have in common? They all form when large rocks are broken into smaller pieces. Small pieces of rock that are moved from place to place are called *sediment*. Sediment can be moved by wind, water, ice, and gravity. ☑

READING CHECK

2. Define What is sediment?

____

____

Sedimentary rocks can be grouped based on what they are made of and how they form. Some sedimentary rocks form when sediment is buried and put under pressure. The pressure causes the sediment pieces to stick together. In time, the sediment hardens and forms a rock. Sedimentary rocks that form when sediment hardens under pressure are called *clastic sedimentary rocks*.

Other sedimentary rocks form when chemicals form mineral crystals. For example, many kinds of chemicals are dissolved in ocean water. If the water evaporates, some of the chemicals will form mineral crystals. Sedimentary rock that forms when minerals that are dissolved in water form crystals are called *chemical sedimentary rocks*.

Critical Thinking

3. Compare How are clastic sedimentary rocks different from organic sedimentary rocks?

____

____

____

____

____

____

Some sedimentary rocks are made of pieces of dead organisms. For example, many animals that live in the ocean have hard shells. When the animals die, their shells fall to the ocean floor. The shells can be buried. In time, pressure can cause the pieces to stick together and form a rock. Sedimentary rocks that form from pieces of dead organisms are called *organic sedimentary rocks*.

Sandstone is a kind of clastic sedimentary rock. It is made of sand that has stuck together and hardened.

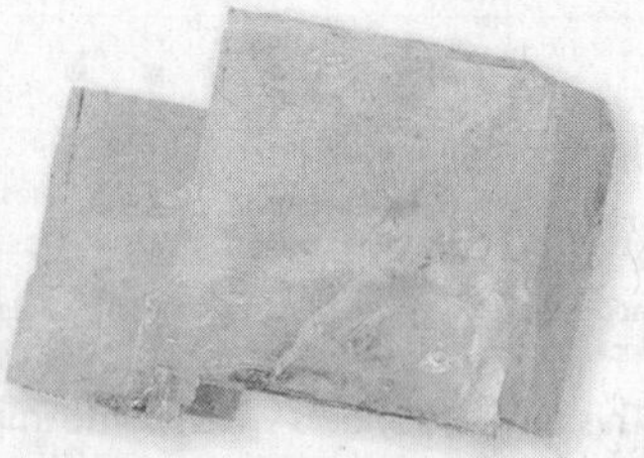

Halite is a kind of chemical sedimentary rock. It forms when seawater evaporates.

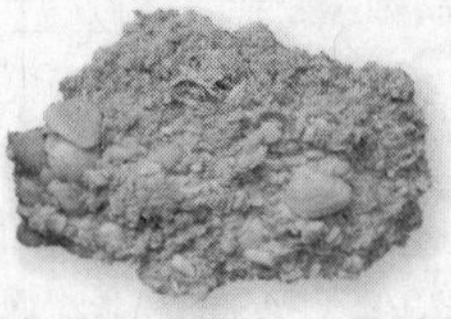

Coquina is a kind of organic sedimentary rock. It is made of the shells of tiny sea creatures.

TAKE A LOOK

4. Describe How does halite form?

____

____

____

Copyright © by Holt, Rinehart and Winston. All rights reserved.

IGNEOUS ROCKS

Igneous rock forms when melted rock cools and hardens. Igneous rocks can be grouped based on the size of the crystals in the rock. When melted rock cools slowly, large crystals can form. Igneous rocks made of large crystals are called *coarse-grained* igneous rocks.

If melted rock cools quickly, only very small crystals can form. Sometimes, the melted rock cools so fast that no crystals can form. Igneous rocks made of tiny crystals or no crystals at all are called *fine-grained* igneous rocks. ☑

Igneous rocks can also be grouped by the types of minerals in them. Some igneous rocks, such as granite, are made of mostly light-colored minerals. Most of these light-colored rocks are found on Earth's crust, in the continents. Other igneous rocks are made of mostly dark-colored minerals. These dark-colored rocks are found mainly in Earth's mantle and in the crust under the oceans.

READING CHECK

5. Identify What is the main factor that affects whether an igneous rock is fine-grained or coarse-grained?

____

____

TAKE A LOOK

6. Infer Where would you be most likely to find granite, in Earth's mantle or Earth's continents? Explain your answer.

____

____

____

____

____

____

Granite is a kind of igneous rock. It has very large crystals, so it is coarse-grained. It is made mostly of light-colored minerals.

Basalt is a kind of igneous rock. It has very small crystals, so it is fine-grained. It is made of dark-colored minerals.

METAMORPHIC ROCKS

Metamorphic rock forms when one kind of rock changes into another kind of rock because of heat or pressure. Most metamorphic rocks form deep within Earth's crust. At these depths, temperature and pressure can be much higher than they are at Earth's surface.

Metamorphic rocks can be put in groups based on how the minerals in them are related. In *foliated* metamorphic rocks, the minerals form bands or stripes within the rock. In *nonfoliated* metamorphic rocks, the minerals do not form bands or stripes. The picture at the top of the next page shows examples of metamorphic rocks.

Copyright © by Holt, Rinehart and Winston. All rights reserved.

Gneiss is a kind of metamorphic rock. The minerals form stripes in the rock, so it is a foliated metamorphic rock.

Marble is a kind of metamorphic rock. The minerals do not form stripes in the rock, so it is a nonfoliated metamorphic rock.

TAKE A LOOK

7. Explain Why is marble considered a nonfoliated metamorphic rock?

____

____

How Do Rocks Change from One Kind to Another?

There are many ways that rocks can change. The **rock cycle** is made up of all of the ways that rocks can change from one kind to another. Some of the processes that can change the form of a rock are melting, cooling, heating, increasing pressure, weathering, and erosion. *Weathering* happens when rock is broken into smaller pieces. *Erosion* happens when sediment is moved from one place to another. ☑

READING CHECK

8. Identify Give three ways that rock can change from one form to another.

____

____

____

HOW ROCK MOVES THROUGH THE ROCK CYCLE

Rocks can take many different paths through the rock cycle. The path that a rock takes depends on the forces that act on the rock. These forces change, depending on where the rock is located. For example, metamorphic rock usually forms deep beneath Earth's surface, where pressure and temperature are high.

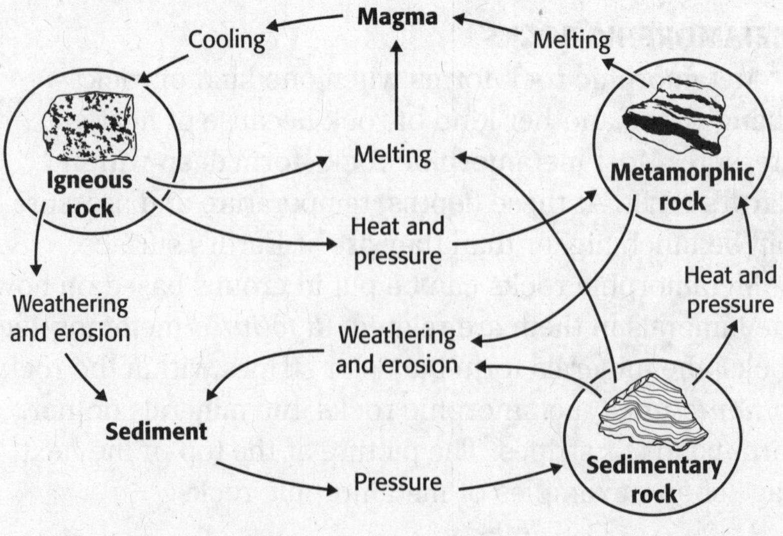
Magma
Cooling
Melting
Igneous rock
Melting
Metamorphic rock
Heat and pressure
Heat and pressure
Weathering and erosion
Weathering and erosion
Sediment
Pressure
Sedimentary rock

TAKE A LOOK

9. Use Models Find two paths through the rock cycle that lead from sedimentary rock to igneous rock. Use a colored pen or marker to trace both paths on the figure.

Copyright © by Holt, Rinehart and Winston. All rights reserved.

How Does Water Move Through Earth's Systems?

The **water cycle** is the constant movement of water between the atmosphere, the land, and the oceans. Most of the energy that drives the water cycle comes from the sun. For example, the sun's energy can turn liquid water in the oceans into water vapor, a gas that enters the atmosphere. The figure below shows some of the ways that water moves through the water cycle.

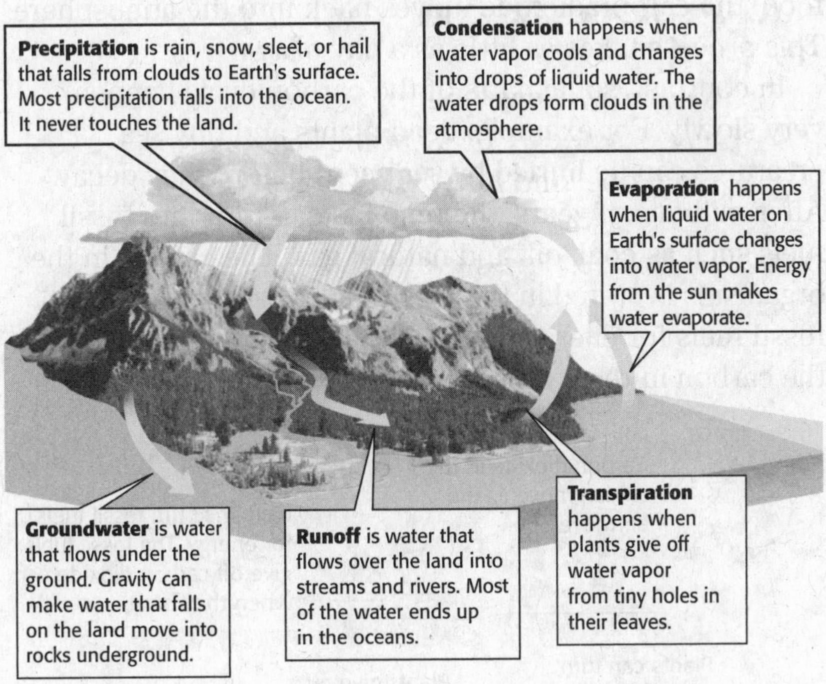
Precipitation is rain, snow, sleet, or hail that falls from clouds to Earth's surface. Most precipitation falls into the ocean. It never touches the land.
Condensation happens when water vapor cools and changes into drops of liquid water. The water drops form clouds in the atmosphere.
Evaporation happens when liquid water on Earth's surface changes into water vapor. Energy from the sun makes water evaporate.
Groundwater is water that flows under the ground. Gravity can make water that falls on the land move into rocks underground.
Runoff is water that flows over the land into streams and rivers. Most of the water ends up in the oceans.
Transpiration happens when plants give off water vapor from tiny holes in their leaves.

How Do Elements Move Through Earth's Systems?

Many chemical elements, such as carbon, nitrogen, and phosphorus, are important for life on Earth. These elements are constantly moving from one place to another. They move between living things and nonliving things through many different processes.

Critical Thinking

10. Predict Consequences Imagine that evaporation stopped happening on Earth. What could happen to the water cycle? Explain your answer.

Say It

Discuss In a small group, talk about the ways that precipitation, runoff, and groundwater can affect people.

Copyright © by Holt, Rinehart and Winston. All rights reserved.

CARBON CYCLE

Carbon is an important part of many of the chemicals that make up living things. However, carbon is not found only in living things. It is also found in the atmosphere, the land, and water. The **carbon cycle** is the way that carbon moves between living things and nonliving things.

Some parts of the carbon cycle happen fairly quickly. For example, plants can turn carbon dioxide from the atmosphere into food. Other organisms get food from plants. When plants and other organisms break down the food, the carbon dioxide moves back into the atmosphere. This process takes as little as a day or two.

In contrast, some parts of the carbon cycle happen very slowly. For example, dead plants and tiny sea creatures can be buried by sediment before they decay. After millions of years, the remains can turn into fossil fuels such as coal, oil, and natural gas. The carbon in the organisms is stored in the fossil fuels. People can burn fossil fuels for energy. When people burn the fossil fuels, the carbon in them moves back into the environment.

CALIFORNIA STANDARDS CHECK

6.5.b Students know matter is transferred over time from one organism to others in the food web and between organisms and the physical environment.

Word Help: transfer
to carry or cause to pass from one thing to another

11. Identify Give one way that carbon can move from the atmosphere into living things. Give one way that carbon can move from living things into the atmosphere?

____

____

____

____

____

____

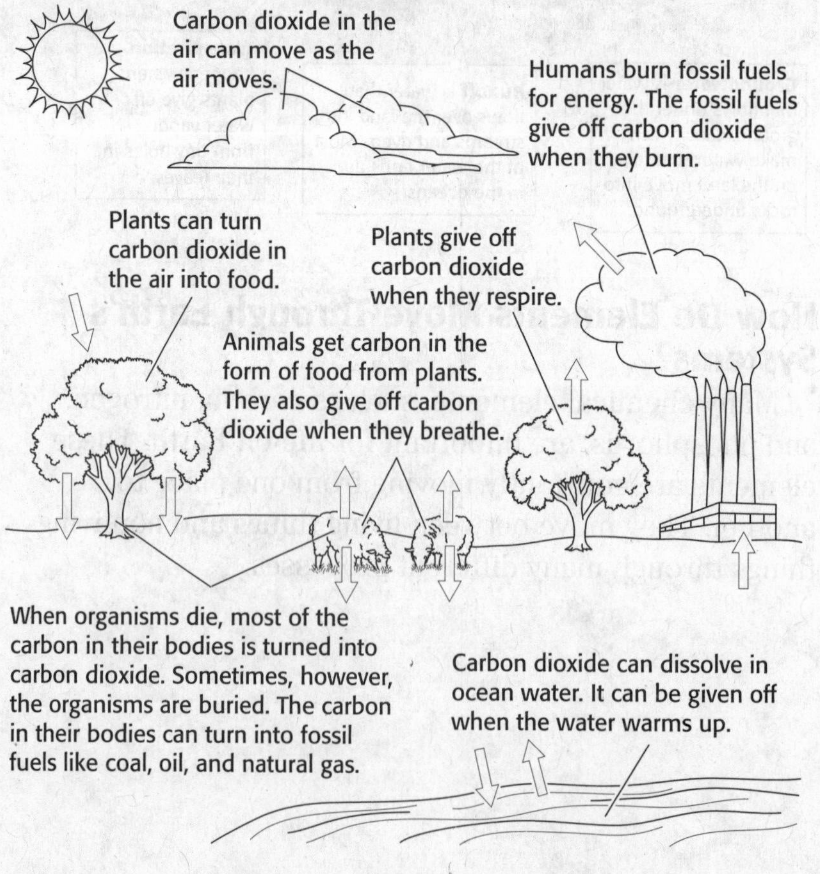
Carbon dioxide in the air can move as the air moves.
Humans burn fossil fuels for energy. The fossil fuels give off carbon dioxide when they burn.
Plants can turn carbon dioxide in the air into food.
Plants give off carbon dioxide when they respire.
Animals get carbon in the form of food from plants. They also give off carbon dioxide when they breathe.
When organisms die, most of the carbon in their bodies is turned into carbon dioxide. Sometimes, however, the organisms are buried. The carbon in their bodies can turn into fossil fuels like coal, oil, and natural gas.
Carbon dioxide can dissolve in ocean water. It can be given off when the water warms up.

TAKE A LOOK

12. Explain How does the carbon in fossil fuels get into the atmosphere?

____

____

____

Copyright © by Holt, Rinehart and Winston. All rights reserved.

NITROGEN CYCLE

The movement of nitrogen through the Earth system is called the **nitrogen cycle**. Nitrogen can be found in several different forms. In the atmosphere, nitrogen is mainly found as a gas. However, most living things cannot use nitrogen gas as a nutrient. Special bacteria change the nitrogen gas into a form of nitrogen that plants can use. Other organisms get their nitrogen from plants.

When organisms die, the nitrogen in them can be returned to the environment. Bacteria can change the nitrogen in dead organisms back into nitrogen gas. The nitrogen gas then can move into the atmosphere. ☑

READING CHECK

13. Explain Why are bacteria important in the nitrogen cycle? Give two reasons.

____

____

____

____

____

____

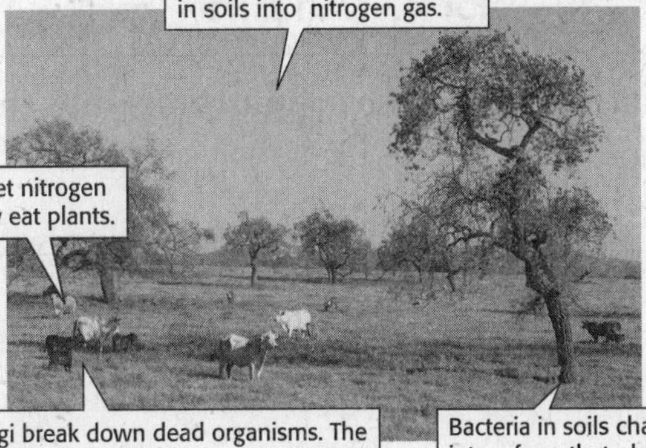
Bacteria can change nitrogen in soils into nitrogen gas.
Animals get nitrogen when they eat plants.
Bacteria and fungi break down dead organisms. The nitrogen in the dead organisms moves into the soil.
Bacteria in soils change nitrogen gas into a form that plants can use.

PHOSPHORUS CYCLE

Like carbon and nitrogen, phosphorus is found in living things. Phosphorus is also found in soil, rock, and water. Plant roots can take up phosphorus from the soil. Animals take in phosphorus when they eat the plants. When the animals and plants die, the phosphorus in their bodies moves back into the soil.

OTHER CYCLES IN NATURE

Rock, water, carbon, nitrogen, and phosphorus move through the Earth in cycles. Many other forms of matter also move in cycles through Earth's systems. In fact, almost all of the elements that living things need to survive are cycled through Earth's systems. When living things die, every element in their bodies is recycled.

Every cycle in nature is connected to every other cycle. For example, water can carry nitrogen and carbon through the environment. Living things are also important in keeping the cycles moving.

Critical Thinking

14. Summarize Why does matter on Earth have to be recycled in order for living things to survive?

____

____

____

____

____

____

Copyright © by Holt, Rinehart and Winston. All rights reserved.

Name ______________________ Class ______________ Date ______________

Section 4 Review

6.4.a, 6.5.a, 6.5.b

SECTION VOCABULARY

carbon cycle the movement of carbon from the nonliving environment into living things and back

nitrogen cycle the process in which nitrogen circulates among the air, soil, water, plants, and animals in an ecosystem

rock cycle the series of processes in which rock forms, changes from one type to another, is destroyed, and forms again by geologic processes

water cycle the continuous movement of water between the atmosphere, the land, and the oceans

1. Describe How does water move from the oceans into groundwater? Use some of the steps in the water cycle in your answer.

____

____

____

2. Identify What is the main feature that scientists use to put rocks into three main groups?

____

____

3. Identify Relationships Give two ways that human beings contribute to the carbon cycle.

____

____

4. Summarize What are the steps in the nitrogen cycle?

____

____

____

____

5. Apply Concepts Give three reasons why living things are important in moving matter through cycles.

____

____

____

Copyright © by Holt, Rinehart and Winston. All rights reserved.

Name ______________________ Class ______________ Date ____________

CHAPTER 4 Material Resources

SECTION 1

Natural Resources

BEFORE YOU READ

After you read this section, you should be able to answer these questions:

- What is the difference between a renewable resource and a nonrenewable resource?
- How can you protect natural resources?

California Science Standards

6.6.b

What Are Earth's Resources?

Earth provides what you need to survive. You breathe air from Earth's atmosphere. You drink water from Earth's rivers, lakes, and aquifers. You eat food from Earth's plants and animals.

A **natural resource** is any material from Earth that is used by people. Air, soil, fresh water, petroleum, rocks, minerals, forests, and wildlife are examples of natural resources. The figure below shows some examples of natural resources. ☑

STUDY TIP

Summarize After you read this section, make a chart giving the definitions of renewable and nonrenewable resources. In the chart, include two examples of each kind of resource.

READING CHECK

1. Define In your own words, write a definition of *natural resource.*

____

____

____

Examples of Natural Resources

Gasoline and plastic are both made from oil that is pumped out of Earth's crust.
These windmills produce electricity from the energy in wind. The energy in wind comes mainly from the sun.
This pile of lumber is made of wood, which comes from trees.

TAKE A LOOK

2. Identify Give two examples of natural resources that are not shown in the figure.

____

____

Copyright © by Holt, Rinehart and Winston. All rights reserved.

What Types of Resources Exist on Earth?

Natural resources can be grouped based on how fast they can be replaced. Some natural resources are nonrenewable. Others are renewable.

NONRENEWABLE RESOURCES

Some resources, such as coal, petroleum, and natural gas, take millions of years to form. A **nonrenewable resource** is a resource that is used much faster than it can be replaced. *Renew* means "to begin again." When nonrenewable resources are used up, people can no longer use them. ☑

READING CHECK

3. Define What is a nonrenewable resource?

____

____

RENEWABLE RESOURCES

Some natural resources, such as trees and fresh water, can grow or be replaced quickly. A **renewable resource** is a natural resource that can be replaced as fast as it is used.

Many renewable resources are only renewable if people do not use them too fast. For example, wood is usually considered a renewable resource. However, if people cut down trees faster than the trees can grow back, wood is no longer a renewable resource. Some renewable resources, such as the sun, will never be used up, no matter how fast people use them.

TAKE A LOOK

4. Explain Describe how some renewable resources can become nonrenewable resources.

____

____

____

____

____

____

Fresh water and trees are both renewable resources. However, they can be used up if people use them too quickly.

Copyright © by Holt, Rinehart and Winston. All rights reserved.

How Can We Protect Natural Resources?

Whether the natural resources you use are renewable or nonrenewable, you should be careful how you use them. In order to *conserve* natural resources, you should try to use them only when you have to. For example, leaving the water running while you are brushing your teeth wastes clean water. Turning the water off while you brush your teeth saves water so that it can be used in the future.

The energy we use to heat our homes, drive our cars, and run our computers comes from natural resources. Most of these resources are nonrenewable. If we use too much energy now, we might use up these resources. Therefore, reducing the amount of energy you use can help to conserve natural resources. You can conserve energy by being careful to use it only when you need to. The table shows some ways you can conserve energy.

Instead of...	You can...
...leaving the lights on all the time	...turn them off when you're not in the room
...running the washing machine when it is only half full	...run it only when it is full
...using a car to travel everywhere	...walk, ride a bike, or use public transportation when you can

Recycling is another important way that you can help to conserve natural resources. **Recycling** means using things that have been thrown away to make new objects. Fewer natural resources are used to make objects from recycled materials than from materials that aren't recycled. Recycling also helps to conserve energy. For example, it takes less energy to recycle an aluminum can than to make a new one. ☑

Conserving resources also means taking care of them even when you are not using them. For example, it is important to keep our drinking water clean. Polluted water can harm the living things, including humans, that need water in order to live.

Critical Thinking

5. Explain Why is it important to conserve all natural resources, even if they are renewable resources?

____
____
____
____
____
____
____

TAKE A LOOK

6. Brainstorm Fill in the blank spaces in the table with some other ways you can conserve natural resources.

☑ READING CHECK

7. Identify How does recycling conserve natural resources?

____
____
____
____
____

Copyright © by Holt, Rinehart and Winston. All rights reserved.

Name ______________________ Class ______________ Date ______________

Section 1 Review

6.6.b

SECTION VOCABULARY

natural resource any natural material that is used by humans, such as water, petroleum, minerals, forests, and animals

nonrenewable resource a resource that forms at a rate that is much slower than the rate at which the resource is consumed

recycling the process of recovering valuable or useful materials from waste or scrap

Wordwise The prefix *re-* means "again." The root *cycl* means "circle" or "wheel."

renewable resource a natural resource that can be replaced at the same rate at which the resource is consumed

1. Identify What is the difference between a renewable resource and a nonrenewable resource?

____

____

2. List Give four ways to conserve natural resources.

____

____

____

____

3. Explain Why is wood usually considered a renewable resource? When would it be considered a nonrenewable resource?

____

____

____

4. Describe What does it mean to conserve natural resources?

____

____

5. Explain Why are coal, oil, and natural gas considered nonrenewable resources, even though they come from living things that can reproduce?

____

____

Copyright © by Holt, Rinehart and Winston. All rights reserved.

Name ____________________ Class __________ Date __________

CHAPTER 4 Material Resources

SECTION 2

Rock and Mineral Resources

BEFORE YOU READ

After you read this section, you should be able to answer these questions:

- What are minerals?
- What are rocks?
- How are rock and mineral resources used?

California Science Standards

6.6.b, 6.6.c

How Are Rocks Different from Minerals?

A **mineral** is a naturally formed, inorganic solid that forms crystals and is always made of the same elements. Let's look at each part of this definition more closely.

- Minerals form naturally. For example, most diamonds form far underground. These diamonds are minerals. However, some diamonds are made by people. These diamonds are not minerals, because they did not form naturally.
- Minerals are solids. For example, frozen water, or ice, in a glacier is considered a mineral because it is solid. However, liquid water in a stream is not a mineral. ☑
- Minerals form by inorganic processes. In other words, they are not formed from living things. For example, coal is not a mineral because it forms from the remains of plants.
- Minerals are crystals. *Crystals* are solids whose particles are lined up in a repeating pattern. For example, volcanic glass is not a mineral because the particles in it are not lined up in a pattern.
- Minerals are always made of the same elements. For example, every sample of the mineral fluorite contains the same elements: calcium and fluorine. In many cases, the elements in minerals are valuable natural resources.

A *rock* is a natural material that makes up most of the solid part of Earth. Most rocks are made from one or more minerals. Rocks can also be made from solid materials that are not minerals. For example, a rock can contain coal or volcanic glass, neither of which are minerals.

STUDY TIP

Describe After you read this section, make a chart showing the uses of different rock and mineral resources.

READING CHECK

1. Explain Why is the salt dissolved in ocean water not considered a mineral?

____

Critical Thinking

2. Compare How are rocks different from minerals?

____

____

____

____

____

Copyright © by Holt, Rinehart and Winston. All rights reserved.

How Do Minerals Form?

Minerals form in many different ways. The table below shows five ways that minerals can form.

Process	Description	Examples of minerals that form this way
Evaporation	When a body of salt water dries up, minerals are left behind.	gypsum, halite
Metamorphism	High temperatures and pressures deep underground can cause the minerals in rock to change into different minerals.	garnet, graphite, magnetite, talc
Deposition	When water carries dissolved materials into lakes and seas, minerals form and sink to the bottom.	calcite, dolomite
Reaction	Water underground can be heated by hot rock. The hot water can dissolve some minerals and deposit other minerals in their place.	gold, copper, sulfur, pyrite, galena
Cooling	Melted rock can cool slowly under Earth's surface. As the melted rock cools, minerals form.	mica, feldspar, quartz

TAKE A LOOK

3. Identify Give three minerals that form by metamorphism and three minerals that form by reaction.

Metamorphism:

____

____

____

Reaction:

____

____

____

How Are Minerals Removed from Earth?

People mine many kinds of rocks and minerals from the ground and make them into objects we need. Some kinds of rocks and minerals have more useful materials in them than others. An **ore** is a rock or mineral that contains enough useful materials for it to be mined at a profit.

There are two ways of removing ores from Earth: surface mining and subsurface mining. The type of mining that people use depends on how close the ore is to the surface. ☑

READING CHECK

4. Explain What determines the type of mining that people use to remove an ore from Earth's crust?

____

____

SURFACE MINING

People use surface mining to remove ores that are near Earth's surface. Open-pit mining is used to remove gold and copper from the ground. Explosives break up the overlying rock and ore. Then, trucks haul the ore from the mine to a processing plant.

Quarries are open mines that are used to remove sand, gravel, and crushed rock. The layers of rock near the surface are removed and used to make buildings and roads.

Strip mines are often used to mine coal. The coal is removed in large pieces. These pieces are called *strips*.

Copyright © by Holt, Rinehart and Winston. All rights reserved.

SUBSURFACE MINING

People use subsurface mining to remove ores that are deep underground. Iron, coal, and salt can be mined in subsurface mines. ☑

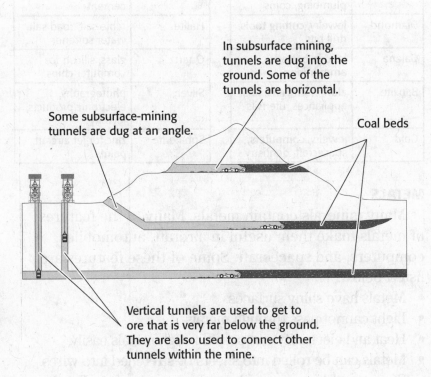

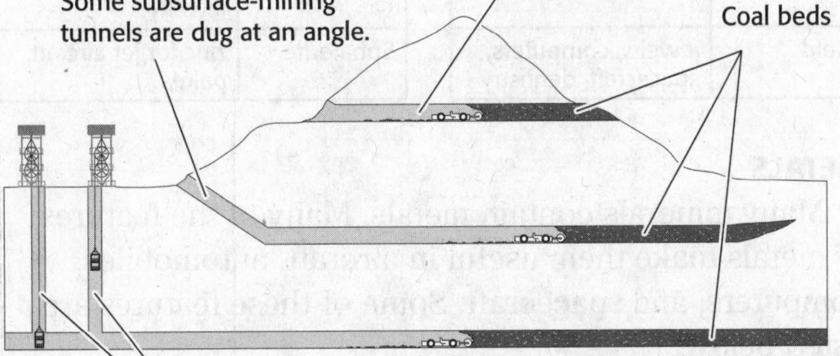
In subsurface mining, tunnels are dug into the ground. Some of the tunnels are horizontal.
Some subsurface-mining tunnels are dug at an angle.
Coal beds
Vertical tunnels are used to get to ore that is very far below the ground. They are also used to connect other tunnels within the mine.

READING CHECK

5. List Give three resources that can be mined using subsurface mining.

____

TAKE A LOOK

6. Identify What are three kinds of tunnels used in subsurface mining?

____

RESPONSIBLE MINING

Mining can help us get the resources we need, but it can also create problems. Mining may destroy or harm the places where plants and animals live. The wastes from mining can be poisonous. They can pollute water. ☑

One way to reduce these problems is to return the land to its original state after mining is finished. This is called *reclamation*. Since the mid-1970s, laws have required the reclamation of land used for mining. Another way to reduce the problems with mining is to use less of the resources that are mined.

READING CHECK

7. Describe What are two problems with mining?

____

Copyright © by Holt, Rinehart and Winston. All rights reserved.

CALIFORNIA STANDARDS CHECK

6.6.c Students know the natural origin of the materials used to make common objects.

8. Identify Give two uses for the mineral silver and two uses for the mineral bauxite.

Silver:

____

____

Bauxite:

____

____

What Are Rocks and Minerals Used For?

The table shows how some common minerals are used.

Mineral	Uses
Copper	electrical wire, plumbing, coins
Diamond	jewelry, cutting tools, drill bits
Galena	lead in batteries, ammunition
Bauxite	aluminum cans, foil, appliances, utensils
Gold	jewelry, computers, spacecraft, dentistry

Mineral	Uses
Gypsum	wallboard, plaster, cement
Halite	table salt, road salt, water softener
Quartz	glass, silicon for computer chips
Silver	photography, electronic products, jewelry
Sphalerite	zinc for jet aircraft, paints

METALS

Many minerals contain metals. Many of the features of metals make them useful in aircraft, automobiles, computers, and spacecraft. Some of these features are listed below.

- Metals have shiny surfaces.
- Light cannot pass through metals.
- Heat and electricity can pass through metals easily.
- Metals can be rolled into sheets or stretched into wires.

Some metals react easily with air and water. For example, iron can react with oxygen in the air to produce rust. Other metals do not react very easily. For example, gold does not react with very many chemicals. Therefore, it is used in many spacecraft.

NONMETALS

Many minerals also contain nonmetals. Some important features of nonmetals are listed below.

- Nonmetals have shiny or dull surfaces.
- Light can pass through some kinds of nonmetals.
- Heat and electricity cannot pass through nonmetals easily.

Nonmetals are very important. For example, the mineral calcite is used to make cement. Sand is made of small pieces of rocks and minerals. It contains silica, which is used to make computer chips and glass.

Say It

Discuss In a small group, talk about some of the ways that metals and nonmetals are used in everyday objects.

Copyright © by Holt, Rinehart and Winston. All rights reserved.

Name ________________ Class ________________ Date ________________

Section 2 Review

6.6.b, 6.6.c

SECTION VOCABULARY

mineral a naturally formed, inorganic solid that has a definite chemical structure	**ore** a natural material whose concentration of economically valuable minerals is high enough for the material to be mined profitably

1. Identify What are five features of minerals?

____

____

____

____

2. Describe Fill in the spaces in the table to describe metals and nonmetals.

Type of material	Main features	Common objects made from it
Metal	has shiny surfaces does not transmit light transmits heat and electricity easily can be rolled into sheets or stretched into wires	
Nonmetal		

3. List What are three ways minerals can form?

____

____

____

4. Apply Concepts Are minerals renewable resources or nonrenewable resources? Explain your answer.

____

____

____

Copyright © by Holt, Rinehart and Winston. All rights reserved.

Name ______________________ Class ______________________ Date ______________________

SECTION 3

Using Material Resources

California Science Standards

6.6.b, 6.6.c

BEFORE YOU READ

After you read this section, you should be able to answer these questions:

- What resources come from Earth?
- What resources come from living things?
- How can we use resources wisely?

STUDY TIP

Ask Questions As you read this section, write down any questions that you think of. Then, talk about your questions with the class.

What Resources Come from Earth?

Earth's resources are found in Earth's atmosphere, crust, and oceans. They also come from the organisms that live on Earth. Earth's resources can be divided into two categories: energy resources and material resources. Energy resources are natural resources that humans use to produce energy. **Material resources** are natural resources that humans eat, drink, or use to make objects.

RESOURCES FROM THE ATMOSPHERE

The air around us contains oxygen gas, which almost all living things need to survive. Other valuable resources are also found in Earth's atmosphere. For example, nitrogen gas in the atmosphere is used to make fertilizer. Argon gas is used in light bulbs. ☑

READING CHECK

1. Identify Give two resources that are found in Earth's atmosphere.

____

____

ROCK AND MINERAL RESOURCES

Rocks and minerals are used to make many of the objects that we use every day. The table below shows some of the ways that people use different minerals.

TAKE A LOOK

2. Infer What is the most likely reason that copper and platinum are used in electrical wires, but salt is not?

____

____

____

Mineral	Uses
Iron	making steel; cars and other vehicles; cooking utensils
Aluminum	cars, buildings, and aircraft; food and drink containers
Copper	electrical wires; computers; money
Platinum	jewelry; cars and other vehicles; electrical wires
Calcite	cement; fertilizer; building stone
Salt	food; clearing ice from roads; source of the element chlorine

The figure on the top of the next page shows how people get salt out of sea water.

Copyright © by Holt, Rinehart and Winston. All rights reserved.

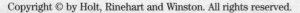

Ocean water has salt in it. Most of the salt is the chemical sodium chloride. In its solid form, sodium chloride forms the mineral halite.

People pump ocean water into large ponds. The water evaporates. The salt is left behind as crystals.

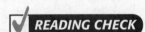

The salt is used for flavoring food and to keep food from going bad. It can also be used to help melt ice on roads.

Math Focus

3. Calculate One kilogram of ocean water has about 35 g of salt in it. One pound of salt has a mass of about 455 g. How many kilograms of ocean water would have to evaporate to produce one pound of salt?

____

PETROLEUM

Petroleum is a mixture of liquid chemicals that is found in Earth's crust. You may know that gasoline and other fuels are made from petroleum. Petroleum can also be used to make many other products, such as tar, wax, and plastic. ☑

READING CHECK

4. Identify List three ways that people use petroleum products in their everyday lives.

____

____

____

What Resources Come from Living Things?

Many of the products we use are made with resources from Earth. Many other products are made with resources from living things.

PLANT RESOURCES

People use resources from plants in many different ways. Farming provides us with plants for food and drinks, and also with food for animals. Cotton produces fibers that can be woven into cloth or braided into ropes or baskets. Trees supply fruit and nut crops, as well as wood for lumber, paper, and fuel. The sap of some trees can be used to make products such as rubber and maple syrup.

Copyright © by Holt, Rinehart and Winston. All rights reserved.

ANIMAL RESOURCES

People also use resources from animals. Animals provide transportation for people and for cargo. They help farmers till the soil and plant crops. People use animals for food and clothing. Fur from sheep, goats, and llamas can be made into fabric. The skin from some animals can be used for leather. Animal wastes can be used for fertilizer and cooking fuel.

CALIFORNIA STANDARDS CHECK

6.6.c Students know the natural origin of the materials used to make common objects.

5. Identify Give three everyday objects that are made from plant resources, and three everyday objects that are made from animal resources.

Plant Resources:

____

____

Animal Resources:

____

____

____

Resources from Plants and Animals

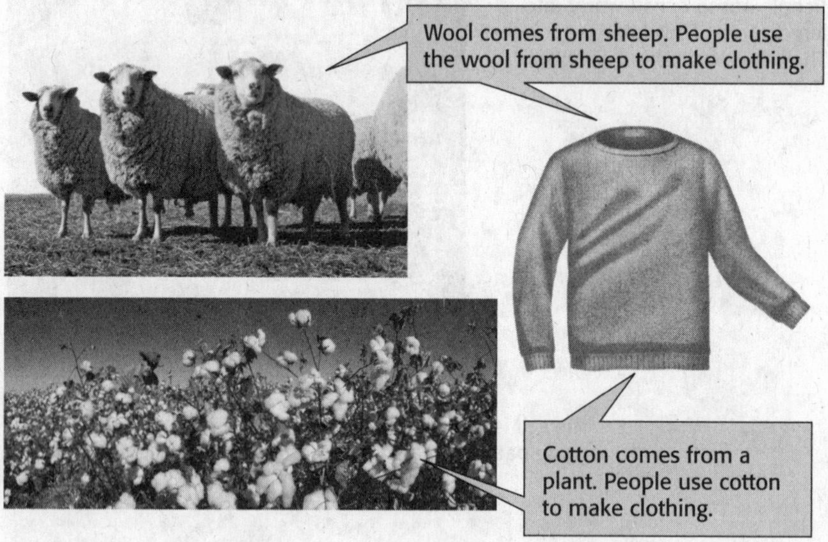
Wool comes from sheep. People use the wool from sheep to make clothing.
Cotton comes from a plant. People use cotton to make clothing.

What Are the Costs of Material Resources?

Using natural resources has many costs. Most natural resources must be removed from rock, water, or the air before they can be used. Then, the resources have to be made into the products we need. The products have to be shipped from place to place. Plant and animal resources may need special care. Using natural resources can harm the environment. ☑

READING CHECK

6. Describe What are two costs related to using material resources?

____

____

ECONOMIC COSTS

It costs money to get natural resources from Earth and from living things. Companies will only make a product if they can sell it for more than it costs them to make it. If it becomes too expensive to get a natural resource for a product, a cheaper resource must be used instead.

The figure on the next page shows how trees can be made into paper. Each step costs money. For example, companies have to pay people to do the work in each step.

Copyright © by Holt, Rinehart and Winston. All rights reserved.

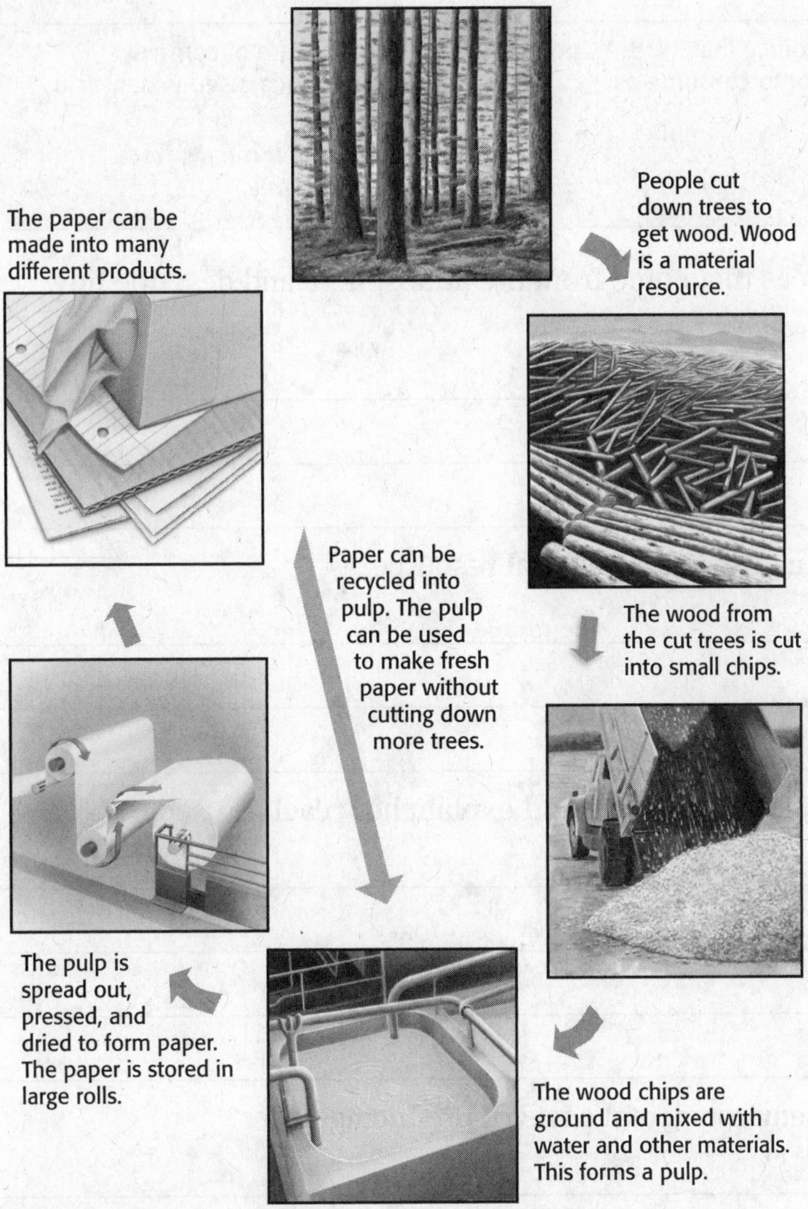
People cut down trees to get wood. Wood is a material resource.
The wood from the cut trees is cut into small chips.
The wood chips are ground and mixed with water and other materials. This forms a pulp.
The pulp is spread out, pressed, and dried to form paper. The paper is stored in large rolls.
The paper can be made into many different products.
Paper can be recycled into pulp. The pulp can be used to make fresh paper without cutting down more trees.

Say It

Discuss In a small group, talk about ways that the costs of using material resources can be reduced.

TAKE A LOOK

7. Explain How does recycling paper reduce the cost of using some material resources?

____

____

____

____

____

____

____

ENVIRONMENTAL COSTS

Often, the price of an item in a store is only based on how much money it costs to make the item. However, using natural resources also has environmental costs. For example, trees are cut down to make paper. The trees may come from old-growth forests. Cutting down trees can harm the places where other plants and animals live. The machines that are used to make the paper can pollute the air and water.

Some communities have passed laws that require companies to protect the environment. These laws can require that companies produce less pollution. They may also require companies to restore the land used for mining or to replant trees that were cut down.

Critical Thinking

8. Analyze Relationships Why does protecting the environment sometimes cause products to cost more money?

____

____

____

____

____

Copyright © by Holt, Rinehart and Winston. All rights reserved.

Name ______________________ Class ______________ Date ______________

Section 3 Review

6.6.b, 6.6.c

SECTION VOCABULARY

material resource a natural resource that humans use to make objects or to consume as food and drink	**petroleum** a liquid mixture of complex hydrocarbon compounds; used widely as a fuel source **Wordwise** The root *petr* means "rock." Another example is *petrify*.

1. Identify Give two resources that come from the atmosphere and describe how each is used.

____

____

____

2. Describe How is petroleum used as a material resource?

____

____

____

3. List Give four rock or mineral resources and explain how each is used.

____

____

____

____

4. Describe What are three environmental costs of producing paper?

____

____

____

5. Identify Give two examples of renewable material resources and two examples of nonrenewable material resources.

____

____

____

____

Copyright © by Holt, Rinehart and Winston. All rights reserved.

Name ______________________ Class ______________ Date ______________

CHAPTER 5 Energy Resources

SECTION 1

Fossil Fuels

California Science Standards

6.3.b, 6.6.a, 6.6.b

BEFORE YOU READ

After you read this section, you should be able to answer these questions:

- What are the different kinds of fossil fuels?
- How do fossil fuels form?
- What are the problems with using fossil fuels?

What Are Fossil Fuels?

How do plants and animals that lived hundreds of millions of years ago affect your life today? If you turned on the lights or traveled to school in a car or bus, you probably used some of the energy from plants and animals that lived long ago.

Energy resources are natural resources that people use to produce energy, such as heat and electricity. Most of the energy we use comes from fossil fuels. A **fossil fuel** is an energy resource made from the remains of plants and tiny animals that lived long ago. The different kinds of fossil fuels are petroleum, coal, and natural gas. ☑

Fossil fuels are an important part of our everyday life. When fossil fuels burn, they release a lot of heat energy. Power plants use the energy to produce the electricity we use. Cars use the energy to move.

However, there are also some problems with using fossil fuels. Fossil fuels are nonrenewable, which means that they cannot be replaced once they have been used. Also, when they burn, they release pollution.

STUDY TIP

Compare In your notebook, make a table to show the similarities and differences between different kinds of fossil fuels.

READING CHECK

1. Identify Where do we get most of the energy we use?

____

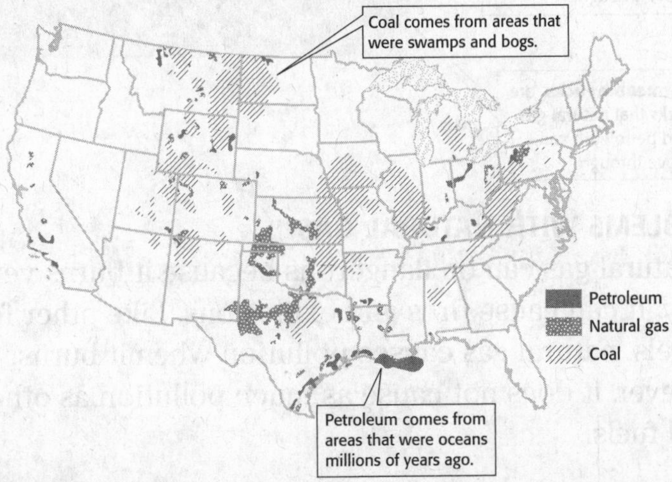

Coal comes from areas that were swamps and bogs.
Petroleum
Natural gas
Coal
Petroleum comes from areas that were oceans millions of years ago.

TAKE A LOOK

2. Describe In general, where are the natural gas deposits in the United States?

____

____

Copyright © by Holt, Rinehart and Winston. All rights reserved.

What Is Natural Gas?

A *hydrocarbon* is a compound that contains the elements carbon, hydrogen, and oxygen. **Natural gas** is a mixture of hydrocarbons that are in the form of gases. Natural gas includes methane, propane, and butane, which can be separated from one another.

Most natural gas is used for heating. Your home may be heated by natural gas. Your kitchen stove might run on natural gas. Some natural gas is used for creating electrical energy, and some cars are able to run on natural gas, too. ☑

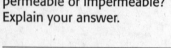

READING CHECK

3. Identify What is most natural gas used for?

____

HOW NATURAL GAS FORMS

When tiny sea creatures die, their remains settle to the ocean floor and are buried in sediment. The sediment slowly becomes rock. Over millions of years, the remains of the sea creatures turn into natural gas.

The formation of natural gas is always happening. Some of the sea life that dies today will become natural gas millions of years from now.

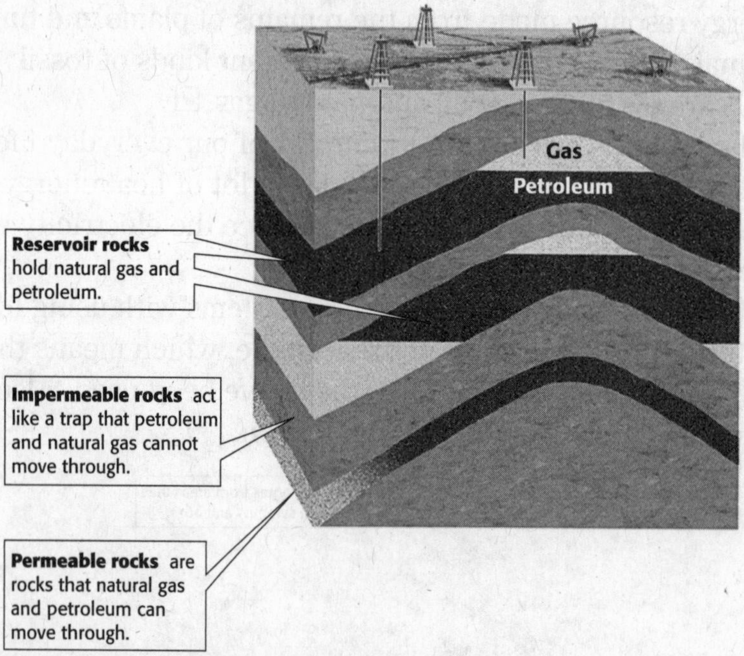
Gas
Petroleum
Reservoir rocks hold natural gas and petroleum.
Impermeable rocks act like a trap that petroleum and natural gas cannot move through.
Permeable rocks are rocks that natural gas and petroleum can move through.

TAKE A LOOK

4. Infer Is reservoir rock permeable or impermeable? Explain your answer.

____

____

____

PROBLEMS WITH NATURAL GAS

Natural gas can be dangerous because it burns very easily. It can cause fires and explosions. Like other fossil fuels, natural gas causes pollution when it burns. However, it does not cause as much pollution as other fossil fuels.

Copyright © by Holt, Rinehart and Winston. All rights reserved.

What Is Petroleum?

Petroleum is a mixture of hydrocarbons that are in the form of liquids. It is also known as crude oil. At a *refinery*, petroleum is separated into many different products, including gasoline, jet fuel, kerosene, diesel fuel, and fuel oil. ☑

Petroleum products provide more than 40% of the world's energy, including fuel for airplanes, trains, boats, ships, and cars. Petroleum is so valuable that it is often called "black gold."

READING CHECK

5. Describe What form is petroleum found in naturally?

____

____

HOW PETROLEUM FORMS

Petroleum forms the same way natural gas does. Tiny sea creatures die and then get buried in sediment, which turns into rock. The organisms eventually become petroleum or natural gas, which is stored in permeable rock within Earth's crust.

PROBLEMS WITH PETROLEUM

Petroleum can be harmful to animals and their environment. For example, in June 2000, the carrier ship *Treasure* sank off the coast of South Africa and spilled more than 400 tons of oil into the ocean. The oil covered penguins and other sea creatures, making it hard for them to swim, breathe, and eat.

Burning petroleum causes smog. **Smog** is a brownish haze that forms when sunlight reacts with pollution in the air. Smog can make it hard for people to breathe. Many cities in the world have problems with smog.

Math Focus

6. Make a Graph Use the information below to show where the world's crude oil comes from.

Middle East: 66%
North America and South America: 15%
Europe and Asia: 12%
Africa: 7%

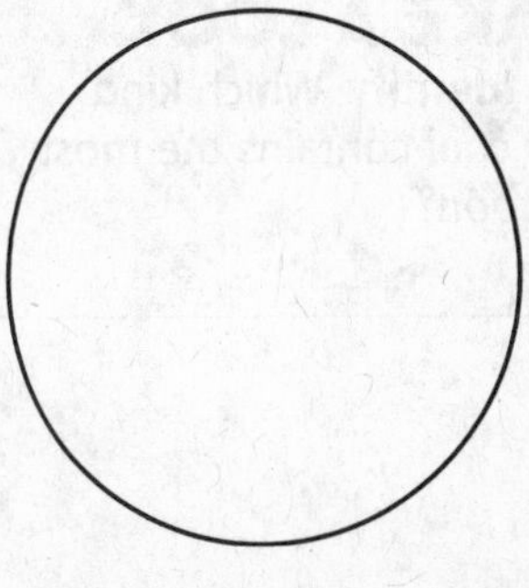

Copyright © by Holt, Rinehart and Winston. All rights reserved.

What Is Coal?

Coal is a solid fossil fuel that is made of partly decayed plant material. Coal was once the main source of energy in the United States. Like other fossil fuels, coal releases heat when it is burned. Many people used to burn coal in stoves to heat their homes. Trains in the 1800s and 1900s were powered by coal-burning steam locomotives. Coal is now used in power plants to make electricity.

HOW COAL FORMS

When swamp plants die, they sink to the bottom of the swamp. If they do not decay completely, coal can start to form. Coal forms in several different stages.

TAKE A LOOK

7. Identify Name the three types of coal.

____

____

____

Stage 1: Formation of Peat
Dead swamp plants that have not decayed can turn into **peat**, a crumbly brown material made mostly of plant material and water. Dried peat is about 60% carbon. In some parts of the world, peat is dried and burned for fuel. Peat is not coal, but it can turn into coal.

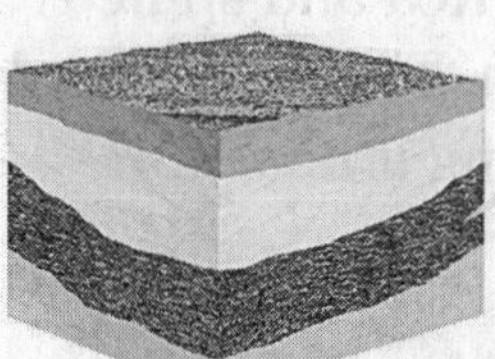

Stage 2: Formation of Lignite
If sediment buries the peat, pressure and temperature increase. The peat slowly changes into a type of coal called **lignite**. Lignite coal is harder than peat. Lignite is about 70% carbon.

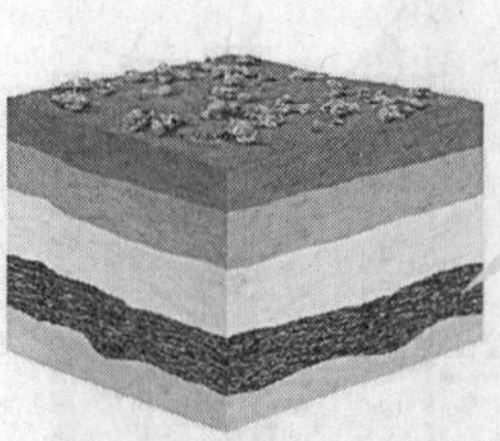

Stage 3: Formation of Bituminous Coal
If more sediment is added, pressure and temperature force more water and gases out of the lignite. Lignite slowly changes into **bituminous** coal. Bituminous coal is about 80% carbon.

TAKE A LOOK

8. Identify Which kind of coal contains the most carbon?

____

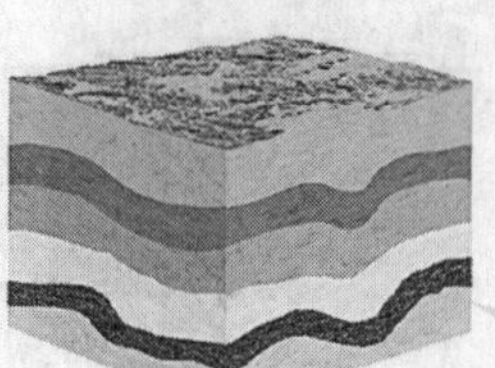

Stage 4: Formation of Anthracite
If more sediment accumulates, temperature and pressure continue to increase. Bituminous coal slowly changes into **anthracite**. Anthracite coal is the hardest type of coal. Anthracite coal is about 90% carbon.

Copyright © by Holt, Rinehart and Winston. All rights reserved.

PROBLEMS WITH COAL

Mining coal can create environmental problems. When coal is mined from Earth's surface, people remove the layers of soil above the coal. This can harm the plants that need soil to grow and the animals that need soil for shelter. If the land is not restored after mining, wildlife habitats can be destroyed for years.

Coal that is on Earth's surface can cause pollution. Water that flows through coal can pick up poisonous metals. That water can then flow into streams and lakes and pollute water supplies.

When coal is burned without pollution controls, sulfur dioxide is released. Sulfur dioxide can combine with the water in the air to produce sulfuric acid. Sulfuric acid is one of the acids in acid precipitation. **Acid precipitation** is rain, sleet, or snow that has a high concentration of acids, often because of air pollutants. Acid precipitation is also called "acid rain." Acid precipitation can harm wildlife, plants, and buildings.

In 1935, this statue had not been damaged by acid precipitation.

By 1994, acid precipitation had caused serious damage to the statue.

CALIFORNIA STANDARDS CHECK

8.3.b Students know different natural energy and material resources, including air, soil, rocks, minerals, petroleum, fresh water, wildlife, and forests, and know how to classify them as renewable or nonrenewable.

Word Help: energy
what makes things happen

Word Help: resource
anything that can be used to take care of a need

9. Classify Are fossil fuels renewable or nonrenewable resources? Explain your answer.

____

____

____

____

____

Critical Thinking

10. Infer What do you think is the reason that fossil fuels are still used today, even though they create many environmental problems?

____

____

____

____

____

Copyright © by Holt, Rinehart and Winston. All rights reserved.

Name ______________________ Class ______________ Date ______________

Section 1 Review

6.3.b, 6.6.a, 6.6.b

SECTION VOCABULARY

acid precipitation rain, sleet, or snow that has a high concentration of acids

coal a fossil fuel that forms underground from partially decomposed plant material

fossil fuel a nonrenewable energy resource formed from the remains of organisms that lived long ago; examples include oil, coal, and natural gas

natural gas a mixture of gaseous hydrocarbons located under the surface of Earth, often near petroleum deposits; used as a fuel

petroleum a liquid mixture of complex hydrocarbon compounds; used widely as a fuel source

Wordwise The root *petr* means "rock." Another example is *petrify*.

smog photochemical haze that forms when sunlight acts on industrial pollutants and burning fuels

1. Compare How are petroleum and natural gas different?

____

____

____

2. Compare Fill in the table to compare the different kinds of fossil fuels.

Kind of fossil fuel	What it is	How it forms
Coal		
	a mixture of gases containing carbon, hydrogen, and oxygen	

3. Summarize What are some of the problems with using fossil fuels for energy?

____

____

____

____

Copyright © by Holt, Rinehart and Winston. All rights reserved.

Name ______________________ Class ______________ Date ______________

CHAPTER 5 Energy Resources

SECTION 2

Alternative Energy

BEFORE YOU READ

After you read this section, you should be able to answer these questions:

- What are some kinds of alternative energy?
- What are the benefits of alternative energy?
- What are the problems with alternative energy?

California Science Standards

6.3.b, 6.6.a, 6.6.b

What Is Alternative Energy?

What would your life be like if you couldn't turn on the lights, microwave your dinner, take a hot shower, or ride the bus to school? We get most of the energy we use for heating and electricity from fossil fuels. However, fossil fuels can be harmful to the environment and to living things. In addition, they are nonrenewable resources, so we cannot replace them when they are used up.

Many scientists are trying to find alternative energy sources. *Alternative energy sources* are sources of energy that are not fossil fuels. Some sources can be converted easily into usable energy. Others are not as easy to use.

STUDY TIP

Compare and Contrast In your notebook, make a chart to show each kind of alternative energy source and its benefits and problems.

What Is Nuclear Energy?

One kind of alternative energy source is nuclear energy. **Nuclear energy** is the energy that is released when atoms come together or break apart. Nuclear energy can be obtained in two main ways: fission and fusion. ☑

READING CHECK

1. List What are the two ways in which nuclear energy is produced?

____

____

FISSION

Fission happens when an atom splits into two or more lighter atoms. Fission releases a large amount of energy. This energy can be used to generate electricity. All nuclear power that people use is generated by fission.

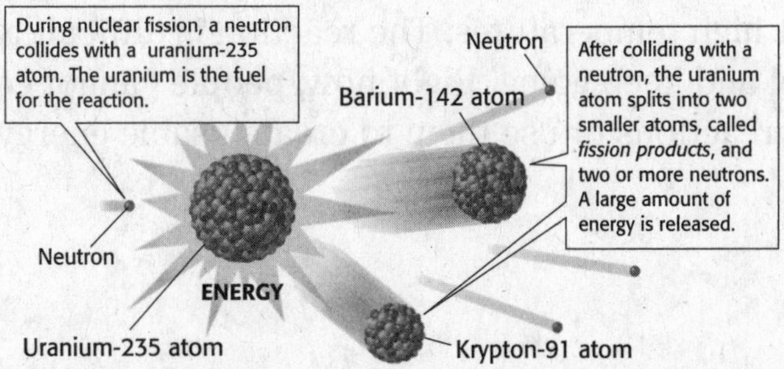

During nuclear fission, a neutron collides with a uranium-235 atom. The uranium is the fuel for the reaction.
Neutron
Barium-142 atom
After colliding with a neutron, the uranium atom splits into two smaller atoms, called *fission products*, and two or more neutrons. A large amount of energy is released.
Neutron
ENERGY
Uranium-235 atom
Krypton-91 atom

TAKE A LOOK

2. Identify What are the fission products in the figure?

____

____

Copyright © by Holt, Rinehart and Winston. All rights reserved.

FISSION'S BENEFITS AND PROBLEMS

One benefit of fission is that it does not cause air pollution. In addition, mining the fuel for nuclear power is less harmful to the environment than mining other energy sources, such as coal. ☑

READING CHECK

3. Explain Why is nuclear energy called a "clean" energy source?

However, nuclear power has several problems. The fission products created in nuclear power plants are dangerous. They must be stored for thousands of years. Nuclear fission plants can release harmful radiation into the environment. Also, nuclear power plants must release extra heat from the fission reaction. This extra heat is not used to make electricity. The extra heat can harm the environment.

FUSION

Fusion happens when two or more atoms join to form a heavier atom. This process occurs naturally in the sun. Fusion releases a lot of energy.

TAKE A LOOK

4. Identify How many protons and how many neutrons are there in the helium-4 nucleus?

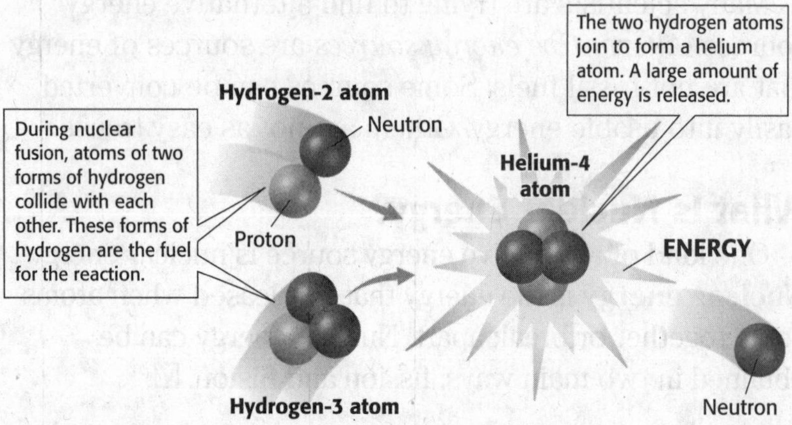
The two hydrogen atoms join to form a helium atom. A large amount of energy is released.
Hydrogen-2 atom
Neutron
During nuclear fusion, atoms of two forms of hydrogen collide with each other. These forms of hydrogen are the fuel for the reaction.
Helium-4 atom
Proton
ENERGY
Hydrogen-3 atom
Neutron

FUSION'S BENEFITS AND PROBLEMS

Fusion has two main benefits. First, fusion does not create a lot of dangerous wastes. Second, the fuels used in fusion are renewable.

The main problem with fusion is that it can take place only at high temperatures. The reaction is difficult to control and keep going. Right now, people cannot control fusion reactions or use them to create usable energy. ☑

READING CHECK

5. Describe What is the main problem with fusion?

Copyright © by Holt, Rinehart and Winston. All rights reserved.

How Can Wind Provide Electricity?

Wind is air that is moving. Moving air contains energy. People can use windmills to turn the energy in wind into electricity. The electricity that is produced by windmills is called **wind power**. Large groups of windmills can make a lot of electricity.

Like all energy sources, wind power has benefits and problems. Since the wind can't be used up, wind energy is renewable. Wind power does not cause air pollution. However, in many areas, the wind isn't strong or regular enough to generate enough electricity for people to use.

Critical Thinking

6. Infer In most cases, people use a large number of windmills to create electricity. What do you think is the reason a lot of windmills are used, instead of just one or two?

This pickup truck shows how large the windmills are.

These windmills near Livermore, California, produce electricity.

TAKE A LOOK

7. Explain Based on what you see in the figure, what do you think is the reason windmills are not used in cities or other crowded areas?

What Are Fuel Cells?

What powers a car? You probably thought of gasoline. But not all cars are powered by gasoline. Some cars are powered by fuel cells. *Fuel cells* change chemical energy into electrical power. The **chemical energy** is released when hydrogen and oxygen react to form water.

Fuel cells have been used in space travel since the 1960s. They have provided space crews with electrical energy and drinking water. Today, fuel cells are used to create electrical energy in some buildings and ships. ☑

The only waste product of fuel cells is water, so they do not create pollution. Fuel cells are also more efficient than engines that use gasoline. However, not very many cars today use fuel cells. The hydrogen and oxygen used in fuel cells can be expensive to make and to store. Many people hope that we will be able to use fuel cells to power cars in the future.

READING CHECK

8. Explain How could fuel cells give space crews electricity and water?

Copyright © by Holt, Rinehart and Winston. All rights reserved.

How Can We Use Sunlight for Power?

Most forms of energy originally come from the sun. For example, the fossil fuels we use today were made from plants. The plants got their energy from the sun. The heat and light that Earth gets from the sun is **solar energy**. This type of energy is a renewable resource.

People can use solar energy to create electricity. *Photovoltaic cells*, or solar cells, can change sunlight into electrical energy. Solar energy can also be used to heat buildings.

Solar energy does not produce pollution and is renewable. The energy from the sun is free. However, some climates don't have enough sunny days to be able to use solar energy all the time. Also, even though sunlight is free, solar cells and solar collectors are expensive to make.

Say It

Share Experiences Have you ever used an object that was powered by sunlight? In a small group, talk about the different ways that sunlight can be used for energy.

How Can Running Water Provide Electricity?

Water wheels have been used since ancient times to help people do work. Today, the energy of falling water is used to generate electrical energy. **Hydroelectric energy** is electrical energy produced from moving water.

Hydroelectric energy causes no air pollution and is considered renewable. Hydroelectric energy is generally not very expensive to produce.

However, hydroelectric energy can be produced only in places that have a lot of fast-moving water. In addition, building a dam and a power plant to generate hydroelectric energy can be expensive. Dams can harm wildlife living in and around the river. Damming a river can cause flooding and erosion.

This dam in California can create electricity because a lot of water moves through it every day.

CALIFORNIA STANDARDS CHECK

6.6.a Students know the utility of energy sources is determined by factors that are involved in converting these sources to useful forms and the consequences of the conversion process.

Word Help: utility
usefulness

Word Help: factor
a condition or event that brings about a result

Word Help: consequence
the effect, or result, of an action or process

Word Help: conversion
a change from one form to another

9. Compare What are hydroelectric energy's problems and benefits?

______

______

______

______

______

Copyright © by Holt, Rinehart and Winston. All rights reserved.

How Can Plants Be Used for Energy?

Plants store energy from the sun. Leaves, wood, and stems contain stored energy. Even the dung of plant-eating animals has a lot of stored energy. These sources of energy are called biomass. **Biomass** is organic matter that can be a source of energy.

Biomass is commonly burned in its solid form to release heat. However, biomass can be changed into a liquid form. The sugar and starch in plants can be made into alcohol and used as fuel. Alcohol can be mixed with gasoline to make a fuel called **gasohol**.

Biomass is not very expensive. It is available almost everywhere. Since biomass grows quickly, it is considered a renewable resource. However, people must be careful not to use up biomass faster than it can grow back.

Critical Thinking

10. Infer What would happen if biomass were used at a faster rate than it was produced?

____

____

____

What Is Geothermal Energy?

Geothermal energy is energy produced by the heat within Earth. This heat makes solid rocks get very hot. If there is any water contained within the solid rock, the water gets hot, too. The hot water can be used to generate electricity and to heat buildings. ☑

Geothermal energy is considered renewable because the heat inside Earth will last for millions of years. Geothermal energy does not create air pollution or harm the environment. However, this kind of energy can be used only where hot rock is near the surface.

READING CHECK

11. List What are two uses for water that has been heated by hot rock?

____

____

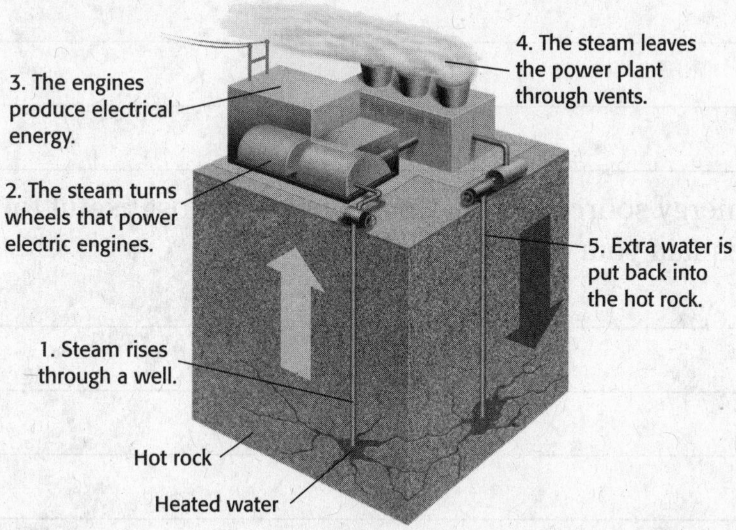
4. The steam leaves the power plant through vents.
3. The engines produce electrical energy.
2. The steam turns wheels that power electric engines.
5. Extra water is put back into the hot rock.
1. Steam rises through a well.
Hot rock
Heated water

TAKE A LOOK

12. Describe On the figure, draw arrows showing the path that the steam takes as it moves through the power plant.

Copyright © by Holt, Rinehart and Winston. All rights reserved.

Name ______________________ Class ______________ Date ______________

Section 2 Review

6.3.b, 6.6.a, 6.6.b

SECTION VOCABULARY

biomass organic matter than can be a source of energy

chemical energy the energy released when a chemical compound reacts to produce new compounds

gasohol a mixture of gasoline and alcohol that is used as a fuel

geothermal energy the energy produced by heat within Earth

hydroelectric energy electrical energy produced by the flow of water

nuclear energy the energy released by a fission or fusion reaction; the binding energy of the atomic nucleus

solar energy the energy received by Earth from the sun in the form of radiation

wind power the use of a windmill to drive an electric generator

1. Explain Why is solar energy considered a renewable resource?

____

____

2. Identify When would biomass *not* be considered a renewable resource?

____

____

3. Apply Concepts Which place is *most likely* to be able to use geothermal energy: a city near a volcano or a city near a waterfall? Explain your answer.

____

____

____

4. Identify Why is wind a useful energy source in some places, but not in others?

____

____

____

5. Analyze Which alternative energy source do you think would be most useful for the place where you live? Explain your answer.

____

____

____

____

____

Copyright © by Holt, Rinehart and Winston. All rights reserved.

SECTION 1
Earth's Structure

BEFORE YOU READ

After you read this section, you should be able to answer these questions:

- What are the layers inside Earth?
- How do scientists study Earth's interior?
- What is the evidence for continental drift?

California Science Standards

6.1.a, 6.1.b

What Is Earth's Interior Made Of?

Earth is made of three main layers. Each layer is made of different materials with different properties. Scientists think about Earth's layers in two ways: by their chemical makeup and by their physical properties.

STUDY TIP

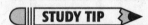

Summarize As you read, make a chart showing the features of Earth's layers. Make sure you include the compositional layers and the physical layers.

COMPOSITIONAL LAYERS

Scientists divide Earth into three layers based on chemical makeup. In the very center of Earth is the core. The **core** is a thick sphere made of iron and nickel. The next layer is the mantle. The **mantle** is a thick layer that is made of hot, solid rock. The outermost layer of Earth is called the **crust**. The crust is the thinnest of Earth's layers.

There are two main kinds of crust: continental crust and oceanic crust. *Continental crust* forms the continents. It is thicker than oceanic crust. Continental crust can be up to 70 km thick. *Oceanic crust* is found beneath the oceans. It contains more iron than continental crust. Most oceanic crust is less than 7 km thick. ☑

READING CHECK

1. Compare How is oceanic crust different from continental crust?

____

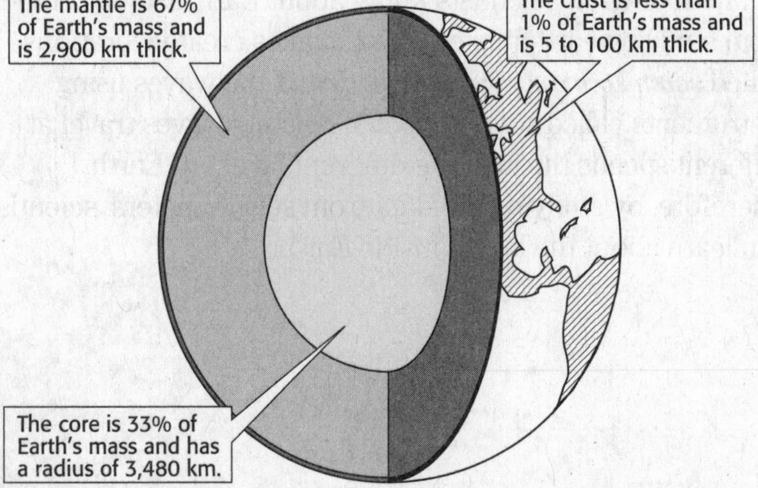
The mantle is 67% of Earth's mass and is 2,900 km thick.
The crust is less than 1% of Earth's mass and is 5 to 100 km thick.
The core is 33% of Earth's mass and has a radius of 3,480 km.

Math Focus

2. Calculate How many times thicker is Earth's mantle than the thickest part of Earth's crust?

____

Copyright © by Holt, Rinehart and Winston. All rights reserved.

EARTH'S PHYSICAL STRUCTURE

Scientists also divide Earth into five layers based on physical properties. The outer layer is the **lithosphere**. It is a cool, rigid layer that includes all of the crust and a small part of the upper mantle. The lithosphere is divided into pieces called *tectonic plates*. These plates move slowly over Earth's surface. ☑

The **asthenosphere** is found beneath the lithosphere. It is a layer of hot, solid rock that flows very slowly, like chewing gum. Beneath the asthenosphere is the *mesosphere*, which is the lower part of the mantle. The mesosphere flows more slowly than the asthenosphere.

There are two physical layers in Earth's core. The outer layer is the *outer core*. It is made of liquid iron and nickel. At the center of Earth is the *inner core*, which is a ball of solid iron and nickel. The inner core is solid because it is under very high pressure.

READING CHECK

3. Define What is the lithosphere?

____

____

____

____

CALIFORNIA STANDARDS CHECK

6.1.b Students know Earth is composed of several layers; a cold, brittle lithosphere; a hot, convecting mantle; and a dense, metallic core.

Word Help: layer
a separate or distinct portion of matter that has thickness

Word Help: core
center

4. Describe What are the five layers of Earth, based on physical properties?

____

____

____

____

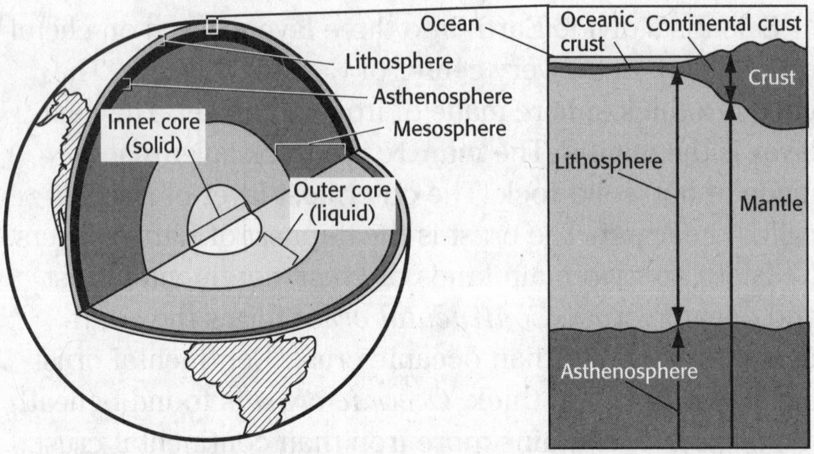
Ocean
Lithosphere
Asthenosphere
Inner core (solid)
Mesosphere
Outer core (liquid)
Oceanic crust
Continental crust
Crust
Lithosphere
Mantle
Asthenosphere

How Do Scientists Study the Inside of Earth?

Much of what scientists know about Earth's layers comes from studying earthquakes. Earthquakes create vibrations called *seismic waves*. Scientists detect the waves using instruments called *seismometers*. Seismic waves travel at different speeds through the different layers of Earth. Therefore, by studying the data from seismometers, scientists can learn about the layers inside Earth.

Copyright © by Holt, Rinehart and Winston. All rights reserved.

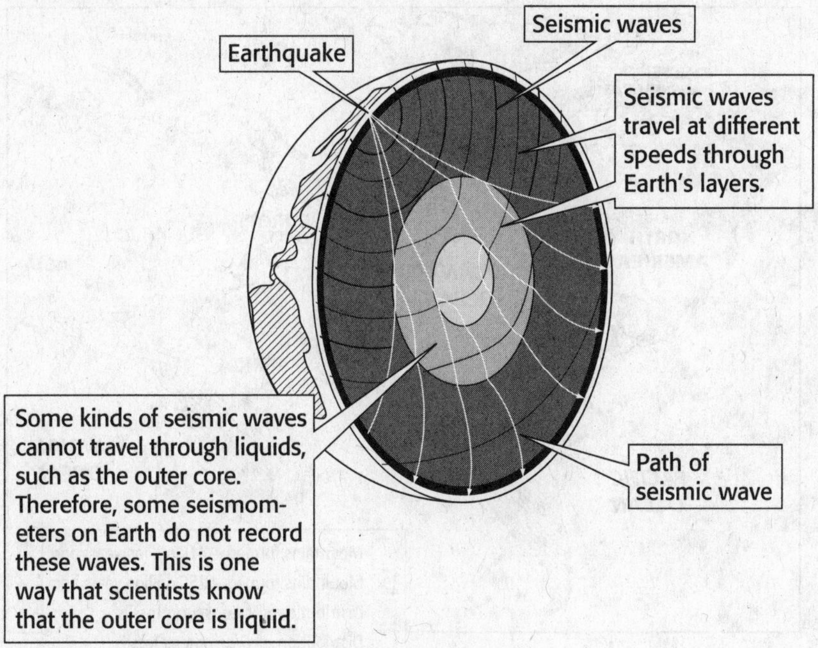
Seismic waves
Earthquake
Seismic waves travel at different speeds through Earth's layers.
Some kinds of seismic waves cannot travel through liquids, such as the outer core. Therefore, some seismometers on Earth do not record these waves. This is one way that scientists know that the outer core is liquid.
Path of seismic wave

TAKE A LOOK

5. Explain What is one way that scientists know the outer core is liquid?

What Is Continental Drift?

Remember that the lithosphere is broken into pieces called tectonic plates that move slowly over Earth's surface. The continents move over Earth's surface along with the plates. Sometimes, the continents are joined together. At other times, such as today, they are broken apart. This slow movement of continents over Earth's surface is called **continental drift**. ☑

READING CHECK

6. Define What is continental drift?

EVIDENCE FOR CONTINENTAL DRIFT

Look at the map of the continents on the next page. Can you see that South America and Africa seem to fit together, like the pieces of a puzzle? Continental drift can explain why this is so. South America and Africa were once part of a single continent. They have since broken apart and moved to their current locations.

Evidence for continental drift can also be found in fossils. For example, many of the same fossils have been found along the matching coastlines of South America and Africa. The organisms that formed these fossils could not have traveled across the Atlantic Ocean. Therefore, the two continents must once have been joined together.

Other evidence comes from similar rock types and mountain ranges on different continents. The locations of similar fossils and mountain ranges are shown on the map on the next page.

Copyright © by Holt, Rinehart and Winston. All rights reserved.

Name ____________ Class ____________ Date ____________

Critical Thinking

7. Infer Which continent was once joined with Greenland? How do you know?

____

____

____

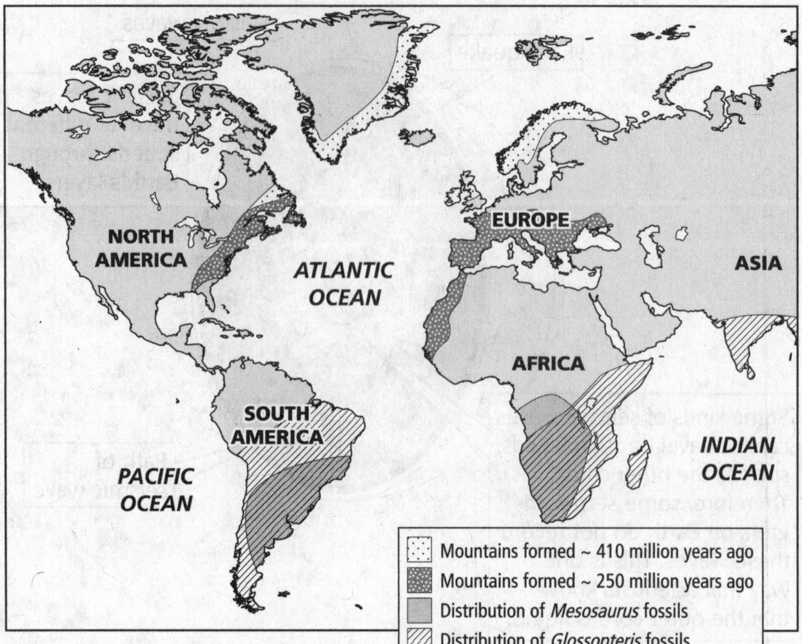
NORTH AMERICA
ATLANTIC OCEAN
EUROPE
ASIA
AFRICA
SOUTH AMERICA
PACIFIC OCEAN
INDIAN OCEAN
Mountains formed ~ 410 million years ago
Mountains formed ~ 250 million years ago
Distribution of *Mesosaurus* fossils
Distribution of *Glossopteris* fossils

Similar fossils and rocks are found on widely separated continents. For example, *Glossopteris* and *Mesosaurus* fossils are found in Africa and in South America. These fossils and rocks indicate that, at one time, all of the continents were joined together.

CALIFORNIA STANDARDS CHECK

6.1.a Students know evidence of plate tectonics is derived from the fit of the continents; the location of earthquakes, volcanoes, and midocean ridges; and the distribution of fossils, rock types, and ancient climatic zones.

Word Help: evidence
information showing whether an idea or belief is true or valid

Word Help: derive
to figure out by reasoning

Word Help: distribution
the relative arrangement of objects or organisms in time or space

8. Explain How do fossils indicate that the continents have moved with time?

____

____

____

____

BREAKUP OF PANGAEA

About 245 million years ago, all of the continents were joined into a single *supercontinent*. This supercontinent was called *Pangaea*. The word *Pangaea* means "all Earth" in Greek. About 200 million years ago, Pangaea began breaking apart. It first separated into two large landmasses called Laurasia and Gondwana. The continents continued to break apart and slowly move to where they are today.

As the continents moved, some of them collided. These collisions produced many of the landforms that we see today, such as mountain ranges and volcanoes.

Copyright © by Holt, Rinehart and Winston. All rights reserved.

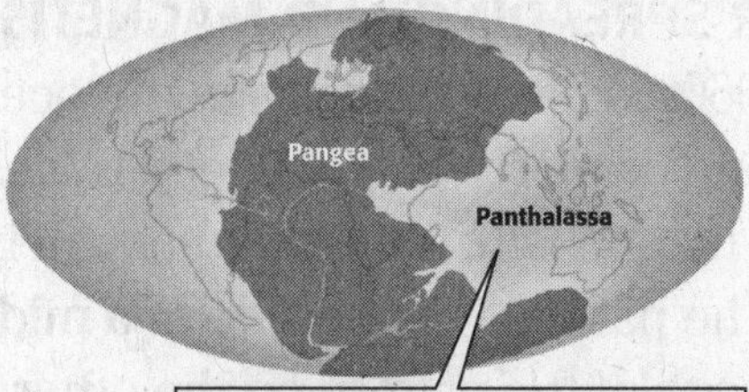
Pangea
Panthalassa

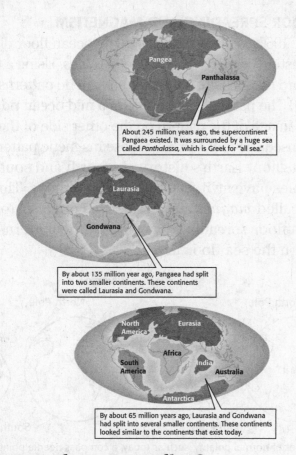

About 245 million years ago, the supercontinent Pangaea existed. It was surrounded by a huge sea called *Panthalassa*, which is Greek for "all sea."

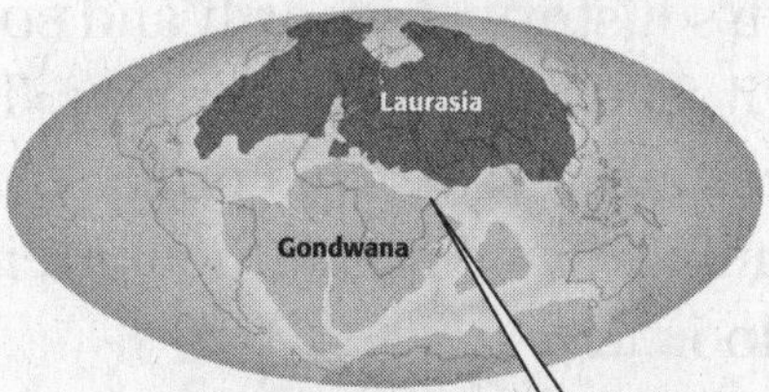
Laurasia
Gondwana

By about 135 million year ago, Pangaea had split into two smaller continents. These continents were called Laurasia and Gondwana.

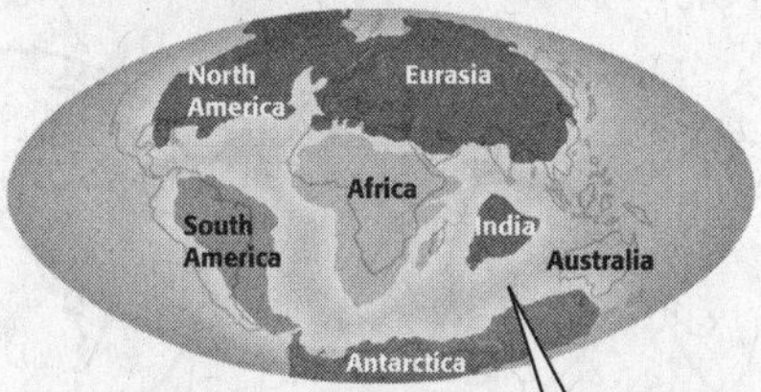
North America
Eurasia
Africa
South America
India
Australia
Antarctica

By about 65 million years ago, Laurasia and Gondwana had split into several smaller continents. These continents looked similar to the continents that exist today.

TAKE A LOOK

9. Describe How were the locations of the continents 65 million years ago different from the locations of the continents today? Give two ways.

____

____

____

____

What Is Sea-Floor Spreading?

Mid-ocean ridges are mountain systems found deep in Earth's oceans. They form a continuous chain that is 50,000 km long. The chain wraps around Earth like the seams of a baseball. Mid-ocean ridges are the sites of intense volcanic activity. ☑

At a mid-ocean ridge, melted rock rises through cracks in the sea floor. As the melted rock cools and hardens, it forms new crust. The newly formed crust pushes the older crust away from the mid-ocean ridge. This process is called **sea-floor spreading**.

READING CHECK

10. Define What is a mid-ocean ridge?

____

____

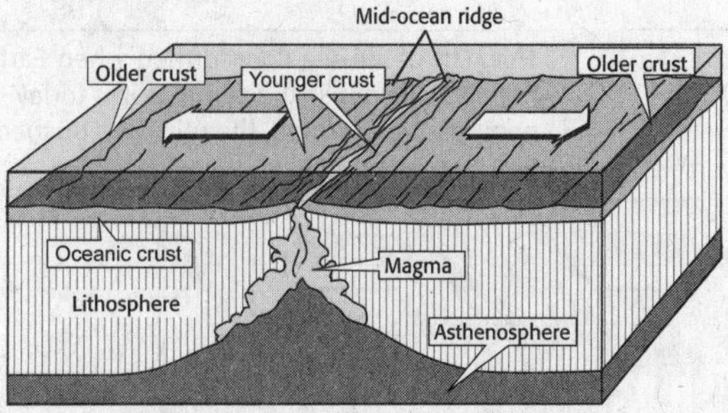
Mid-ocean ridge
Older crust
Younger crust
Older crust
Oceanic crust
Magma
Lithosphere
Asthenosphere

Sea-floor spreading produces new oceanic lithosphere at mid-ocean ridges. The oldest oceanic crust is found far from the ridges, and the youngest crust is found very close to the ridges.

TAKE A LOOK

11. Identify Does the oceanic lithosphere get older or younger as you move closer to the mid-ocean ridge?

____

Copyright © by Holt, Rinehart and Winston. All rights reserved.

SEA-FLOOR SPREADING AND MAGNETISM

In the 1960s, scientists studying the ocean floor discovered an interesting property of mid-ocean ridges. Using a tool that can record magnetism, they found magnetic patterns on the sea floor! The pattern on one side of a mid-ocean ridge was a mirror image of the pattern on the other side of the ridge. What caused the rocks to have these magnetic patterns?

Throughout Earth's history, the north and south magnetic poles have switched places many times. This process is called *magnetic reversal.* This process, together with sea-floor spreading, can explain the patterns of magnetism on the sea floor. ☑

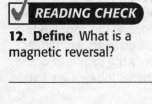

READING CHECK

12. Define What is a magnetic reversal?

____

____

____

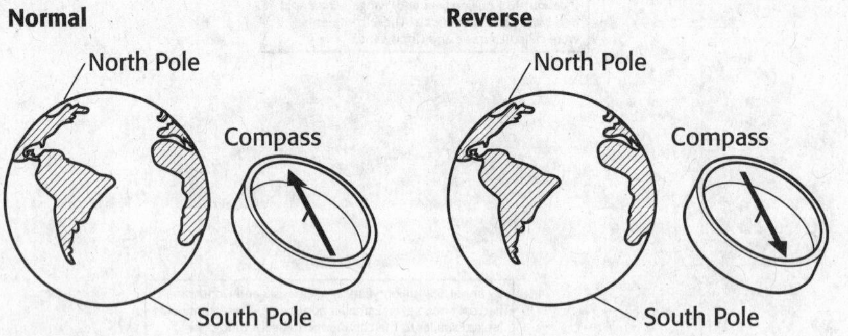
Normal
North Pole
Compass
South Pole
Reverse
North Pole
Compass
South Pole

During times of normal polarity, such as today, a compass needle points toward the North Pole. During times of reverse polarity, a compass needle points toward the South Pole.

As ocean crust forms from melted rock, magnetic minerals form. These minerals act as compasses. As they form, they line up with Earth's magnetic north pole. When the melted rock cools, the minerals are stuck in place.

After Earth's magnetic field reverses, these minerals point to Earth's magnetic south pole. However, new rock that forms will have minerals that point to the magnetic north pole. Therefore, the ocean floor contains "stripes" of rock whose magnetic minerals point to the north or south magnetic poles.

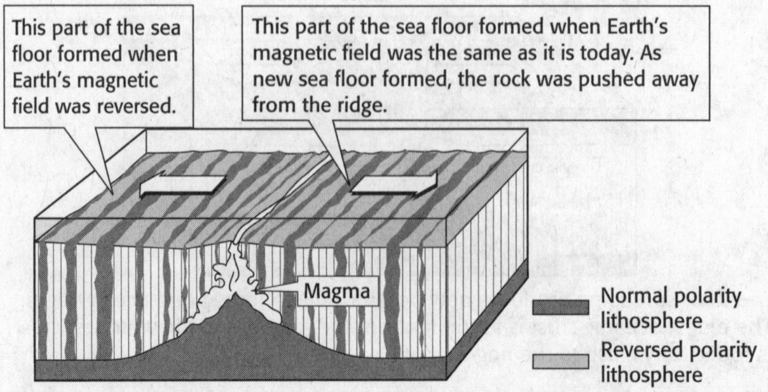
This part of the sea floor formed when Earth's magnetic field was reversed.
This part of the sea floor formed when Earth's magnetic field was the same as it is today. As new sea floor formed, the rock was pushed away from the ridge.
Magma
Normal polarity lithosphere
Reversed polarity lithosphere

TAKE A LOOK

13. Describe How are the "stripes" of magnetism on each side of the ridge related?

____

____

Copyright © by Holt, Rinehart and Winston. All rights reserved.

Name ______________________ Class ______________ Date ______________

Section 1 Review

6.1.a, 6.1.b

SECTION VOCABULARY

asthenosphere the soft layer of the mantle on which the tectonic plates move

continental drift the hypothesis that states that the continents once formed a single landmass, broke up, and drifted to their present locations

core the central part of the Earth below the mantle

crust the thin and solid outermost layer of the Earth above the mantle

lithosphere the solid, outer layer of Earth that consists of the crust and the rigid upper part of the mantle

Wordwise The root *lith* means "stone."

mantle the layer of rock between the Earth's crust and core

sea-floor spreading the process by which new oceanic lithosphere (sea floor) forms as magma rises to Earth's surface and solidifies at a mid-ocean ridge

1. Describe What are the features of each of the three compositional layers of Earth?

____

____

____

2. Identify Give three pieces of evidence that support the idea of continental drift.

____

____

____

3. Compare How is the crust different from the lithosphere? How are they the same?

____

____

____

____

4. Explain How do the parallel magnetic "stripes" near mid-ocean ridges form?

____

____

____

____

____

____

Copyright © by Holt, Rinehart and Winston. All rights reserved.

SECTION 2

The Theory of Plate Tectonics

California Science Standards

6.1.b, 6.1.c, 6.1.e, 6.4.c

BEFORE YOU READ

After you read this section, you should be able to answer these questions:

- What is the theory of plate tectonics?
- What are the three types of tectonic plate boundaries?

What Is the Theory of Plate Tectonics?

As scientists learned more about sea-floor spreading and magnetic reversals, they formed a theory to explain how continents move. This theory, known as plate tectonics, explains how many of the features on Earth's surface form.

The theory of **plate tectonics** states that Earth's lithosphere is broken into many pieces that move slowly over the asthenosphere. These pieces are called **tectonic plates**. There are 12 large plates and many smaller ones. ☑

Tectonic plates can contain different kinds of lithosphere. Some plates contain mostly oceanic lithosphere. Others contain mostly continental lithosphere. Some contain both continental and oceanic lithosphere. The figure below shows Earth's tectonic plates.

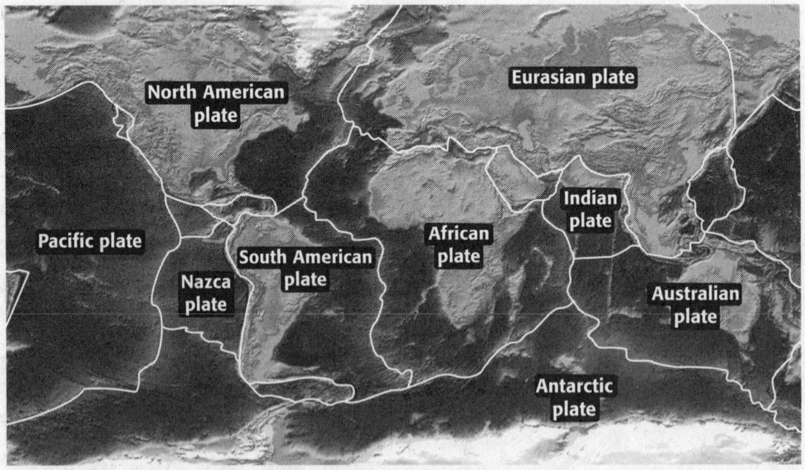
North American plate
Eurasian plate
Indian plate
Pacific plate
African plate
South American plate
Nazca plate
Australian plate
Antarctic plate

Tectonic plates move very slowly—only a few centimeters per year. Scientists can detect this motion only by using special equipment, such as global positioning systems (GPS). This equipment is sensitive enough to pick up even small changes in a continent's location.

STUDY TIP

Compare As you read, make a table showing the features of the three kinds of plate boundaries.

READING CHECK

1. Define What are tectonic plates?

____

____

____

TAKE A LOOK

2. Identify Give the name of one plate that contains mostly oceanic lithosphere and of one plate that contains mostly continental lithosphere.

Oceanic: ______________

Continental: ______________

Copyright © by Holt, Rinehart and Winston. All rights reserved.

STRUCTURE OF A TECTONIC PLATE

The tectonic plates that make up the lithosphere are like pieces of a giant jigsaw puzzle. The figure below shows what a single plate might look like it if were separated from the other plates. Notice that the plate contains both continental and oceanic crust. It also contains some mantle material.

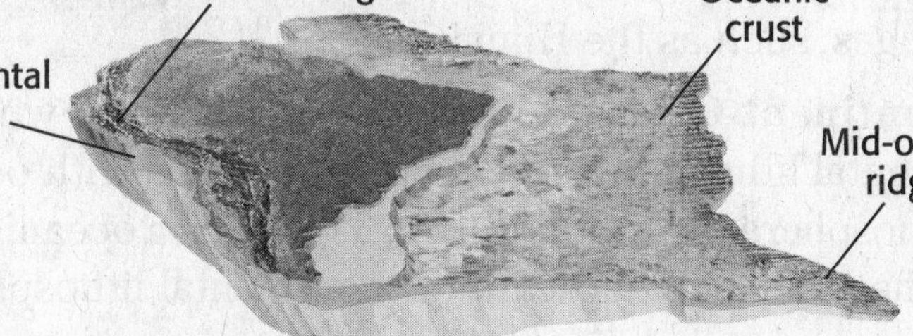
Andes mountain range
Oceanic crust
Continental crust
Mid-ocean ridge
Mantle

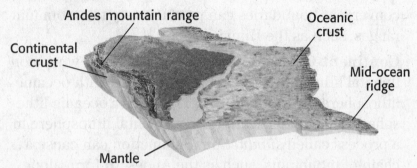

This figure shows what the South American plate might look like if it were lifted off the asthenosphere. Notice that the plate is thickest where it contains continental crust and thinnest where it contains oceanic crust.

TAKE A LOOK

3. Compare Which type of crust is thicker, oceanic crust or continental crust?

____

What Happens Where Tectonic Plates Touch?

The places where tectonic plates meet are called *boundaries*. Earthquakes and volcanoes are more common at tectonic plate boundaries than at other places on Earth. Other features, such as mid-ocean ridges and ocean trenches, are found only at plate boundaries.

There are three types of boundaries between plates: divergent boundaries, convergent boundaries, and transform boundaries. The kinds of features that are found at a plate boundary depend on what kind of plate boundary it is.

DIVERGENT BOUNDARIES

A *divergent boundary* forms where plates are moving apart. Most divergent boundaries are found beneath the oceans. Mid-ocean ridges form at these divergent boundaries. Because the plates are pulling away from each other, cracks form in the lithosphere. Melted rock can rise through these cracks. When the melted rock cools and hardens, it becomes new lithosphere. ☑

READING CHECK

4. Describe What features are found at most divergent boundaries?

____

Copyright © by Holt, Rinehart and Winston. All rights reserved.

CONVERGENT BOUNDARIES

A *convergent boundary* forms where plates are moving together. There are three different types of convergent boundaries:

- **Continent-Continent Boundaries** These form when continental lithosphere on one plate collides with continental lithosphere on another plate. Continent-continent convergent boundaries can produce very tall mountain ranges, such as the Himalayas.
- **Continent-Ocean Boundaries** These form when continental lithosphere on one plate collides with oceanic lithosphere on another plate. The denser oceanic lithosphere sinks underneath the continental lithosphere in a process called *subduction*. Subduction can cause a chain of mountains, such as the Andes, to form along the plate boundary.
- **Ocean-Ocean Boundaries** These form when oceanic lithosphere on one plate collides with oceanic lithosphere on another plate. One of the plates subducts beneath the other. A series of volcanic islands, called an *island arc*, can form along the plate boundary.

Critical Thinking

5. Infer Why do continent-continent convergent boundaries produce very tall mountain ranges?

____

____

____

____

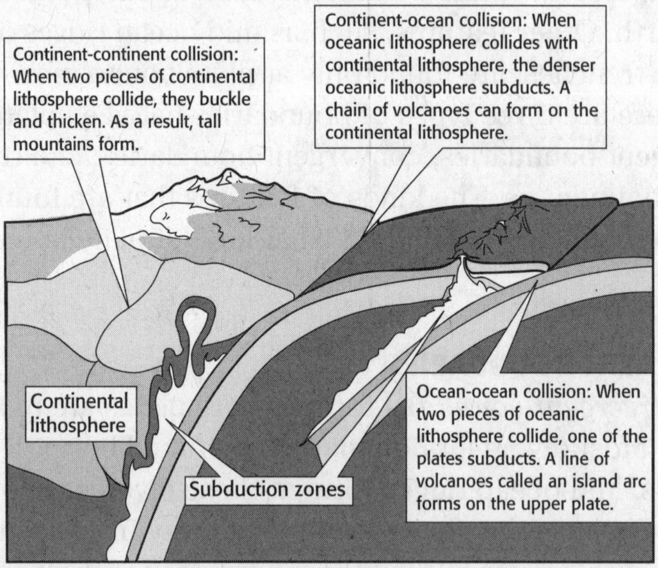
Continent-continent collision: When two pieces of continental lithosphere collide, they buckle and thicken. As a result, tall mountains form.
Continent-ocean collision: When oceanic lithosphere collides with continental lithosphere, the denser oceanic lithosphere subducts. A chain of volcanoes can form on the continental lithosphere.
Continental lithosphere
Subduction zones
Ocean-ocean collision: When two pieces of oceanic lithosphere collide, one of the plates subducts. A line of volcanoes called an island arc forms on the upper plate.

TAKE A LOOK

6. Identify At what two types of convergent boundaries do subduction zones form?

____

____

TRANSFORM BOUNDARIES

A *transform boundary* forms where plates slide past each other horizontally. Most transform boundaries are found near mid-ocean ridges. The ridges are broken into *segments*, or pieces. Transform boundaries separate the segments from one another.

Copyright © by Holt, Rinehart and Winston. All rights reserved.

Why Do Tectonic Plates Move?

Scientists do not know for sure what causes tectonic plates to move. However, they have three main hypotheses to explain plate movements: convection, slab pull, and ridge push.

Scientists used to think that convection in the mantle was the main force that caused plate motions. Remember that *convection* happens when matter carries heat from one place to another. Convection happens in the mantle as rock heats up and expands. As it expands, it becomes less dense and rises toward Earth's surface. ☑

As the hot material rises, cold, dense lithosphere sinks into the mantle at subduction zones. The rising hot material and the sinking cold material form *convection currents*. Until the 1990s, many scientists thought that these convection currents pulled the tectonic plates over Earth's surface.

Today, most scientists think that slab pull is the main force that causes plate motions. During subduction, oceanic lithosphere at the edge of a plate sinks into the mantle. The oceanic lithosphere sinks because it is colder and denser than the mantle. As the lithosphere at the edge of the plate sinks, it pulls the rest of the plate along with it. This process is called *slab pull*.

Another possible cause of plate motions is ridge push. At mid-ocean ridges, new oceanic lithosphere forms. This new lithosphere is warmer and less dense than the older lithosphere farther from the ridge. Therefore, it floats higher on the asthenosphere than the older lithosphere. As gravity pulls the new lithosphere down, the plate slides away from the mid-ocean ridge. This process is called *ridge push*.

Driving Force	Description
Slab pull	Cold, sinking lithosphere at the edges of a tectonic plate pulls the rest of the plate across Earth's surface.
Ridge push	Gravity pulls newly formed lithosphere downward and away from the mid-ocean ridge. The rest of the plate moves because of this force.
Convection currents	Convection currents are produced when hot material in the mantle rises toward the surface and colder material sinks. The currents pull the plates over Earth's surface.

READING CHECK

7. Define What is convection?

____

____

____

Critical Thinking

8. Compare How is slab pull different from ridge push?

____

____

____

____

____

Copyright © by Holt, Rinehart and Winston. All rights reserved.

Name ______________________ Class ______________ Date ______________

Section 2 Review

6.1.b, 6.1.c, 6.1.e, 6.4.c

SECTION VOCABULARY

plate tectonics the theory that explains how large pieces of the Earth's outermost layer, called tectonic plates, move and change shape	**tectonic plate** a block of lithosphere that consists of the crust and the rigid, outermost part of the mantle

1. Define Write your own definition for *tectonic plate.*

2. Identify What are the three main types of plate boundaries?

3. Describe How fast do tectonic plates move?

4. List Give three processes that may cause tectonic plates to move.

5. Explain Why does oceanic lithosphere sink beneath continental lithosphere at convergent boundaries?

6. Describe What is a transform boundary?

7. Identify Give two features that are found only at plate boundaries, and give two features that are found most commonly at plate boundaries.

Copyright © by Holt, Rinehart and Winston. All rights reserved.

SECTION 3

Deforming Earth's Crust

BEFORE YOU READ

After you read this section, you should be able to answer these questions:

- What happens when rock is placed under stress?
- What are three kinds of faults?
- How do mountains form?

California Science Standards

6.1.a, 6.1.d

What Is Deformation?

In the left-hand figure below, the girl is bending the spaghetti slowly and gently. The spaghetti bends, but it doesn't break. In the right-hand figure, the girl is bending the spaghetti quickly and with a lot of force. Some of the pieces of spaghetti have broken.

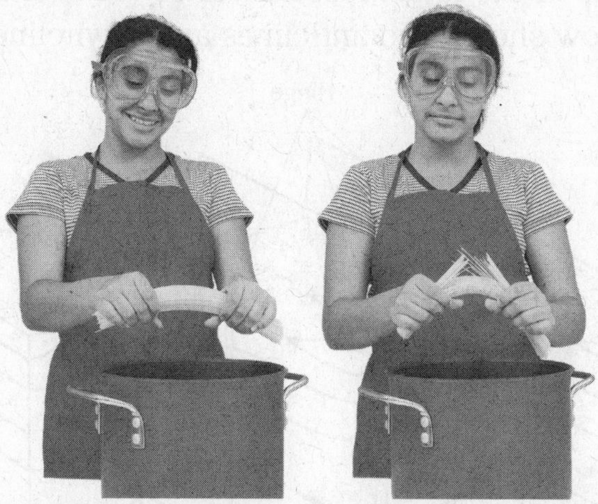

How can the same material bend in one situation but break in another? The answer is that the stress on the material is different in each case. *Stress* is the amount of force per unit area that is placed on an object. ☑

In the left-hand picture, the girl is slowly putting a small amount of stress on the spaghetti. The spaghetti bends under the small amount of stress. In the right-hand picture, the girl is quickly putting a lot of stress on the spaghetti. The spaghetti breaks under this large amount of stress.

Rocks can also bend or break under stress. When a rock is placed under stress, it *deforms*, or changes shape. When a small amount of stress is put on a rock slowly, the rock can bend. However, if the stress is very large or is applied quickly, the rock can break.

STUDY TIP

Learn New Words As you read, underline words that you don't understand. When you learn what they mean, write the words and their definitions in your notebook.

READING CHECK

1. Define What is stress?

____

____

Copyright © by Holt, Rinehart and Winston. All rights reserved.

What Happens When Rock Layers Bend?

Folding happens when rock layers bend under stress. Folding causes rock layers to look bent or buckled. The bends are called *folds*.

Scientists assume that all rock layers start out as horizontal layers. Therefore, when scientists see a fold, they know that deformation has happened. ☑

All folds have a hinge and two limbs. *Limbs* are the sides of a fold. The *hinge* is the place where the two limbs meet.

READING CHECK

2. Explain How do folds indicate that deformation has happened?

TYPES OF FOLDS

Two of the most common types of folds are synclines and anticlines. In a *syncline*, the oldest rocks are found on the outside of the fold. Most synclines are U-shaped. In an *anticline*, it is the youngest rocks that are found on the outside of the fold. Most anticlines are ∩-shaped. The figure below shows two anticlines and a syncline.

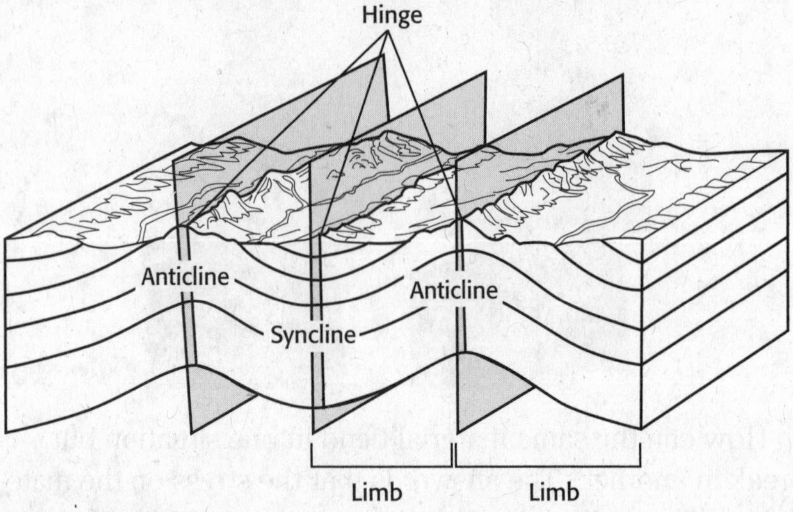
Hinge
Anticline
Anticline
Syncline
Limb
Limb

TAKE A LOOK

3. Color The rock layers in the picture are oldest at the bottom and youngest at the top. Color each rock layer with a different color to help you see where the oldest and youngest rocks are in anticlines and synclines.

SHAPES OF FOLDS

Folds can have many shapes. *Symmetrical* folds, like the ones in the figure above, have two limbs that are bent the same amount. *Asymmetrical* folds have limbs that are bent by different amounts. For example, an *overturned fold* has one limb that is tilted more than 90 degrees. In a *recumbent* fold, the limbs are bent so much that it looks like the fold is lying on its side.

Copyright © by Holt, Rinehart and Winston. All rights reserved.

What Happens When Rock Layers Break?

When rock is put under so much stress that it can no longer bend, it may break. The crack that forms when rocks break and move past each other is called a **fault**. The blocks of rock that are on either side of the fault are called *fault blocks*. When fault blocks move suddenly, they can cause earthquakes. ☑

When a fault forms at an angle, one fault block is called the *hanging wall*, and the other is called the *footwall*. The figure below shows the difference between the hanging wall and the footwall.

Scientists classify faults by how the fault blocks have moved along the fault. There are three main kinds of faults: normal faults, reverse faults, and strike-slip faults.

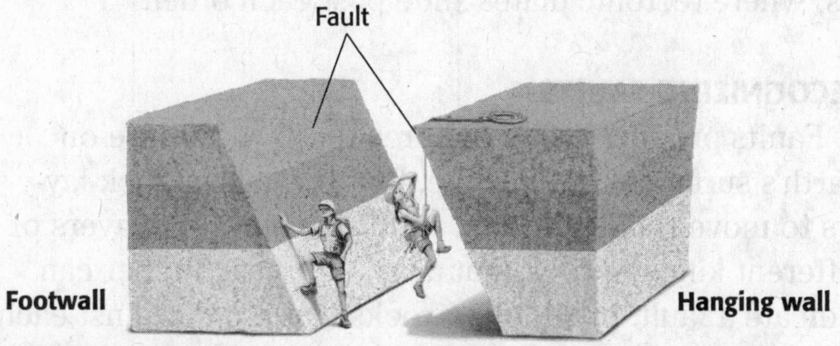
Fault
Footwall
Hanging wall

The footwall is the fault block that is below the fault. The hanging wall is the fault block that is above the fault.

NORMAL FAULTS

In a *normal fault*, the hanging wall moves down, or the footwall moves up, or both. Normal faults form where tectonic plate motions cause tension. *Tension* is stress that pulls rock apart. Therefore, normal faults are common along divergent boundaries, where Earth's crust stretches. The figure at the top of the next page shows a normal fault. ☑

REVERSE FAULTS

In a *reverse fault*, the hanging wall moves up, or the footwall moves down, or both. Reverse faults form where tectonic plate motions cause compression. *Compression* is stress that pushes rock together. Therefore, reverse faults are common along convergent boundaries, where two plates collide. The figure on the top of the next page shows a reverse fault.

✓ **READING CHECK**

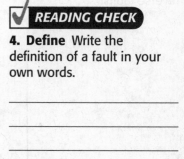

4. Define Write the definition of a fault in your own words.

____
____
____

TAKE A LOOK

5. Compare How is the hanging wall different from the footwall?

____
____

✓ **READING CHECK**

6. Explain Why are normal faults common along divergent boundaries?

____
____
____

Copyright © by Holt, Rinehart and Winston. All rights reserved.

TAKE A LOOK

7. Identify Label the hanging walls and the footwalls on the normal and reverse faults.

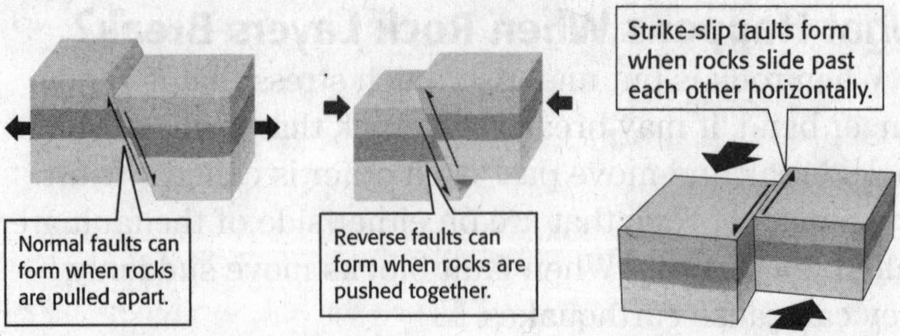

Normal faults can form when rocks are pulled apart.

Reverse faults can form when rocks are pushed together.

Strike-slip faults form when rocks slide past each other horizontally.

STRIKE-SLIP FAULTS

In a *strike-slip fault*, the fault blocks move past each other horizontally. Strike-slip faults form when rock is under shear stress. *Shear stress* is stress that pushes different parts of the rock in different directions. Therefore, strike-slip faults are common along transform boundaries, where tectonic plates slide past each other.

RECOGNIZING FAULTS

Faults produce many features that you can see on Earth's surface. For example, faulting causes rock layers to move relative to one another. Therefore, layers of different kinds of rock that are next to each other can indicate a fault. In addition, rocks can grind against each other when a fault forms. This grinding can produce grooves and scrapes on the surface of the rock. ☑

Some faults are many kilometers long. These faults can produce very large features on Earth's surface. For example, streams may flow in different directions when they cross a fault. Structures made by people, such as fences and roads, can also be disturbed by a fault.

Another feature that can indicate faulting is a scarp. A *scarp* is a row of cliffs that forms on a fault. Scarps form when one fault block is raised above the other fault block, as in a reverse fault. Some scarps are very small, but others may be tens of meters high.

READING CHECK

8. Describe Give two features that can indicate the presence of a fault.

____

____

____

____

Copyright © by Holt, Rinehart and Winston. All rights reserved.

How Do Mountains Form?

As tectonic plates move over Earth's surface, the edges of the plates grind against each other. This produces a lot of stress in Earth's lithosphere. Over very long periods of time, the movements of the plates can form mountains. Mountains can form in three main ways: through folding, faulting, and volcanism.

Critical Thinking

9. Apply Concepts Why does it take a very long time for most mountains to form?

____

____

____

____

____

FOLDED MOUNTAINS

Folded mountains form when rock layers are squeezed together and pushed upward. Folded mountains usually form at convergent boundaries, where continents collide. For example, the Appalachian Mountains formed hundreds of millions of years ago when North America collided with Europe.

FAULT-BLOCK MOUNTAINS

Fault-block mountains form when tension makes the lithosphere break into many normal faults. Along these faults, pieces of the lithosphere drop down compared with other pieces. This produces fault-block mountains. The Tetons are an example of a fault-block mountain range. ☑

READING CHECK

10. Identify What kind of stress forms fault-block mountains?

____

VOLCANIC MOUNTAINS

Volcanic mountains form when melted rock erupts onto Earth's surface. Most major volcanic mountains are found at convergent boundaries.

Volcanic mountains can form on land or on the ocean floor. Volcanoes on the ocean floor can grow so tall that they rise above the surface of the ocean. These volcanoes form islands, such as the Hawaiian Islands.

Most of Earth's active volcanoes are concentrated around the edge of the Pacific Ocean. This area is known as the *Ring of Fire*.

Type of Mountain	Description
Folded	
Fault-block	
Volcanic	

TAKE A LOOK

11. Describe Fill in the table with the features of each kind of mountain. Include where the mountains form and what they are made of.

Copyright © by Holt, Rinehart and Winston. All rights reserved.

Name ______ Class ______ Date ______

Section 3 Review

6.1.a, 6.1.d

SECTION VOCABULARY

fault a break in a body of rock along which one block slides relative to another	**folding** the bending of rock layers due to stress

1. **Compare** How are folding and faulting similar? How are they different?

2. **Describe** Fill in the spaces in the table to describe the three main kinds of faults.

Kind of fault	Description	Kind of stress that produces it
Normal		
	Hanging wall moves up; footwall moves down.	
		shear stress

3. **Identify** Name two kinds of asymmetrical folds.

4. **Explain** Why are strike-slip faults common at transform boundaries?

5. **Infer** Why are fault-block mountains probably uncommon at transform boundaries?

6. **Define** What is the Ring of Fire?

Copyright © by Holt, Rinehart and Winston. All rights reserved.

Name ______________________ Class ______________ Date ____________

CHAPTER 6 Plate Tectonics

SECTION 4

California Geology

BEFORE YOU READ

After you read this section, you should be able to answer these questions:

- How have plate movements affected California?
- What features in California are the result of plate movements?

California Science Standards

6.1.f

How Has Plate Tectonics Affected California?

Much of California's landscape is a result of plate tectonics. Compression along the plate boundaries has produced many of the rugged mountains in California. The steep, rocky coastlines in central and northern California are also a result of compression.

The main plate boundary in California, the San Andreas fault system, runs in a northwest-southeast direction. Many major river valleys and mountain ranges run parallel to this fault, as you can see in the figure below. The Transverse Ranges in southern California are an exception to this trend. The motion along the faults in southern California has caused these mountains to run in an east-west direction. ☑

STUDY TIP

Organize As you read this section, create a time line of the geologic events that are discussed.

READING CHECK

1. Describe In which direction do most of the geologic features in California run?

____

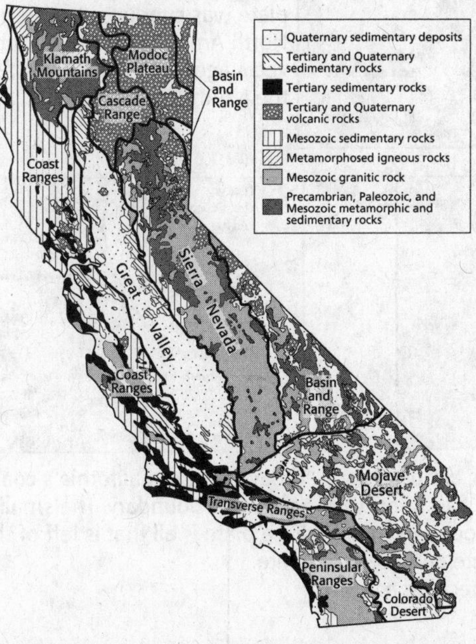

Klamath Mountains
Modoc Plateau
Basin and Range
Cascade Range
Coast Ranges
Great Valley
Sierra Nevada
Coast Ranges
Basin and Range
Mojave Desert
Transverse Ranges
Peninsular Ranges
Colorado Desert
Quaternary sedimentary deposits
Tertiary and Quaternary sedimentary rocks
Tertiary sedimentary rocks
Tertiary and Quaternary volcanic rocks
Mesozoic sedimentary rocks
Metamorphosed igneous rocks
Mesozoic granitic rock
Precambrian, Paleozoic, and Mesozoic metamorphic and sedimentary rocks

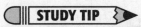

This geologic map of California shows many of the features of the California landscape. Notice that many of the features run in a northwest-southeast direction, parallel to the San Andreas fault system.

TAKE A LOOK

2. Identify What is the main kind of rock that is found in the Coast Ranges?

____

Copyright © by Holt, Rinehart and Winston. All rights reserved.

How Did California Form?

The area that we know as California has been an active plate boundary for more than 225 million years. Before then, the western coast of North America stopped at Nevada and the eastern deserts of California. Most of what is now California was then either part of a distant plate or did not yet exist.

When Pangaea began to break up, the North American plate moved west. The western edge of North America became an active convergent boundary. A long period of subduction began. This subduction lasted from about 160 million to 100 million years ago. It started to "build" the area that we know today as California. ☑

READING CHECK

3. Identify What kind of plate boundary started to form California when Pangaea began to break up?

____

ANCIENT PLATE BOUNDARIES

Three major tectonic plates have affected California's geology: the North American, the Farallon, and the Pacific plates. The interactions of these plates produced many of the features that are found in California today. The figures below show how these plates have interacted over time.

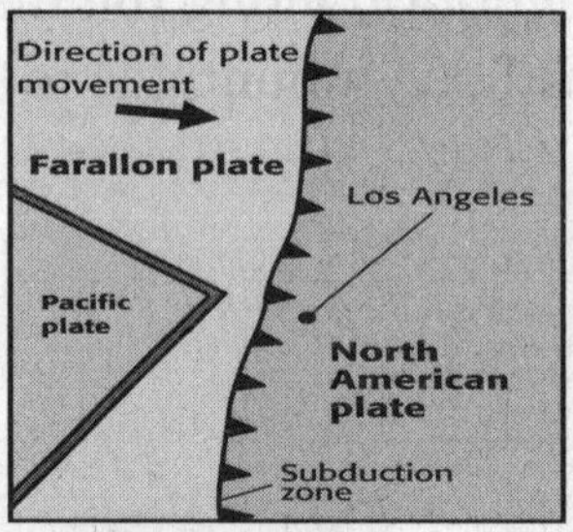
Direction of plate movement
Farallon plate
Los Angeles
Pacific plate
North American plate
Subduction zone

About 30 million years ago, the Farallon plate was subducting underneath the North American plate. About 25 million years ago, the Pacific plate first touched the North American plate.

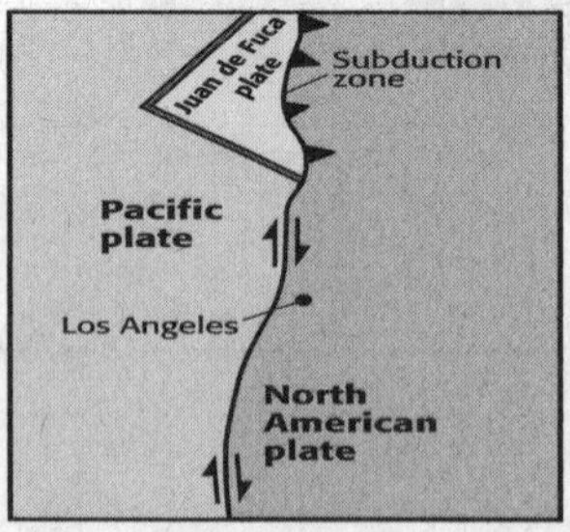
Juan de Fuca plate
Subduction zone
Pacific plate
Los Angeles
North American plate

By 10 million years ago, almost all of the Farallon plate had been subducted. The plate boundary between the Pacific plate and the North American plate became a transform boundary.

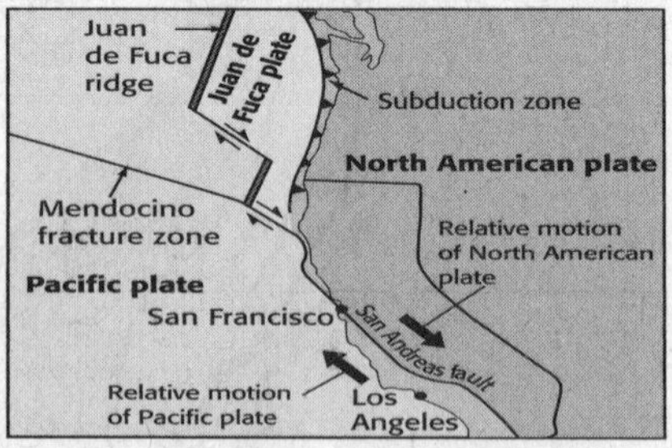
Juan de Fuca ridge
Juan de Fuca plate
Subduction zone
North American plate
Mendocino fracture zone
Relative motion of North American plate
Pacific plate
San Francisco
San Andreas fault
Relative motion of Pacific plate
Los Angeles

Today, most of California's coastline is a transform boundary. The small Juan de Fuca plate is all that is left of the Farallon plate.

TAKE A LOOK

4. Identify What is the only remaining piece of the Farallon plate?

____

Copyright © by Holt, Rinehart and Winston. All rights reserved.

How Did California's Volcanic Mountains Form?

As an oceanic plate subducts, it heats up and releases water into the mantle rock around it. The mantle rock begins to melt. The melted rock rises toward the surface, where it can erupt to form mountains. The subduction of the Farallon plate produced California's well-known volcanic mountain ranges, the Sierra Nevada and the Cascade Mountains.

Say It

Investigate and Share Learn more about how one of California's mountain ranges formed. Share what you learn with a small group or with your class.

SIERRA NEVADA MOUNTAINS

As the Farallon plate subducted beneath the North American plate, large amounts of melted rock were produced. Some of the melted rock rose to the surface and erupted to form volcanoes. However, most of the melted rock remained below Earth's surface. Between about 210 million and 100 million years ago, it cooled and hardened below the ground.

Math Focus

5. Calculate How long did it take for the melted rock to cool and harden?

____

As the melted rock cooled and hardened, it produced a large mass of granite below the ground. This mass of granite is called the *Sierra Nevada batholith*. A **batholith** is a large mass of igneous rock that forms underground.

The Sierra Nevada batholith formed hundreds of millions of years ago. Over time, the rock above the batholith was removed by weathering and erosion. Today, only the batholith remains. It is exposed at the surface in the Sierra Nevada mountain range.

The rocks in the Sierra Nevada mountains formed when melted rock cooled and hardened over hundreds of millions of years. This cooled rock makes up the Sierra Nevada batholith, which was once located far underground. Today, weathering and erosion have removed the overlying rock. Only the rock of the batholith is left.

TAKE A LOOK

6. Describe Why is the Sierra Nevada batholith found above the ground today, even though it formed far beneath the ground?

____

____

____

Copyright © by Holt, Rinehart and Winston. All rights reserved.

CALIFORNIA STANDARDS CHECK

6.1.f. Students know how to explain major features of California geology (including mountains, faults, volcanoes) in terms of plate tectonics.

Word Help: major
of great importance or large scale

Word Help: feature
the shape or form of a thing; characteristic

7. Identify Name four mountain ranges in California that formed as a result of plate tectonics.

____

____

____

____

CASCADE MOUNTAINS

Today, the Juan de Fuca plate is subducting beneath the North American plate. This convergent boundary stretches from northern California to British Columbia, Canada. It is known as the *Cascadia subduction zone.*

Many active volcanoes are found along the Cascadia subduction zone. These volcanoes form the Cascade Mountains. They include Mount St. Helens, in Washington, and Mount Shasta and Lassen Peak, in California.

How Did Other Mountains in California Form?

When a tectonic plate subducts, parts of the plate can be scraped off and become attached to the plate above it. This process is called *accretion.* Accretion can form mountain chains parallel to the plate boundary.

The chunks of Earth's crust that are added to the edge of the continent are called **accreted terranes**. Geologists can identify accreted terranes because the rocks in the terranes are different from the surrounding rocks. For example, rocks from an accreted terrane may be a different age. They may be made of rocks from deep in the ocean or even of rocks from the mantle. In addition, many accreted terranes are surrounded by faults.

The Coast and Transverse mountain ranges in California probably formed by accretion. The Central Valley, Los Angeles Basin, and Ventura Basin separate some of these mountain ranges.

GOLD ORES FROM ACCRETION

Most of the California's gold is found in the western foothills along the Sierra Nevada range. The rocks that contain the gold are thought to be accreted terranes that originally formed near volcanic vents in the ocean.

Type of mountain	How it forms	Examples in California
Volcanic	Magma forms at a convergent plate boundary. It erupts to form volcanoes or cools to form batholiths.	
Accreted terrane		Coast Ranges, Transverse Ranges

TAKE A LOOK

8. Describe Fill in the blank spaces in the table.

Copyright © by Holt, Rinehart and Winston. All rights reserved.

How Do Transform Faults Affect California?

The San Andreas fault system in California is the most famous transform boundary in the world. The San Andreas is the dividing line between the Pacific plate to the west and the North American plate to the east. Most of California is located on the North American plate. The San Andreas fault system is more than 1,000 km long. It includes many smaller faults that branch off or join the main fault. ☑

READING CHECK

9. Identify Which two plates are separated by the San Andreas fault system?

____

____

PLATE MOVEMENT ALONG THE FAULT SYSTEM

Scientists consider the San Andreas fault to be the boundary between the Pacific and North America plates. However, not all of the plate motion takes place on the San Andreas fault itself. For example, in southern California, most plate motion happens on other faults in the fault system. Therefore, it is better to think of the boundary between the Pacific and North American plates as a region, rather than as a single line.

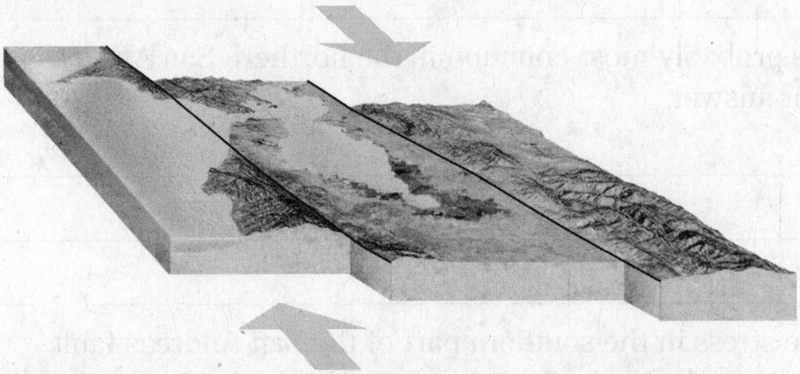

Movement in the San Andreas fault system takes place along several faults. Therefore, the boundary between the Pacific plate and the North American plate can be hard to define.

TAKE A LOOK

10. Explain Why can it be hard to define the exact boundary between the Pacific plate and the North American plate?

____

____

____

The Pacific and North American plates have been moving along the San Andreas fault system for about 25 million years. Over the last 16 million years, the total movement along the fault has been about 315 km.

In southern California, the San Andreas fault system bends to the east. Because of this bend, the Pacific and North American plates are colliding as they slide past each other. The collision produces a lot of compression. This compression causes the crust to be pushed up. The San Bernardino and San Gabriel mountains are some of the mountains that have been produced by this compression.

Copyright © by Holt, Rinehart and Winston. All rights reserved.

Name ______________________ Class ______________ Date ______________

Section 4 Review

6.1.f

SECTION VOCABULARY

accreted terrane a piece of lithosphere that becomes part of a larger landmass when tectonic plates collide at a convergent boundary	**batholith** a large mass of igneous rock in Earth's crust that, if exposed at the surface, covers an area of at least 100 km²

1. Define Write your own definition for *accreted terrane.*

____

____

____

2. Identify What kind of plate boundary is found along most of the coast of California?

____

3. Explain How are Lassen Peak and Mount Shasta related to subduction?

____

____

____

4. Infer What kind of fault is probably most common in the northern San Andreas fault system? Explain your answer.

____

____

____

5. Apply Concepts How is the stress in the southern part of the San Andreas fault system different from the stress in the northern part?

____

____

6. Describe How have the tectonic plate boundaries in California changed over the last 225 million years?

____

____

____

Copyright © by Holt, Rinehart and Winston. All rights reserved.

CHAPTER 7 Earthquakes

SECTION 1

What Are Earthquakes?

California Science Standards

6.1.a, 6.1.d, 6.1.e, 6.3.a

BEFORE YOU READ

After you read this section, you should be able to answer these questions:

- Where do most earthquakes happen?
- What makes an earthquake happen?
- What are seismic waves?

What Is an Earthquake?

Have you ever been in an earthquake? An **earthquake** is a movement or shaking of the ground. Earthquakes happen when huge pieces of the Earth's crust move suddenly and give off energy. This energy travels through the ground and makes it move.

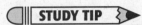

STUDY TIP

Outline As you read, create an outline of this section. Use the headings in the section to guide you.

Where Do Most Earthquakes Happen?

Most earthquakes happen at places where two tectonic plates touch. Tectonic plates are always moving. Where two plates touch, they may be moving toward each other, moving away from each other, or sliding past each other.

The movements of the plates can break the rocky crust in many places. A place where the crust is broken is called a *fault*. Earthquakes happen when rock breaks and slides along a fault. ☑

READING CHECK

1. Define What is a fault?

____

____

A place in the crust where a lot of faults are found is called a *fault zone*. Most fault zones, such as the San Andreas Fault Zone, are found at plate boundaries. But some fault zones are in the middle of plates.

Earthquakes and Plate Boundaries

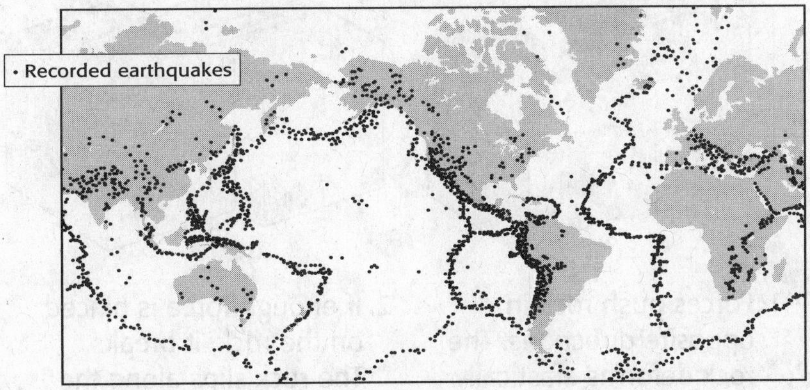
• Recorded earthquakes

TAKE A LOOK

2. Infer Use the earthquake locations to help you figure out where the tectonic plate boundaries are on the map. Use a colored pen or marker to draw tectonic plate boundaries on the map.

Copyright © by Holt, Rinehart and Winston. All rights reserved.

Why Do Earthquakes Happen?

As tectonic plates move, pressure builds up on the rock near the edges of the plates. When a rock is put under pressure, it deforms, or changes shape. Rock can deform in two main ways.

One kind of deformation is called plastic deformation. *Plastic deformation* happens when rock bends and folds like clay. When the pressure is taken away, the rock stays folded.

TAKE A LOOK

3. Explain How do you know that the rock layers in the figure were once under a lot of pressure?

____

____

Folded Layers of Rock

A second kind of deformation is called elastic deformation. *Elastic deformation* happens when rock bends a little bit under pressure but returns to its original size and shape when the pressure stops.

Earthquakes happen when rock breaks under pressure. When the rock breaks, it snaps back to almost the same size and shape as before the earthquake. This snap back is called **elastic rebound**. When the rock breaks and rebounds, it gives off energy. This energy causes the ground to shake.

CALIFORNIA STANDARDS CHECK

6.1.e Students know major geologic events, such as earthquakes, volcanic eruptions, and mountain building, result from plate motions.

Word Help: major
of great importance or large scale

4. Explain How does the movement of tectonic plates cause earthquakes?

____

____

____

____

____

Elastic Rebound

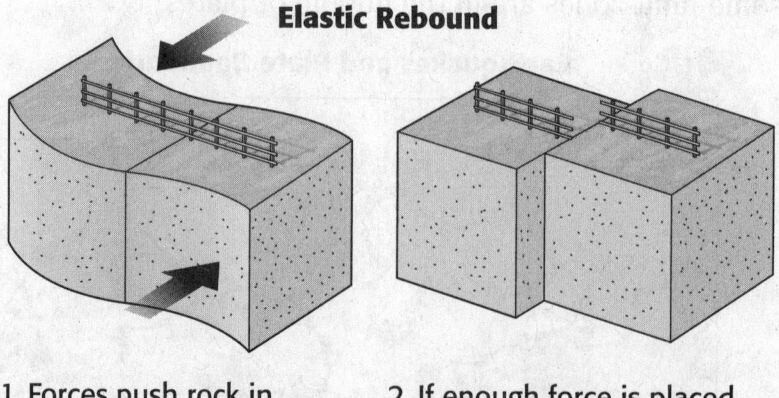

1. Forces push rock in opposite directions. The rock deforms elastically. It does not break.
2. If enough force is placed on the rock, it breaks. The rock slips along the fault. Energy is released.

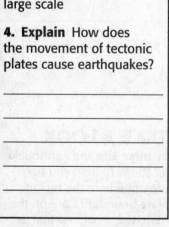

Copyright © by Holt, Rinehart and Winston. All rights reserved.

How Do Earthquakes Happen at Divergent Boundaries?

A *divergent boundary* is a place where two tectonic plates are moving away from each other. As the plates pull apart, the crust stretches. The crust breaks into pieces along faults. The pieces are called *fault blocks* because they are blocks of rock that move along faults. ☑

Most of the crust at divergent boundaries is thin, so the earthquakes tend to be shallow. Most earthquakes at divergent boundaries are no more than 20 km deep. ☑

READING CHECK

5. Define What is a fault block?

____

____

READING CHECK

6. Explain Why do most earthquakes at divergent boundaries happen at shallow depths?

____

____

Earthquakes at Divergent Boundaries

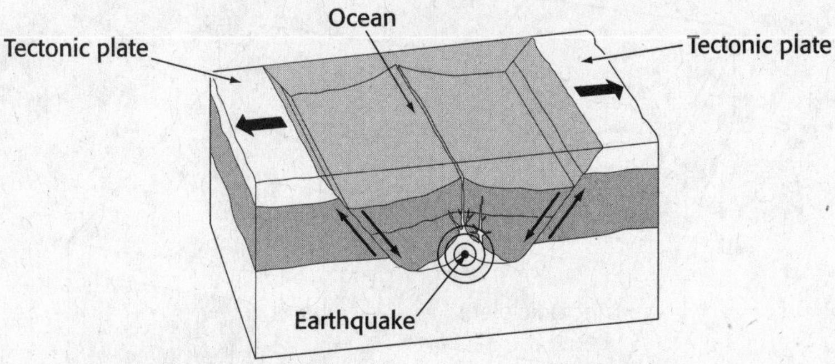
Ocean
Tectonic plate
Tectonic plate
Earthquake

How Do Earthquakes Happen at Convergent Boundaries?

A *convergent boundary* is a place where two tectonic plates collide. When two plates come together, both plates may crumple up to form mountains. Or one plate can *subduct*, or sink, underneath the other plate and into the mantle.

The earthquakes that happen at convergent boundaries can be very strong because so much pressure is produced. When one plate subducts under another, earthquakes can happen inside the sinking plate at depths of up to 700 km. ☑

READING CHECK

7. Explain Why are most earthquakes at convergent boundaries very strong?

____

____

Earthquakes at Convergent Boundaries

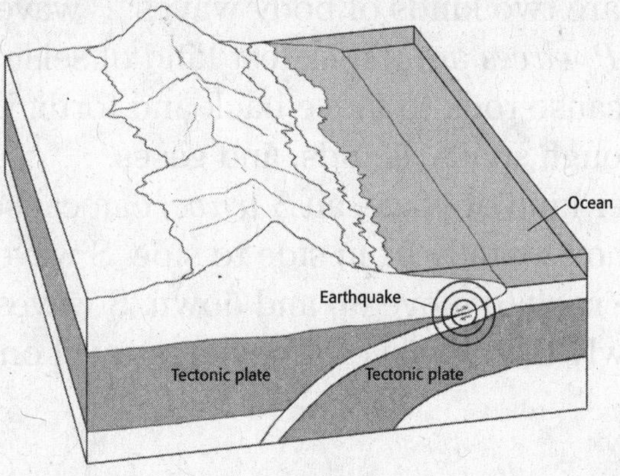
Ocean
Earthquake
Tectonic plate
Tectonic plate

TAKE A LOOK

8. Identify Draw arrows on the figure to show the directions in which the two tectonic plates are moving.

Copyright © by Holt, Rinehart and Winston. All rights reserved.

How Do Earthquakes Happen at Transform Boundaries?

A *transform boundary* is a place where two tectonic plates slide past each other. As the plates move past each other, pressure builds up on the rock. Eventually, the rock breaks along a fault.

The rock will break only if it is brittle. Rocks far below Earth's surface are generally not brittle. Therefore, most earthquakes at transform boundaries happen in the upper 50 km of the crust. ☑

READING CHECK

9. Identify How deep can earthquakes occur at transform boundaries?

____

Earthquakes at Transform Boundaries

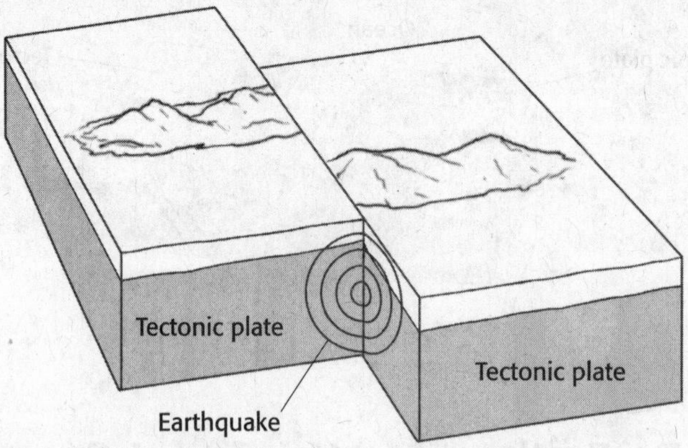
Tectonic plate
Tectonic plate
Earthquake

TAKE A LOOK

10. Identify Draw arrows showing the directions that the tectonic plates in the figure are moving.

How Does Earthquake Energy Travel?

When an earthquake occurs, a lot of energy is given off. This energy travels through the Earth in the form of waves called **seismic waves**.

There are two kinds of seismic waves. *Body waves* are seismic waves that travel through the inside of Earth to the surface. *Surface waves* are seismic waves that travel through the top part of Earth's crust. ☑

READING CHECK

11. List What are the two kinds of seismic waves?

____

____

BODY WAVES

There are two kinds of body waves: P waves and S waves. *P waves* are the fastest kind of seismic wave. P waves cause rock to move back and forth. P waves can move through solids, liquids, and gases.

Another kind of wave, an *S wave*, can cause the rock to move horizontally from side to side. S waves can also cause the rock to move up and down. S waves travel more slowly than P waves. They can travel only through solids.

Copyright © by Holt, Rinehart and Winston. All rights reserved.

SURFACE WAVES

Surface waves travel along the top of Earth's crust. Only the very top part of the crust moves when a surface wave passes. Surface waves travel much more slowly than body waves, but they cause a lot more damage. ☑

READING CHECK

12. Compare Which kind of seismic waves travel the most slowly?

____

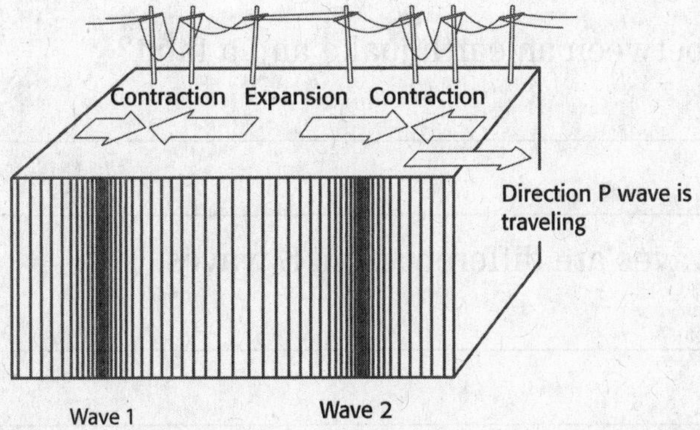
Contraction
Expansion
Contraction
Direction P wave is traveling
Wave 1
Wave 2

P waves are body waves that squeeze and stretch rock.

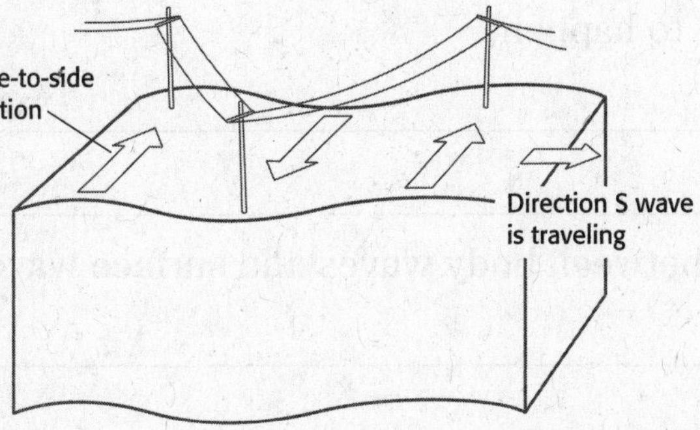
Side-to-side motion
Direction S wave is traveling

S waves are body waves that can move rock from side to side.

TAKE A LOOK

13. Compare How are the motions of P waves and S waves different?

____

____

____

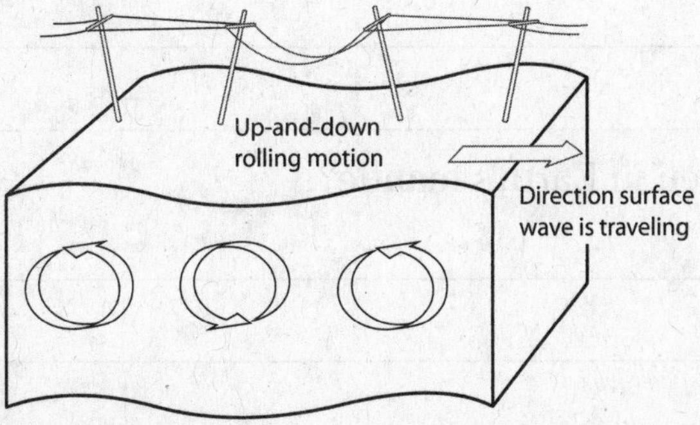
Up-and-down rolling motion
Direction surface wave is traveling

Surface waves can move the ground up and down in a circular motion.

Critical Thinking

14. Infer What do you think is the reason that surface waves usually cause the most damage?

____

____

____

____

Copyright © by Holt, Rinehart and Winston. All rights reserved.

Name ______________________ Class ______________ Date ______________

Section 1 Review

6.1.a, 6.1.d, 6.1.e, 6.3.a

SECTION VOCABULARY

earthquake a movement or trembling of the ground that is caused by a sudden release of energy when rocks along a fault move **elastic rebound** the sudden return of elastically deformed rock to its undeformed shape	**seismic wave** a wave of energy that travels through the Earth, away from an earthquake in all directions

1. Compare What is the difference between an earthquake and a fault?

____

____

2. Compare Give two ways that P waves are different from S waves.

____

____

3. Identify Where do most earthquakes happen?

____

4. Describe What causes an earthquake to happen?

____

____

5. Explain What is the main difference between body waves and surface waves?

____

____

6. Apply Concepts Why are some earthquakes stronger than others?

____

____

____

7. Infer Why do few earthquakes happen in Earth's mantle?

____

____

____

Copyright © by Holt, Rinehart and Winston. All rights reserved.

Name ______________________ Class ______________ Date ______________

CHAPTER 7 Earthquakes

SECTION 2

Earthquake Measurement

BEFORE YOU READ

After you read this section, you should be able to answer these questions:

- How do scientists know exactly where an earthquake happened?
- How are earthquakes measured?
- What affects the intensity of an earthquake?

California Science Standards

6.1.g

How Do Scientists Study Earthquakes?

Scientists who study earthquakes use an important tool called a seismometer. A *seismometer* records the vibrations that are caused by seismic waves. When the waves from an earthquake reach a seismometer, it records them as lines on a chart called a *seismogram*.

Remember that earthquakes happen when rock in Earth's crust breaks. The rock breaks in just one area, even though the earthquake can sometimes be felt many miles away.

The place inside the Earth where the rock first breaks is called the earthquake's **focus**. The place on Earth's surface that is right above the focus is called the **epicenter**. Scientists can use seismograms to find the epicenter of an earthquake. ☑

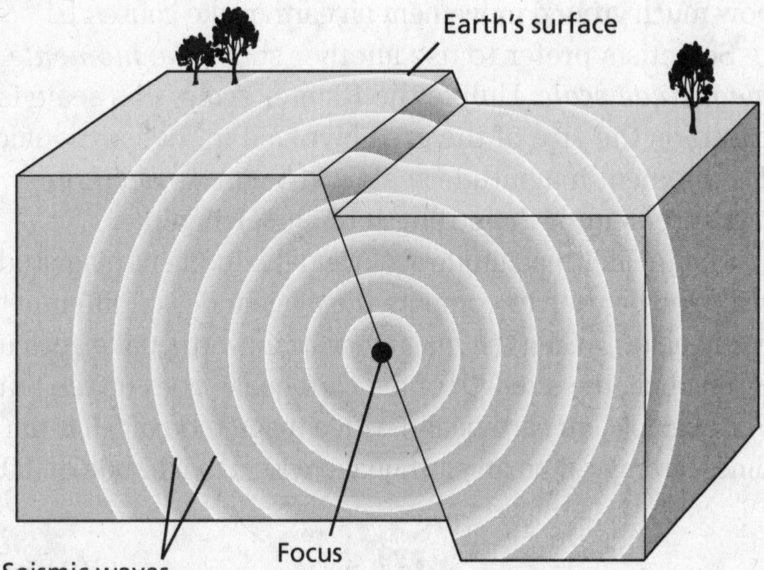

Earth's surface
Seismic waves
Focus

STUDY TIP

Ask Questions Read this section quietly to yourself. As you read, write down questions that you have. Discuss your questions in a small group.

READING CHECK

1. Explain What is the difference between the epicenter of an earthquake and the focus of the earthquake?

____

____

____

____

TAKE A LOOK

2. Identify On the figure, mark the epicenter of the earthquake with a star.

Copyright © by Holt, Rinehart and Winston. All rights reserved.

How Do Scientists Know Where an Earthquake Happened?

Scientists can use seismograms to figure out how far each seismometer is from the epicenter of an earthquake. Then they can find the epicenter by using the steps below.

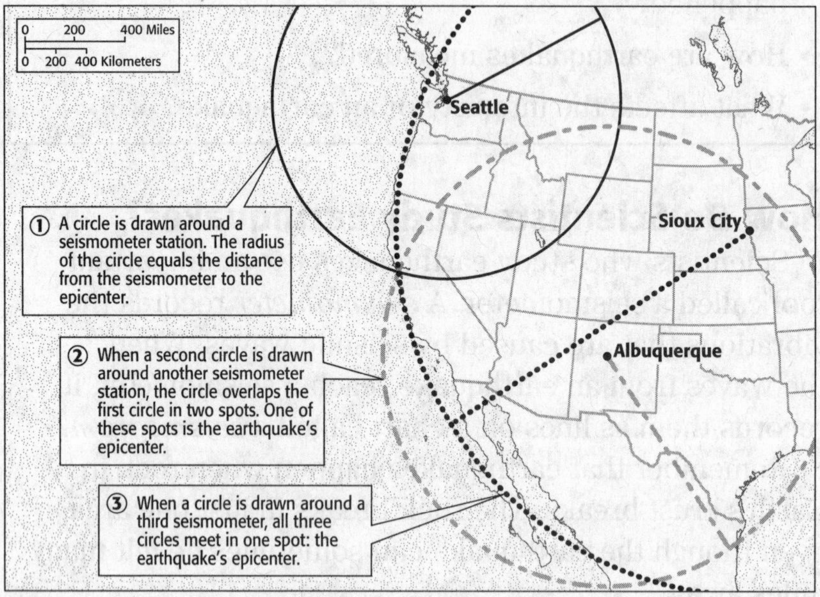
0 200 400 Miles
0 200 400 Kilometers
Seattle
Sioux City
Albuquerque
① A circle is drawn around a seismometer station. The radius of the circle equals the distance from the seismometer to the epicenter.
② When a second circle is drawn around another seismometer station, the circle overlaps the first circle in two spots. One of these spots is the earthquake's epicenter.
③ When a circle is drawn around a third seismometer, all three circles meet in one spot: the earthquake's epicenter.

TAKE A LOOK

3. Identify On the map, draw a star at the earthquake's epicenter.

What Is Earthquake Magnitude?

The strength of an earthquake is called its **magnitude**. Scientists use two different scales to measure magnitude. One scale is called the Richter magnitude scale. It measures how much ground movement an earthquake causes. ☑

Scientists prefer to use another scale, *the moment magnitude scale*. Unlike the Richter scale, this scale is based on the size of the area involved in the earthquake. The moment magnitude scale gives a more accurate measurement of the strength of an earthquake.

The Richter magnitude scale and the moment magnitude scale can be used to compare the magnitudes of different earthquakes. When the magnitude of an earthquake goes up by one unit, the strength of the earthquake goes up ten times. For example, an earthquake with a magnitude of 5.0 is ten times stronger than an earthquake with a magnitude of 4.0.

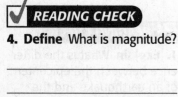

✓ **READING CHECK**

4. Define What is magnitude?

____

____

Math Focus

5. Multiply How much stronger is a magnitude 6.0 earthquake than a magnitude 4.0 earthquake? Show your work.

____

____

____

____

Copyright © by Holt, Rinehart and Winston. All rights reserved.

What Is Earthquake Intensity?

The **intensity** of an earthquake is a measure of how much damage it caused and how much it was felt by people. Scientists in the United States use the Modified Mercalli intensity scale to compare the intensities of different earthquakes. ☑

Intensity	Description of effects
I	generally is not felt by people
II	is felt by only a few people
III	is felt by most people indoors
IV	is felt by many people; causes dishes to rattle
V	is felt by nearly everyone; breaks or tips over some objects
VI	is felt by everyone; causes some damage to buildings
VII	causes some damage to many buildings
VIII	causes a lot of damage to ordinary buildings
IX	causes much damage to earthquake-resistant buildings
X	destroys many buildings; bends railroad tracks
XI	destroys almost all buildings; breaks bridges
XII	causes total destruction

✓ ***READING CHECK***

6. Define What is intensity?

____

____

____

The effects of an earthquake can be very different from place to place. One earthquake can have many different intensity values, even though it has only one magnitude. Many things can affect the intensity of an earthquake.

Factor	How it affects intensity
Size of the earthquake	Intensity is higher in larger earthquakes.
Distance from the epicenter	Areas that are closer to the epicenter have higher intensities.
Local geology	Areas with hard rock in the ground have lower intensities than areas on soft or wet soil.
Building features	Short buildings made of wood or steel are not damaged as much as taller buildings made of brick or concrete.

Critical Thinking

7. Explain Why is it that an earthquake can have many intensity values, even though it has only one magnitude?

____

____

____

____

____

____

____

____

INTENSITY MAPS

Scientists can measure the intensities of an earthquake in different areas. They can use the measurements to make an intensity map of an earthquake.

An intensity map shows which areas had the most damage from the earthquake. By looking at intensity maps for past earthquakes, scientists can figure out which areas are likely to be damaged in future earthquakes.

Copyright © by Holt, Rinehart and Winston. All rights reserved.

Name ________________ Class ________ Date ________

Section 2 Review

6.1.g

SECTION VOCABULARY

epicenter the point on Earth's surface directly above an earthquake's starting point, or focus **focus** the point along a fault at which the first motion of an earthquake occurs	**intensity** in Earth science, the amount of damage caused by an earthquake **magnitude** a measure of the strength of an earthquake

1. Explain What is the difference between a seismometer and a seismogram?

____

____

____

2. Compare Fill in the chart below to show the differences between an earthquake's magnitude and its intensity.

Feature	What does it measure?	What scale is used to measure it?
Magnitude		
Intensity		

3. Describe How is the focus of an earthquake related to the epicenter?

____

____

____

4. Infer An area that is far from the epicenter of an earthquake generally has a lower intensity than an area closer to the epicenter. Why is this so?

____

____

____

5. Apply Concepts Which city would most likely be the epicenter of an earthquake: San Francisco, California, or St. Paul, Minnesota? Explain your answer.

____

____

____

Copyright © by Holt, Rinehart and Winston. All rights reserved.

Name ______________________ Class ______________ Date ______________

CHAPTER 7 Earthquakes

SECTION 3

Earthquakes and Society

BEFORE YOU READ

After you read this section, you should be able to answer these questions:

- Can scientists predict when earthquakes will happen?
- Why do some buildings survive earthquakes better than others?
- How can you prepare for an earthquake?

California Science Standards

6.1.g, 6.2.d

What Is Earthquake Hazard?

Earthquake hazard is a measure of how likely it is that an area will have a damaging earthquake in the future. Scientists figure out the earthquake-hazard level by looking at the history of earthquakes in an area. Areas that have had many large earthquakes in the past have higher earthquake-hazard levels than areas without many earthquakes.

STUDY TIP

Concept Map As you read, underline the main ideas in this section. After you have read the section, create a Concept Map using your underlined ideas.

Earthquake Hazard Map of the Continental United States

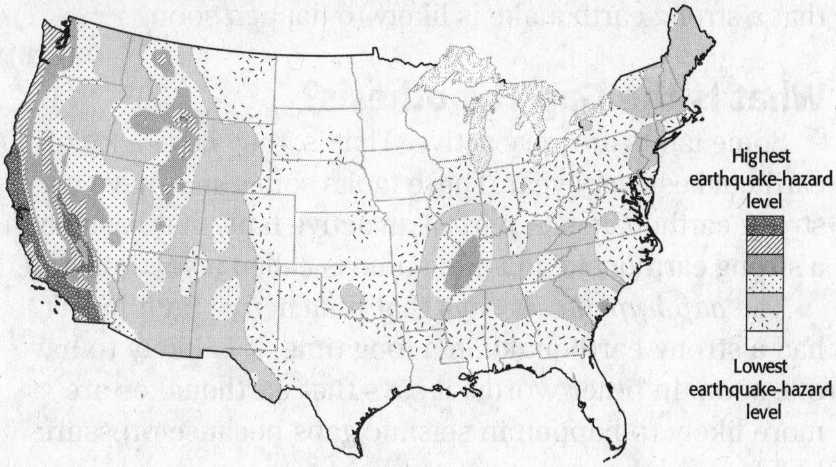

Highest
earthquake-hazard
level
Lowest
earthquake-hazard
level

TAKE A LOOK

1. Identify Which parts of California have the highest earthquake-hazard level?

____

____

Look at the map in the figure above. Notice that California has the highest earthquake-hazard level in the country. The San Andreas Fault Zone runs through most of California, and a lot of earthquakes happen there. As a result, California has a high earthquake-hazard level.

Copyright © by Holt, Rinehart and Winston. All rights reserved.

Can Scientists Predict Earthquakes?

Scientists can't predict earthquakes. But by looking at how often earthquakes have happened in the past, they can estimate where and when an earthquake will happen.

Look at the table below. It shows the number of earthquakes of different sizes that happen every year. There are a lot more weak earthquakes than strong earthquakes every year.

Kind of earthquake	Magnitude on the Richter scale	Average number every year
Great	8.0 and higher	1
Major	7.0 to 7.9	18
Strong	6.0 to 6.9	120
Moderate	5.0 to 5.9	800
Light	4.0 to 4.9	6,200
Minor	3.0 to 3.9	49,000
Very minor	2.0 to 2.9	365,000

Math Focus

2. Calculate About how many times more light earthquakes than strong earthquakes happen every year?

____

Scientists can guess when an earthquake will happen by looking at how many earthquakes have happened in the past. For example, if only a few strong earthquakes have happened recently in a fault zone, scientists can predict that a strong earthquake is likely to happen soon.

What Is the Gap Hypothesis?

Some faults are very active. That is, they have a lot of earthquakes every year. These faults sometimes have very strong earthquakes. A part of an active fault that hasn't had a strong earthquake in a long time is called a **seismic gap**.

The *gap hypothesis* says that if an active fault hasn't had a strong earthquake in a long time, it is likely to have one soon. In other words, it says that earthquakes are more likely to happen in seismic gaps because pressure builds up in these places.

Critical Thinking

3. Apply Concepts What do you think is the reason that strong earthquakes are more likely to happen in seismic gaps?

____

____

____

____

____

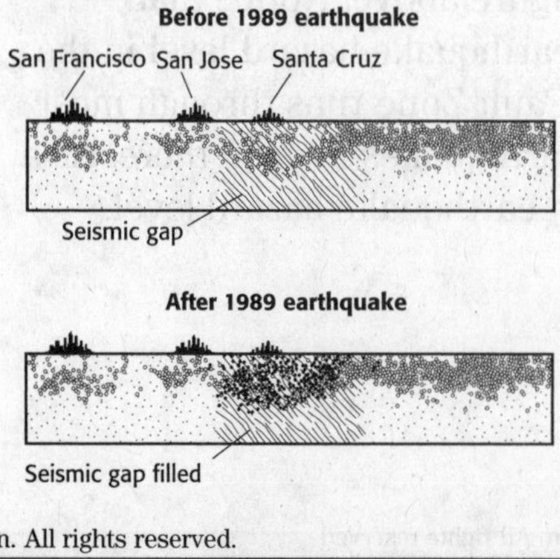
Before 1989 earthquake
San Francisco
San Jose
Santa Cruz
Seismic gap
After 1989 earthquake
Seismic gap filled

○ **Earthquakes before the 1989 earthquake** In 1988, seismologists found a seismic gap in the San Andreas Fault. The pictures show a side view of the fault and the locations of earthquakes on the fault.

• **1989 earthquake and aftershocks** Seismologists predicted that there was a 30% chance of a strong earthquake in the seismic gap within 30 years. In 1989, a magnitude 6.0 earthquake happened at Loma Prieta, California. Many seismologists think that the gap hypothesis helped to predict this earthquake.

Copyright © by Holt, Rinehart and Winston. All rights reserved.

How Do Earthquakes Affect Buildings?

Have you ever seen pictures of a city after a strong earthquake? You may have noticed that some buildings survive without very much damage. Other buildings, however, are totally destroyed.

Buildings that are damaged in an earthquake can be dangerous to the people in or near them. The damage can also be useful, though. Engineers can study it to learn how to make buildings that can better survive earthquakes.

Critical Thinking

4. List Give three factors that can affect how much a building will be damaged in an earthquake.

____

____

____

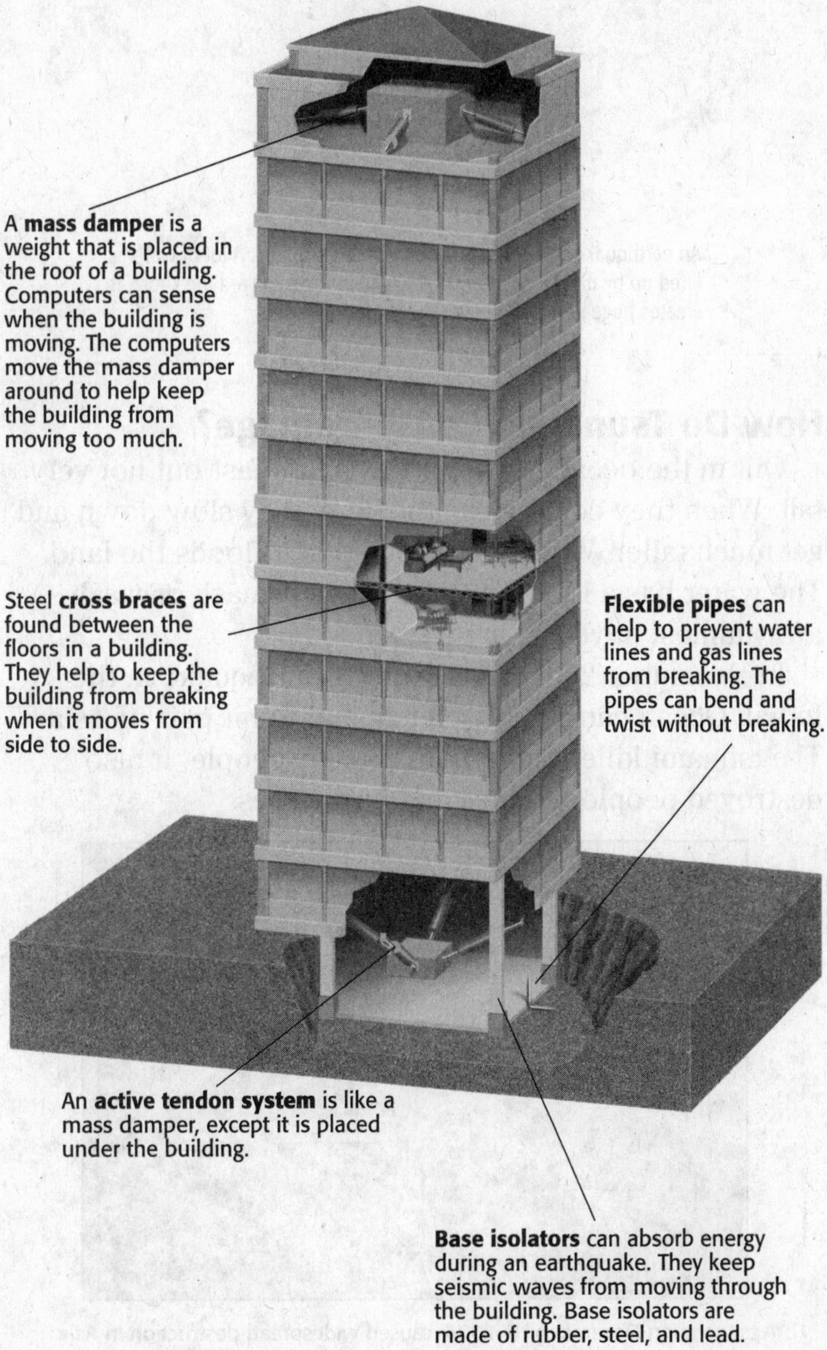
A **mass damper** is a weight that is placed in the roof of a building. Computers can sense when the building is moving. The computers move the mass damper around to help keep the building from moving too much.
Steel **cross braces** are found between the floors in a building. They help to keep the building from breaking when it moves from side to side.
Flexible pipes can help to prevent water lines and gas lines from breaking. The pipes can bend and twist without breaking.
An **active tendon system** is like a mass damper, except it is placed under the building.
Base isolators can absorb energy during an earthquake. They keep seismic waves from moving through the building. Base isolators are made of rubber, steel, and lead.

TAKE A LOOK

5. Compare How is a mass damper different from an active tendon system?

____

____

____

____

Copyright © by Holt, Rinehart and Winston. All rights reserved.

What Are Tsunamis?

Earthquakes can affect water as well as land. A **tsunami** is a series of very long water waves. Tsunamis are often called "tidal waves." Tsunamis can form when a strong earthquake happens on the sea floor. They can travel up to 800 km/h, from one side of the ocean to the other. ☑

READING CHECK

6. Identify What is one thing that can cause a tsunami?

____

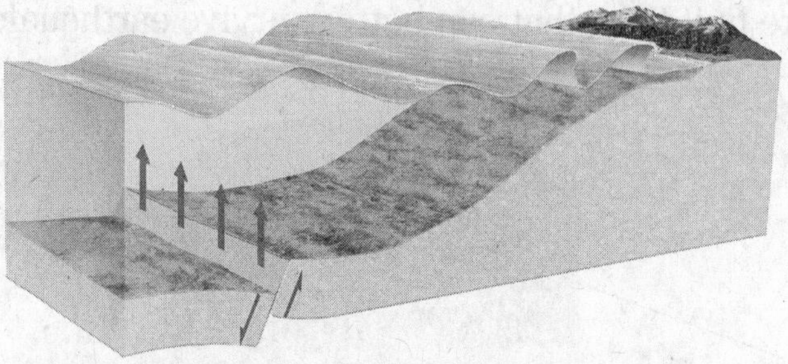

An earthquake on the ocean floor causes a tsunami. A lot of water is lifted up by the earthquake. As the water moves toward the shore, it creates huge waves that flood the land.

TAKE A LOOK

7. Compare What happens to the size of the waves as a tsunami moves closer to shore?

____

How Do Tsunamis Cause Damage?

Out in the ocean, tsunami waves are fast but not very tall. When they come close to shore, they slow down and get much taller. When a tsunami hits, it floods the land. The water has a lot of energy and can smash or wash away almost anything in its path.

In December 2004, a very strong earthquake in the Indian Ocean caused a huge tsunami to hit part of Asia. The tsunami killed more than 280,000 people. It also destroyed people's homes and businesses.

CALIFORNIA STANDARDS CHECK

6.2.d Students know earthquakes, volcanic eruptions, landslides, and floods change human and wildlife habitats.

8. Infer How can a tsunami change wildlife habitats?

____

____

____

A tsunami on December 26, 2004, caused widespread destruction in Asia.

Copyright © by Holt, Rinehart and Winston. All rights reserved.

How Can You Prepare for an Earthquake?

If you live in a place with a high earthquake-hazard level, you and your family should have an earthquake plan. If you plan ahead and practice your plan, you will be better prepared for an earthquake. ☑

READING CHECK

9. Explain Why is it important to make and practice an earthquake plan?

How Can You Make an Earthquake Plan?

YOUR SAFE HOME

Put heavy things near the floor so that they do not fall during an earthquake. Keep electrical wires and other things that can start a fire away from things that can burn.

SAFE PLACES IN YOUR HOME

Make sure you know a safe place in each room in your home. Safe places are areas that are far from windows and heavy objects that can fall or break. ☑

READING CHECK

10. Identify Think of your home. Write down a safe place in your home where you can go during an earthquake.

A PLAN TO MEET OTHERS

Talk to your family, friends, or neighbors and set up a place where you all will meet after an earthquake. If you all know where to meet each other, it will be easy to make sure that everyone is safe.

AN EARTHQUAKE KIT

Your earthquake kit should have things that you may need after an earthquake. Remember that your electricity may not work after an earthquake. It may be difficult to find clean water to drink.

What Should Be in an Earthquake Kit?
Water
A fire extinguisher
A small radio with batteries
Medicines
Food that won't go bad
A flashlight with batteries
Extra batteries
A first-aid kit

TAKE A LOOK

11. List List four foods that are good to put in an earthquake kit.

Copyright © by Holt, Rinehart and Winston. All rights reserved.

What Should You Do During an Earthquake?

If you are inside when an earthquake happens, crouch or lie face down under a table or a desk. Make sure you are far from any windows or heavy objects that can break or fall. Cover your head with your hands. ☑

READING CHECK

12. List Look around your classroom. List two safe places that you can go in case of an earthquake.

____

____

If you are outside during an earthquake, lie face down on the ground. Make sure you are far from buildings, power lines, and trees. Cover your head with your hands.

If you are in a car or bus, you should ask the driver to stop. Everyone should stay inside the car or bus until the earthquake is over.

If you are near the shore when an earthquake happens, you should try to get to high ground in case there is a tsunami. If there is no high ground, try to get to the top of a building or tree.

Say It

Share Experiences Have you ever been in an earthquake or heard about an earthquake that happened somewhere else? In a small group, talk about what it was like.

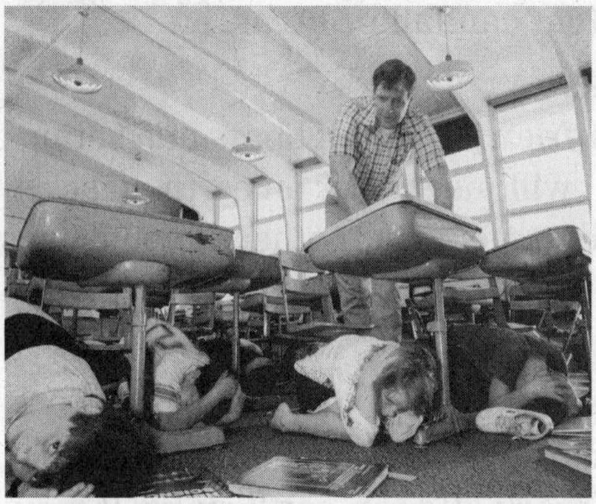

If you are inside during an earthquake, you should crouch under a table or a desk.

What Should You Do After an Earthquake?

Being in an earthquake can be scary. After an earthquake happens, people are often confused about what happened. They may not know what to do or where to go. If you can stay calm, you can help to keep yourself and others safe.

After an earthquake, look around you. If you are near something dangerous, like a power line or broken glass, get away as quickly as you can. Never go into a building after an earthquake until a parent, teacher, police officer, or firefighter tells you it is safe. ☑

READING CHECK

13. Identify Who should you ask if you want to know whether it is safe to go back into a building after an earthquake?

____

____

____

Always remember that there could be aftershocks. Aftershocks are weak earthquakes that sometimes happen after a large earthquake. Even though they are weaker than the main earthquake, they can still be strong and damaging.

Copyright © by Holt, Rinehart and Winston. All rights reserved.

Name ______________________ Class ______________ Date ______________

Section 3 Review

6.1.g, 6.2.d

SECTION VOCABULARY

seismic gap an area along a fault where relatively few earthquakes have occurred recently but where strong earthquakes have occurred in the past	**tsunami** a large ocean wave caused by an earthquake or other disturbance such as a volcanic eruption or a meteorite impact

1. Identify What is the relationship between earthquakes and tsunamis?

____

____

2. Graphic Organizer Fill in the chart below to show what you should do during an earthquake.

If you are...	Then you should...
...inside a building,	
	...lie face down on the ground with your hands on your head, far from power lines or fire hazards.
...in a car or bus,	
...at the beach,	

3. Identify How do engineers know how to make a building more likely to survive an earthquake?

____

____

4. Identify Relationships What is the relationship between the strength of an earthquake and how often it occurs?

____

5. Infer In most cases, you should stay inside a car or a bus in an earthquake. When might it be better to leave a car or bus during an earthquake?

____

____

____

Copyright © by Holt, Rinehart and Winston. All rights reserved.

Name ______________________ Class ______________ Date ______________

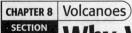

CHAPTER 8 Volcanoes

SECTION 1
Why Volcanoes Form

California Science Standards

6.1.d, 6.1.e

BEFORE YOU READ

After you read this section, you should be able to answer these questions:

- How do volcanoes form?
- What causes magma to form?
- Where do most volcanoes form?

What Is a Volcano?

STUDY TIP

Summarize in Pairs Read this section quietly to yourself. With a partner, take turns explaining what you learned. Work together to figure out things that you or your partner didn't understand.

When you think of a volcano, what comes into your mind? Most people think of a steep mountain with smoke coming out. In fact, a **volcano** is any place where gas, ash, and melted rock come out of the ground. A volcano can be a tall mountain, or it can be a small hole in the ground.

How Do Volcanoes Form?

At many tectonic plate boundaries, the rock that is just below the surface melts. The melted rock, or **magma**, is less dense than the solid rock, so it rises toward the surface. When the magma gets to Earth's surface, it erupts and forms a volcano. Whether or not a rock melts depends on three things: temperature, pressure, and fluid. ☑

READING CHECK

1. List What three things control whether or not a rock will melt?

____

TEMPERATURE

If a rock gets hot enough, parts of it will melt. On Earth's surface, few places are hot enough for rocks to melt. Deep underground, temperatures can be very high. In many cases, however, heat alone is not enough to make a rock melt. In order for the rock to melt, the pressure on it must drop, or water or other fluids must be added to it.

PRESSURE

READING CHECK

2. Explain Why does pressure have to be low in order for magma to form?

____

____

____

Magma takes up more space than solid rock. This means that rock must be able to expand in order to melt. In many places deep underground, hot rock cannot melt because it is under too much pressure. Near Earth's surface, however, the pressure is low because not much rock is pushing down from above. As solid rock rises to the surface, the decreasing pressure allows the rock to melt. ☑

Copyright © by Holt, Rinehart and Winston. All rights reserved.

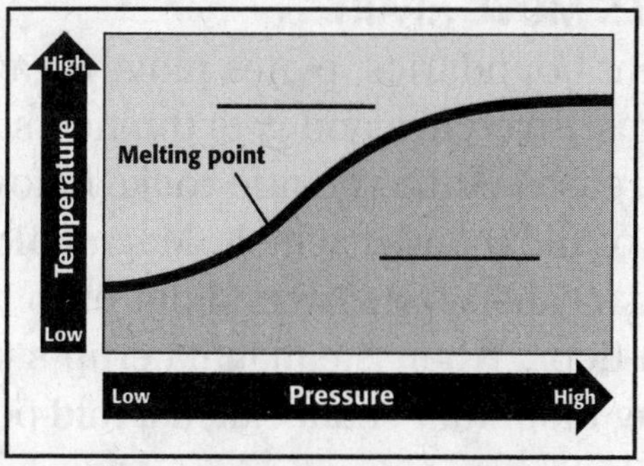
High
Temperature
Low
Melting point
Low
Pressure
High

TAKE A LOOK
3. Identify The graph shows what happens to rock as pressure and temperature change. Fill in the blank spaces on the graph with the following words: melted rock and solid rock.

FLUIDS

Rock that has fluids like water in it can melt more easily than dry rock. Therefore, when fluids are added to rocks that are already very hot, the rocks can melt. The illustration below shows how this can happen.

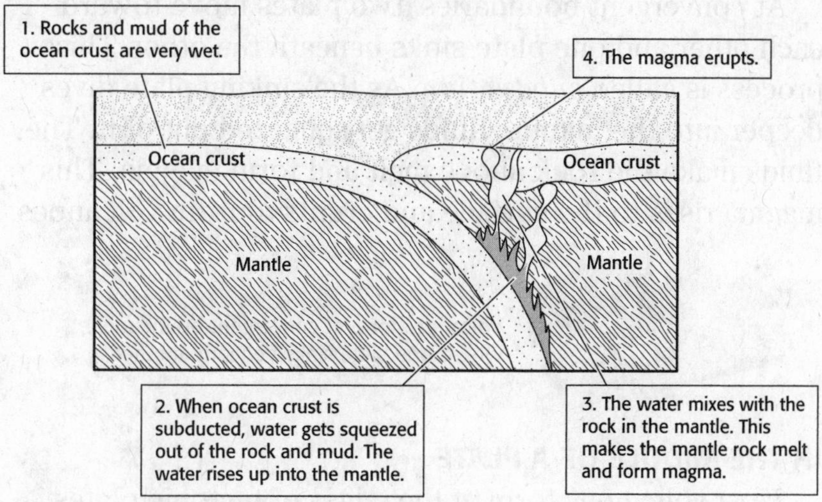
1. Rocks and mud of the ocean crust are very wet.
4. The magma erupts.
Ocean crust
Ocean crust
Mantle
Mantle
2. When ocean crust is subducted, water gets squeezed out of the rock and mud. The water rises up into the mantle.
3. The water mixes with the rock in the mantle. This makes the mantle rock melt and form magma.

TAKE A LOOK
4. Identify Draw a volcano in the proper place on the diagram.

Where Do Volcanoes Form?

Look at the map below. It shows the locations of many of the volcanoes in the world. Can you see that most volcanoes are found in chains? The chains of volcanoes seem to break up the Earth's crust into pieces. The pieces are *tectonic plates*. Most volcanoes form at tectonic plate boundaries. ☑

READING CHECK
5. Identify Where do most volcanoes form?

____

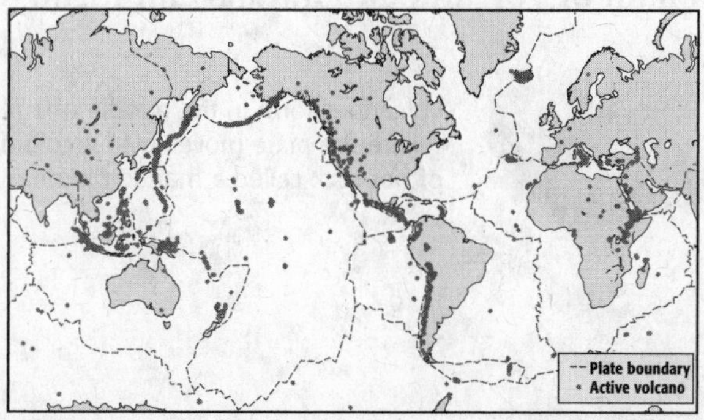
Plate boundary
Active volcano

Copyright © by Holt, Rinehart and Winston. All rights reserved.

WHERE PLATES MOVE APART

At divergent boundaries, plates move away from each other. The crust stretches and gets thinner, so the pressure decreases on the mantle rocks below. This causes parts of the mantle to melt. Magma flows up through cracks called *fissures* and out onto the land or on the ocean floor. When the magma erupts underwater, it forms a long mountain chain called a mid-ocean ridge. ☑

READING CHECK

6. Explain What causes the rocks in the mantle at divergent boundaries to melt?

____

____

Volcanoes form at divergent boundaries, where tectonic plates pull apart.

WHERE PLATES COME TOGETHER

At convergent boundaries, two plates move toward each other and one plate sinks beneath the other. This process is called *subduction*. As the sinking plate dives deeper into the mantle, fluids get squeezed out of it. The fluids make the rock above melt and form magma. This magma rises to the surface and erupts to form volcanoes.

TAKE A LOOK

7. Explain Draw arrows on the divergent boundary and convergent boundary to show which ways the plates are moving.

Volcanoes form at convergent boundaries, where one plate sinks beneath another.

IN THE MIDDLE OF A PLATE

Most volcanoes form at the edges of tectonic plates, but some form in the middle. These volcanoes form where columns of hot rock called *mantle plumes* rise up through the mantle and crust. When the hot rock gets near the surface, it can melt and then erupt to form a volcano.

Areas on Earth's surface that have volcanoes like these are called *hot spots*. As the plate moves over a mantle plume, a chain of volcanic mountains can form.

Critical Thinking

8. Infer Hawaii is a chain of volcanoes in the middle of the Pacific Ocean. How did the Hawaiian island chain form?

____

____

____

____

____

Volcanoes form in the middle of a plate where the plate moves over a column of hot rock called a mantle plume.

Copyright © by Holt, Rinehart and Winston. All rights reserved.

Section 1 Review

6.1.d, 6.1.e

SECTION VOCABULARY

magma liquid rock produced under Earth's surface; igneous rocks can form when magma cools and solidifies	**volcano** a vent or fissure in the Earth's surface through which magma and gases are expelled

1. Explain How are volcanoes and magma related?

2. List Name three things that can cause rock to melt and form magma.

3. Describe What causes magma to form at a divergent plate boundary?

4. Describe What causes magma to form at a convergent plate boundary?

5. Compare How are hot spot volcanoes different from volcanoes in other places?

6. Apply Concepts Why do most volcanoes form at plate boundaries?

7. Explain Give one reason that a very hot rock might not melt.

Copyright © by Holt, Rinehart and Winston. All rights reserved.

Name ______________________ Class ______________ Date ______________

CHAPTER 8 Volcanoes

SECTION 2

Types of Volcanoes

California Science Standards

6.1.d, 6.1.e

BEFORE YOU READ

After you read this section, you should be able to answer these questions:

- Why are there different kinds of volcanoes?
- What are the parts of a volcano?
- How do types of lava differ?

STUDY TIP

Color When you come to an illustration, use crayons or colored pencils to color it in. Use a different color for each part of the drawing. Include a legend.

Why Are Volcanoes Different from Each Other?

You have probably noticed that not all volcanoes look the same. Some are tall and steep. Others are low and wide. The type of volcano that forms depends on the type of lava it is made of.

Lava is magma that has reached Earth's surface. Lava at different places on the Earth is made up of different materials. In some places, lava is thin and runny, like water. In other places, lava is thick and stiff. Because there are different types of lava, there are different types of volcanoes.

CALIFORNIA STANDARDS CHECK

6.1.d Students know that earthquakes are sudden motions along breaks in the crust called faults and that volcanoes and fissures are locations where magma reaches the surface.

1. Describe Use the illustration to describe how magma inside a volcano becomes a lava flow outside.

____

____

____

____

____

____

____

____

What Are the Parts of a Volcano?

There are some features that most volcanoes have in common: a magma chamber, vents, and lava flows. In addition, most volcanoes experience earthquakes.

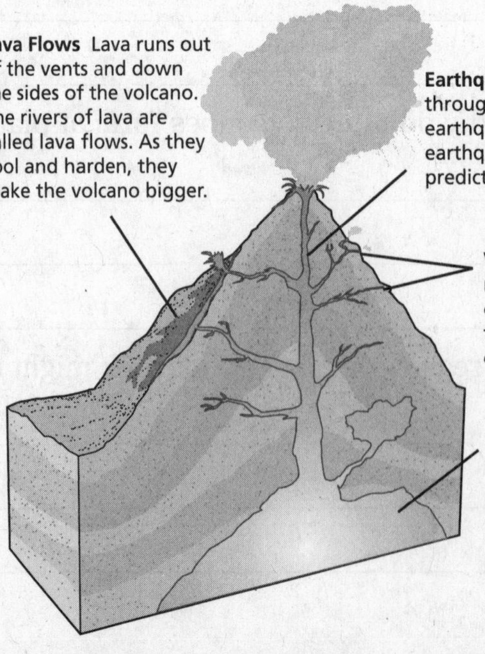

Lava Flows Lava runs out of the vents and down the sides of the volcano. The rivers of lava are called lava flows. As they cool and harden, they make the volcano bigger.
Earthquakes When magma rises through the crust, it can cause small earthquakes. By studying the earthquakes, scientists can sometimes predict when a volcano will erupt.
Vents When the magma reaches the surface, it erupts out of holes or openings in the volcano. These holes are called vents.
Magma Chamber Magma moves from the mantle into a pocket under the surface called a magma chamber. When the magma chamber is full, magma rises through the crust and erupts out of the volcano.

Copyright © by Holt, Rinehart and Winston. All rights reserved.

Name ______________________ Class ______________ Date ______________

What Kind of Lava Forms at Hot Spots?

Most of the magma that erupts at hot spots forms when parts of Earth's mantle melt. Magma that comes from the mantle is mafic. **Mafic** is a word that describes rock, magma, or lava that has a lot of iron and magnesium but not much silica in it. Most mafic rocks are dark in color. ☑

READING CHECK

2. Define What is mafic rock?

____

____

Aa is lava that forms a thick, brittle crust as it cools. The crust is torn into sharp pieces as lava moves underneath it.

Pahoehoe is lava that forms a thin, flexible crust as it cools. The crust wrinkles as the lava moves underneath it.

Blocky lava is cool, stiff lava that does not travel very far from the volcano. Blocky lava usually oozes from a volcano and forms piles of rocks with sharp edges.

Pillow lava is lava that erupts under water. As it cools, it forms rounded lumps that look like pillows. Pillow lava is common at hot spots and divergent boundaries.

What Kinds of Volcanoes Form at Hot Spots?

Because mafic lava is so runny, it can travel a long distance before it hardens. As a result, mafic volcanoes tend to be very wide and not very steep. Mafic volcanoes are rarely explosive. ☑

READING CHECK

3. Explain Why are mafic volcanoes so wide?

____

____

____

SHIELD VOLCANOES

If mafic lava erupts from the same place for thousands of years, lava flows build up. A wide, tall mountain called a *shield volcano* can form. From the side, a shield volcano looks like a knight's shield. Most hot-spot volcanoes are shields for two reasons: mafic lava is very runny, and the volcano erupts over and over, for a very long period of time.

Lava flows

Mauna Kea is a shield volcano in Hawaii. Its base is on the sea floor, but its top pokes above the ocean. If you measured the height of Mauna Kea from the sea floor all the way to the top of the volcano, it would be taller than Mount Everest!

Shield volcanoes form when many layers of mafic lava build up over time.

TAKE A LOOK

4. Identify On the illustration, draw and label arrows pointing to the magma chamber and the vents.

Copyright © by Holt, Rinehart and Winston. All rights reserved.

What Kinds of Volcanoes Form at Divergent Boundaries?

Like the lava at hot spots, most of the lava that erupts at divergent boundaries comes from the mantle. As the tectonic plates move apart, part of the mantle melts. Magma rises up and fills the space between the plates. Some of it erupts on the surface. Like hot-spot lava, the lava at divergent boundaries is mafic. ☑

READING CHECK

5. Compare How is lava at divergent boundaries similar to lava at hot spots?

____

____

MID-OCEAN RIDGE

Most divergent boundaries are on the ocean floor. When the mafic lava erupts, it forms a long chain of undersea volcanoes known as the *mid-ocean ridge.*

Although mafic lava is thin and runny, it does not travel very far underwater, because it hardens quickly. The reason mid-ocean ridges are wide and low is because the plates are moving apart. As soon as lava erupts and hardens, it begins moving away from the vent.

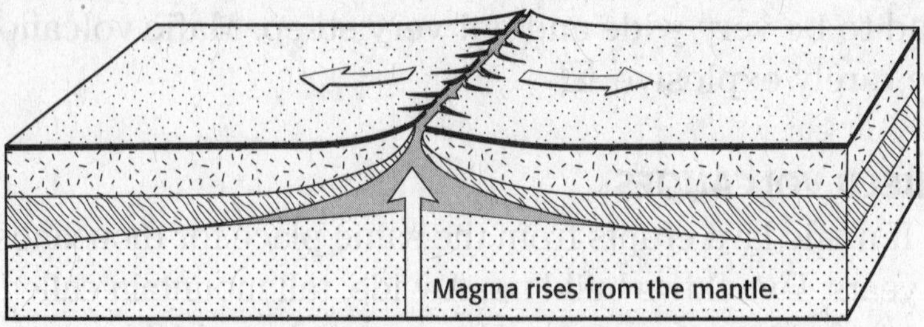
Magma rises from the mantle.

At mid-ocean ridges, magma moves upward from the mantle. The magma erupts as lava. It then cools and hardens to form new ocean floor. The plates keep moving apart and more lava erupts. This process is called sea-floor spreading.

ICELAND

Iceland is part of the mid-ocean ridge that runs down the center of the Atlantic Ocean. But it is also a hot spot. There is a mantle plume beneath the mid-ocean ridge under Iceland. Iceland is not deep underwater like the rest of the mid-ocean ridge because there are so many volcanic eruptions.

Critical Thinking

6. Infer Iceland has an area of 100,000 km^2. The Big Island of Hawaii is only 10,000 km^2. Explain why Iceland is so much bigger.

____

____

____

____

The Krafla volcano in Iceland has erupted 29 times in the last few thousand years.

Copyright © by Holt, Rinehart and Winston. All rights reserved.

What Kind of Lava Forms at Convergent Boundaries?

The lava that erupts at convergent boundaries comes from melted crust as well as melted mantle. Some of the lava is mafic, but some of it is felsic. **Felsic** is a word that describes rocks, magma, and lava that have a lot of silica but not very much iron and magnesium in them. Most felsic rocks are light in color. Felsic magma and lava are thick and sticky. ☑

READING CHECK

7. Compare Name one way that the lava at convergent boundaries is different from the lava at hot spots.

____

____

What Kinds of Volcanoes Form at Convergent Boundaries?

Have you ever shaken a can of soda and then opened it? The foamy soda probably exploded out of the can. Felsic magma does the same thing when it gets near Earth's surface.

The low pressure near the surface causes the gas and water trapped in the magma to form bubbles. The bubbles cannot get out because the magma is thick and sticky. When the pressure inside the bubbles gets too high, the magma explodes. ☑

READING CHECK

8. Explain Why does felsic magma explode when it gets near the surface?

____

____

____

____

When a volcano explodes, a lot of material is given off. Gas, ash, and even large chunks called *pyroclastic rocks* can be blown out of the volcano. Ash and dust can race down the side of a volcano like a river, forming a *pyroclastic flow*. Pyroclastic flows are very dangerous. They can be as hot as 700°C and can move at 200 km/h. A pyroclastic flow can destroy everything in its path.

Volcanic bombs are large blobs of lava that harden in the air.

Lapilli are small bits of lava that harden before they hit the ground. Lapilli are usually about the size of pebbles.

Volcanic ash forms when gases trapped in magma or lava form bubbles. When the bubbles explode, they create millions of tiny pieces.

Volcanic blocks are large pieces of solid rock that come out of a volcano.

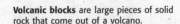

Math Focus

9. Multiply How fast can pyroclastic flows move in miles per hour?
1 km = 0.62 mi

____

Copyright © by Holt, Rinehart and Winston. All rights reserved.

Pyroclastic flows associated with the 1991 eruption of Mount Pinatubo in the Philippines had temperatures that reached 750°C.

TAKE A LOOK

10. Apply Concepts At what kind of tectonic plate boundary do you think the Philippines are located?

______

COMPOSITE VOLCANOES

Many of the volcanoes that form at convergent boundaries are composite volcanoes. *Composite volcanoes* are made of layers of different materials. Some layers are made of mafic lava flows. Other layers are made of ash, dust, and other pyroclastic materials. Composite volcanoes are also called *stratovolcanoes* because they are made of strata, or layers. ☑

Unlike shield volcanoes, most composite volcanoes have very steep sides. Composite volcanoes can erupt many times, but there may be hundreds of years between eruptions.

READING CHECK

11. Explain Why are composite volcanoes also called stratovolcanoes?

______

______

______

Mount Fuji is a composite volcano in Japan. It has erupted many times.

TAKE A LOOK

12. Explain How can you tell that Mount Fuji is a composite volcano rather than a shield volcano?

______

______

______

Copyright © by Holt, Rinehart and Winston. All rights reserved.

CINDER CONES

Sometimes lava and ash come out of small vents and spray out into the air. Pieces of lava harden in the air and then rain down onto the ground. These hardened pieces of lava are called *cinders*. The cinders and ash build up around the vent and form a small, steep volcano called a *cinder cone*.

Many cinder cones form at convergent boundaries, but they can form at hot spots also. In fact, some shield volcanoes have cinder cones on their sides. Unlike shield volcanoes and composite volcanoes, most cinder cones erupt only once. ☑

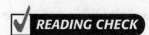

READING CHECK

13. Explain Why are cinder cones smaller than shield volcanoes?

____

____

____

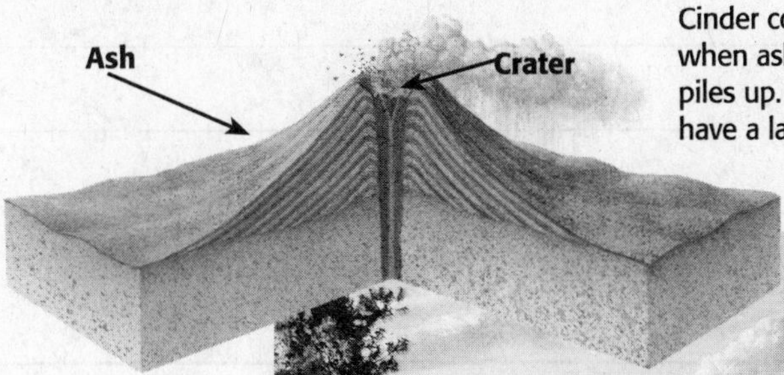
Ash
Crater

Cinder cone volcanoes form when ash from an eruption piles up. Most are small. Some have a large crater at the top.

Paracutín is a cinder cone volcano in Mexico. It formed in a farmer's cornfield in 1943. It erupted for nine years. At the end of all that time, it was only 400 m tall.

Copyright © by Holt, Rinehart and Winston. All rights reserved.

Name ______________________ Class ______________ Date ______________

Section 2 Review

6.1.d, 6.1.e

SECTION VOCABULARY

felsic describes magma, lava, or igneous rock that is rich in feldspars and silica and that is generally light in color

lava magma that flows onto Earth's surface; the rock that forms when lava cools and solidifies

mafic describes magma, lava, or igneous rock that is rich in magnesium and iron and that is generally dark in color

1. Compare What is the difference between lava and magma?

____

____

2. Explain Why are there different types of volcanoes?

____

____

____

3. Contrast Fill in the chart below to show the differences between mafic and felsic lava.

Type of lava	Where it forms	What it is like	How it erupts
mafic	hot spots and divergent boundaries		flows in rivers
felsic		thick and sticky	

4. Explain Why are shield volcanoes wider than composite volcanoes and not as steep?

____

____

5. Describe Relationships Describe one way that earthquakes and volcanoes are related.

____

____

____

Copyright © by Holt, Rinehart and Winston. All rights reserved.

Name ______________________ Class ______________ Date ______________

CHAPTER 8 Volcanoes

SECTION 3

Effects of Volcanic Eruptions

BEFORE YOU READ

After you read this section, you should be able to answer these questions:

- How do volcanic eruptions affect humans and wildlife?
- How can a volcanic eruption affect global temperature?
- What are the benefits of volcanic eruptions?

California Science Standards

6.2.d

What Are Some Harmful Effects of Volcanoes?

A volcanic eruption can hurt or kill humans and other living things. A volcano's blast can destroy buildings and trees. An eruption can also cause large changes in the environment.

Pyroclastic flows can race down the side of a volcano and burn or bury everything in their path. Snow on top of a tall volcano can melt and cause dangerous floods. Volcanic ash can mix with water and create fast-moving mudflows called **lahars**.

Ash from a volcano can make it hard for people and animals to breathe. It can also damage or ruin crops by covering their leaves. Ash and gas from a volcano can mix into the air and travel all around the world. The ash in the air can block sunlight and make the world cooler for several years.

STUDY TIP

List As you read, make a list of the effects of volcanic eruptions. Using a colored pencil, underline harmful effects in red and helpful effects in green.

TAKE A LOOK

1. Identify What is a likely harmful effect of the volcanic eruption shown in the picture?

____

____

Copyright © by Holt, Rinehart and Winston. All rights reserved.

What Are Some Helpful Effects of Volcanoes?

Although volcanoes can cause damage, they can also be very useful. For example, volcanic rocks, like other rocks, break down to form soil. The soil that forms from volcanic rocks contains nutrients that many plants need in order to grow. Many farmers rely on volcanic soil to grow enough food. ☑

READING CHECK

2. Explain Why is volcanic soil good for growing plants?

____

____

RENEWABLE ENERGY

Volcanoes can also give people a clean, renewable energy source. When magma rises toward the surface, it heats up the rocks around it. Water underground heats up when it flows through cracks in the hot rock.

People can pump the hot water out of the rocks and use it to make electricity or to heat buildings. This kind of energy is called *geothermal energy*.

There is a very large geothermal energy power plant in Santa Rosa, California, called the Geysers. The Geysers is the largest geothermal energy plant in the world! San Francisco gets some of its electricity from the Geysers.

Critical Thinking

3. Apply Concepts Explain why a divergent boundary, such as the one in Iceland, would be a good source of geothermal energy.

____

____

____

USES FOR VOLCANIC MATERIALS

Rocks from volcanoes can be used to build large buildings. The early Romans used volcanic ash from Mount Vesuvius to make cement. Some of the buildings they made are still standing today!

Volcanic materials can be used for many other things. Volcanic ash is used in cat litter, soaps, cleaners, and polishes. Pumice, a volcanic rock, can be added to soil to help air and water move through it more easily. Pumice is even used to filter drinking water.

Say It

Break Down Words In a group, break down the word *geothermal* into two parts. Discuss what you think the two parts mean. If you are having trouble, look them up in a dictionary.

Copyright © by Holt, Rinehart and Winston. All rights reserved.

Name ______________________ Class ______________ Date ______________

Section 3 Review

6.2.d

SECTION VOCABULARY

lahar a mudflow that forms when volcanic ash and debris mix with water during a volcanic eruption	

1. Describe How can a lahar affect wildlife?

____

____

____

2. Compare Fill in the chart below with examples of helpful and harmful effects of volcanic eruptions.

Type of effect	Examples
Helpful	
Harmful	

3. Explain How can a volcanic eruption in one part of the world cause the temperature to change in another part of the world?

____

____

____

4. Describe How did the early Romans use volcanic ash?

____

____

5. Explain What can geothermal energy be used for?

____

____

____

Copyright © by Holt, Rinehart and Winston. All rights reserved.

SECTION 1

Weathering

California Science Standards

6.2.a, 6.2.b

BEFORE YOU READ

After you read this section, you should be able to answer these questions:

- What is weathering?
- What causes mechanical weathering?
- What causes chemical weathering?

What Is Weathering?

STUDY TIP

Compare Make a chart showing the ways that mechanical weathering and chemical weathering can happen.

Have you ever wondered how large rocks turn into smaller rocks? **Weathering** is the process in which rocks break down. There are two main kinds of weathering: mechanical weathering and chemical weathering.

What Is Mechanical Weathering?

Mechanical weathering happens when rocks are broken into pieces by physical means. There are many ways that mechanical weathering can happen. ☑

READING CHECK

1. Define What is mechanical weathering?

____

____

ICE WEDGING

Ice wedging is one kind of mechanical weathering. *Ice wedging* breaks rock into pieces during cycles of freezing and thawing.

The cycle of ice wedging starts when water seeps into cracks in a rock. As the water freezes, it expands. The ice pushes against the sides of the cracks. This causes the cracks to widen. When the ice melts, the water seeps further into the cracks. As the cycle repeats, the cracks get bigger. Finally, the rock breaks apart.

Critical Thinking

2. Infer Would ice wedging happen if water did not expand as it froze? Explain your answer.

____

____

____

____

Ice Wedging

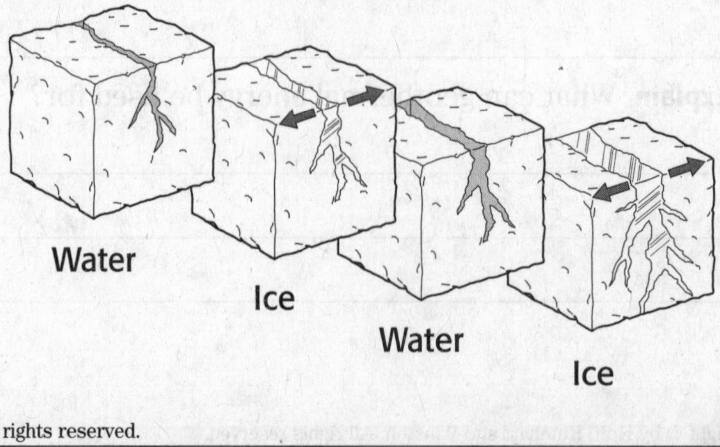
Water
Ice
Water
Ice

Copyright © by Holt, Rinehart and Winston. All rights reserved.

ABRASION

As you scrape a large block of chalk against a board, tiny pieces of the chalk rub off on the board. The large piece of chalk wears down and becomes smaller. The same process happens with rocks. **Abrasion** is a kind of mechanical weathering that happens when rocks are worn away by contact with other rocks. Abrasion happens whenever one rock hits another. Water, wind, and gravity can cause abrasion.

Water can cause abrasion by moving rocks and causing them to hit each other. The rocks in this river are rounded because of abrasion.

Wind can cause abrasion when it blows sand against rocks. This rock has been shaped by blowing sand.

Gravity can cause abrasion by making rocks grind against each other as they slide downhill. As the rocks grind against each other, they are broken into smaller pieces.

EXFOLIATION

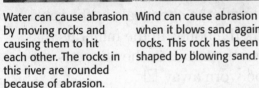

Abrasion or other kinds of weathering can remove layers of rock from a large body of rock. When the layers of rock are removed, they no longer press down on the rest of the rock. Therefore, there is less pressure on the rest of the rock. The lower pressure on the rest of the rock allows it to expand.

When the rock expands, sheets of rock can peel away from the surface of the rock. **Exfoliation** is a kind of mechanical weathering that happens when sheets of rock peel away from a larger rock because pressure is removed.

PLANT GROWTH

Have you ever seen sidewalks and streets that are cracked because of tree roots? As plants grow, their roots get larger. The roots can make cracks in rock wider. In time, an entire rock can split apart. Roots don't grow fast, but they are very powerful!

CALIFORNIA STANDARDS CHECK

6.2.a Students know water running downhill is the dominant process in shaping the landscape, including California's landscape.

Word Help: dominant
having the greatest effect

Word Help: process
a set of steps, events, or changes

3. Explain How does running water cause abrasion?

____

____

____

____

____

Say It

Explore Meanings In a small group, look up the word *exfoliate* in the dictionary. Talk about where the word comes from and the different ways it can be used.

Copyright © by Holt, Rinehart and Winston. All rights reserved.

ANIMALS

Did you know that earthworms cause a lot of weathering? They tunnel through the soil and move pieces of rock around. This motion breaks some of the rocks into smaller pieces. It also exposes more rock surfaces to weathering. ☑

Any animal that burrows in the soil causes mechanical weathering. Ants, worms, mice, coyotes, and rabbits are just a few of the animals that can cause weathering. The mixing and digging that animals do can also cause chemical weathering, another kind of weathering.

READING CHECK

4. Describe How can earthworms cause weathering?

____

____

____

____

What Is Chemical Weathering?

Rocks can be broken down by chemical means. **Chemical weathering** happens when rocks break down because of chemical reactions.

Water, acids, and air can all cause chemical weathering. They react with the chemicals in the rock. The reactions can break the bonds in the minerals that make up the rock. When the bonds in the minerals are broken, the rock can be worn away. ☑

READING CHECK

5. List What are three things that can cause chemical weathering?

____

____

____

WATER

If you drop a sugar cube into a glass of water, the sugar cube will dissolve after a few minutes. In a similar way, water can dissolve some of the chemicals that make up rocks. Even very hard rocks, such as granite, can be broken down by water. However, this process may take thousands of years or more.

Chemical Weathering in Granite

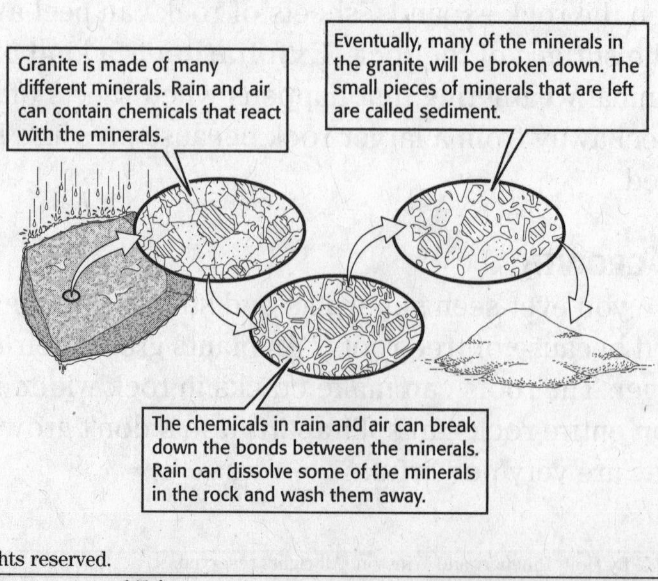
Granite is made of many different minerals. Rain and air can contain chemicals that react with the minerals.
Eventually, many of the minerals in the granite will be broken down. The small pieces of minerals that are left are called sediment.
The chemicals in rain and air can break down the bonds between the minerals. Rain can dissolve some of the minerals in the rock and wash them away.

TAKE A LOOK

6. Infer What do you think is the reason it takes a very long time for granite to break down?

____

____

____

____

____

Copyright © by Holt, Rinehart and Winston. All rights reserved.

Name ______________________ Class ______________ Date ____________

ACID PRECIPITATION

Precipitation, such as rain and snow, always contains a little bit of acid. However, sometimes precipitation contains more acid than normal. Rain, sleet, or snow that contains more acid than normal is called **acid precipitation.**

Acid precipitation forms when small amounts of certain gases mix with water in the atmosphere. The gases come from natural sources, such as volcanoes. They are also produced when people burn fossil fuels, such as coal and oil. ☑

The acids in the atmosphere fall back to the ground in rain and snow. Acids can dissolve materials faster than plain water can. Therefore, acid precipitation can cause very rapid weathering of rock.

READING CHECK

7. Identify What are two sources of the gases that produce acid precipitation?

____

____

ACIDS IN GROUNDWATER

In some places, water flows through rock underground. This water, called *groundwater*, may contain weak acids. When the groundwater touches rock, a chemical reaction takes place. The chemical reaction dissolves the rock. Over a long period of time, huge caves can form where rock has been dissolved.

This cave was formed when acids in groundwater dissolved the rock.

TAKE A LOOK

8. Explain Caves like the one in the picture are not found everywhere. What do you think controls where a cave forms?

____

____

____

Copyright © by Holt, Rinehart and Winston. All rights reserved.

ACIDS FROM LIVING THINGS

All living things make weak acids in their bodies. When the living things touch rock, some of the acids are transferred to the surface of the rock.

The acids react with chemicals in the rock and weaken it. The different kinds of mechanical weathering can more easily remove rock in these weakened areas. ☑

The rock may also crack in the weakened areas. Even the smallest crack can expose more of the rock to both mechanical weathering and chemical weathering.

READING CHECK

9. Explain How can acids from living things cause mechanical weathering?

____

____

____

____

AIR

Have you ever seen a rusted car or building? Metal reacted with something to produce rust. What did the metal react with? In most cases, the answer is air.

The oxygen in the air can react with many metals. These reactions are examples of a kind of chemical weathering called *oxidation*. This common form of chemical weathering is what causes rust. Rocks can rust if they have a lot of iron in them.

Many people think that rust forms only when metal gets wet. In fact, oxidation can happen even without any water around. However, when water is present, oxidation happens much more quickly.

Oxidation can cause rocks to weaken. Oxidation changes the metals in rocks into different chemicals. These chemicals are more easily broken down than the metals were.

Factor	How does it cause chemical weathering?
Water	
Acid precipitation	
Acids in groundwater	
Acids from living things	
Air	

TAKE A LOOK

10. Describe Fill in the blank spaces in the table to describe how different factors cause chemical weathering.

Copyright © by Holt, Rinehart and Winston. All rights reserved.

Name ______________________ Class ______________ Date ____________

Section 1 Review

6.2.a, 6.2.b

SECTION VOCABULARY

abrasion the grinding and wearing away of rock surfaces through the mechanical action of other rock or sand particles

acid precipitation rain, sleet, or snow that contains a high concentration of acids

chemical weathering the process by which rocks break down as a result of chemical reactions

exfoliation the process by which sheets of rock peel away from a large body of rock because pressure is removed

mechanical weathering the process by which rocks break down into smaller pieces by physical means

weathering the natural process by which atmospheric and environmental agents, such as wind, rain, and temperature changes, disintegrate and decompose rocks

1. List What are three things that can cause abrasion?

____

____

____

2. Explain Fill in the spaces to show the steps in the cycle of ice wedging.

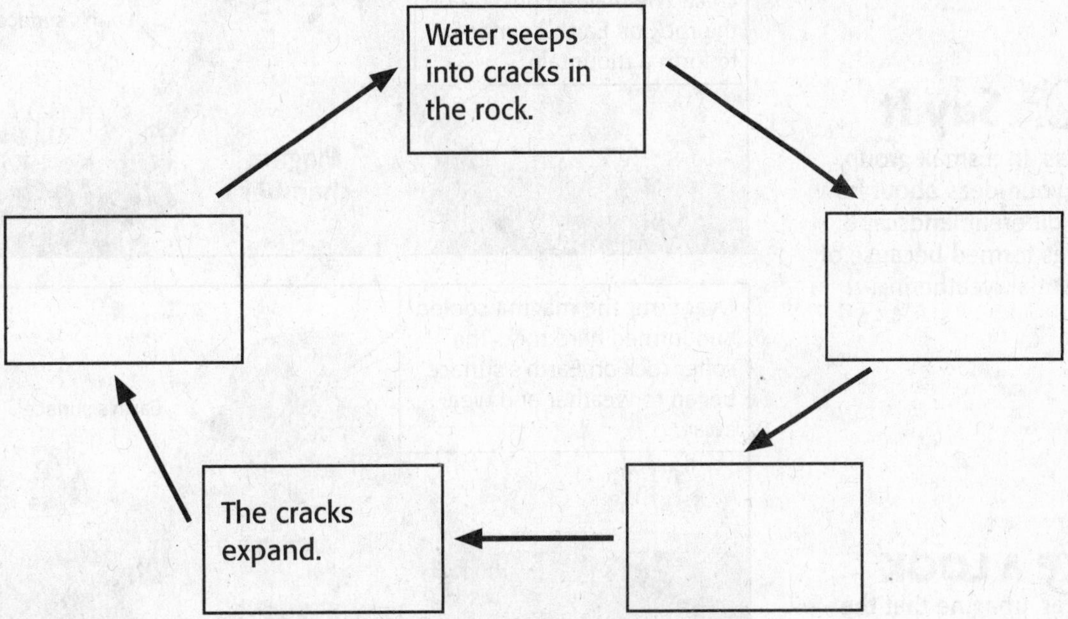

Water seeps into cracks in the rock.
The cracks expand.

3. Identify How can acids cause chemical weathering?

____

____

____

Copyright © by Holt, Rinehart and Winston. All rights reserved.

SECTION 2

Rates of Weathering

California Science Standards

6.2.c

BEFORE YOU READ

After you read this section, you should be able to answer these questions:

- What is differential weathering?
- What factors affect how fast rock weathers?

STUDY TIP

Apply Concepts After you read this section, think about how the different factors that control weathering can affect objects that you see every day.

What Is Differential Weathering?

Hard rocks, such as granite, weather more slowly than softer rocks, such as limestone. **Differential weathering** happens when softer rock weathers away and leaves harder, more resistant rock behind. The figures below show an example of how differential weathering can shape the landscape.

Say It

Discuss In a small group, share your ideas about how some different landscape features formed because of differential weathering.

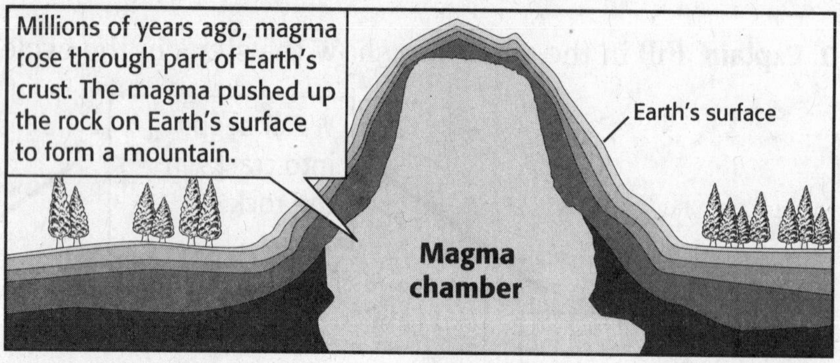
Millions of years ago, magma rose through part of Earth's crust. The magma pushed up the rock on Earth's surface to form a mountain.
Earth's surface
Magma chamber

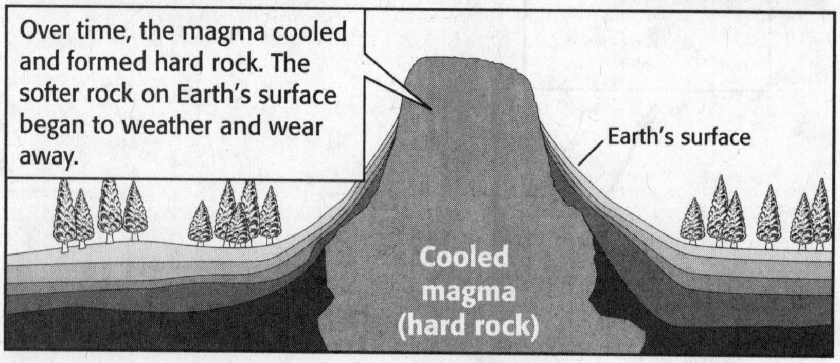
Over time, the magma cooled and formed hard rock. The softer rock on Earth's surface began to weather and wear away.
Earth's surface
Cooled magma (hard rock)

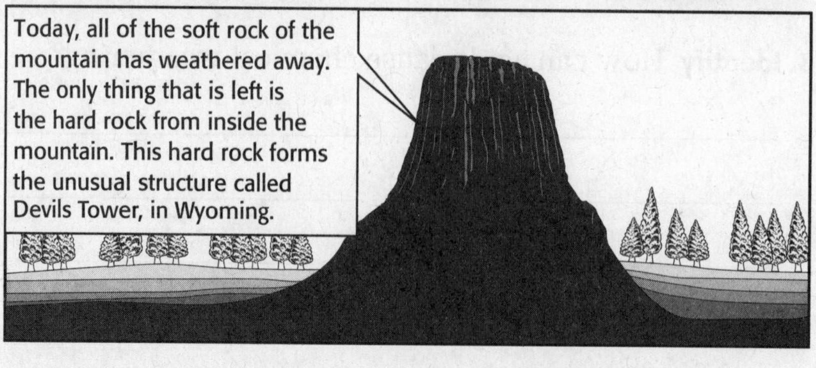
Today, all of the soft rock of the mountain has weathered away. The only thing that is left is the hard rock from inside the mountain. This hard rock forms the unusual structure called Devils Tower, in Wyoming.

TAKE A LOOK

1. Infer Imagine that the rock on the outside of the mountain was much harder than the cooled magma. How might Devils Tower look today?

____

____

____

____

Copyright © by Holt, Rinehart and Winston. All rights reserved.

How Does a Rock's Surface Area Affect Weathering?

Most chemical weathering takes place only on the outer surface of a rock. Therefore, rocks with a lot of surface area weather faster than rocks with a little surface area. However, if a rock has a large surface area as well as a large volume, it takes longer for the rock to wear down. The figure below shows how the surface area and volume of a rock affect how fast it wears down.

Imagine a rock in the shape of a cube. Each side of the rock is 4 m long. The volume of the cube is 4 m × 4 m × 4 m = 64 m³. Each face of the rock has an area of 4 m × 4 m = 16 m². Because there are eight faces on the rock, it has a surface area of

8 × 16 m² = ______ m².

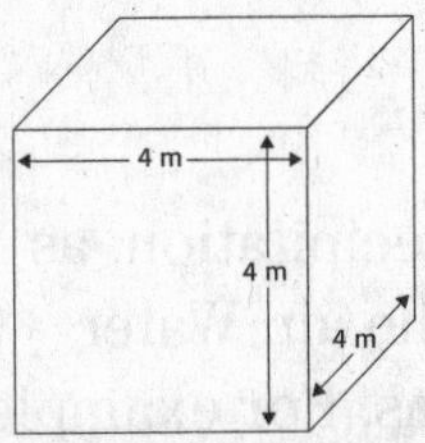
4 m
4 m
4 m

Cube with 4 m sides: volume = 64 m³, surface area = ______ m²

Now imagine eight smaller rocks that are shaped like cubes. Each small rock's side is 2 m long. The volume of each small rock is 2 m × 2 m × 2 m = 8 m³. The total volume of all 8 rocks is 8 × 8 m³ = 64 m³, the same as the large rock. Each face of each small rock has an area of 2 m × 2 m = 4 m². Each small rock has eight faces, and there are eight rocks total. Therefore, the total surface area of all eight small rocks is 8 × 8 × 4 m² = ______ m². This is twice as big as the surface area of the large rock.

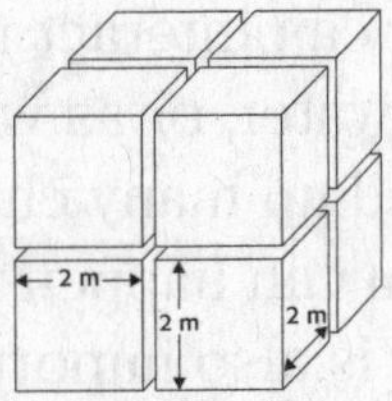
2 m
2 m
2 m

Eight cubes with 2 m sides: volume = 64 m³, surface area = ______ m²

Over time, the rock weathers. Its volume and surface area get smaller.

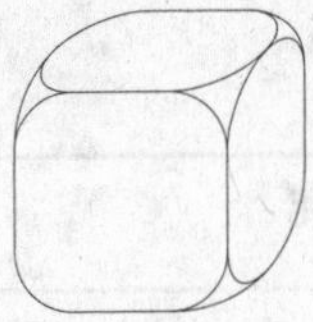

In the same amount of time, the smaller rocks weather more and become much smaller. They lose a larger fraction of their volume than the larger rock.

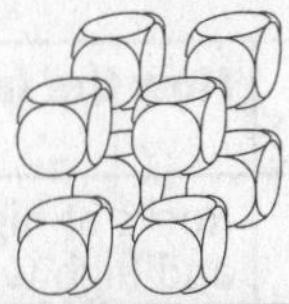

More time passes. The large rock is weathered even more. It is now much smaller than it was before it was weathered.

In the same amount of time, the small rocks have completely worn away. They took much less time to wear away than the large rock. Even though their volume was the same, they had more surface area than the large rock. The large surface area allowed them to wear away more quickly.

Critical Thinking

2. Explain Why does most chemical weathering happen only on the outer surface of a rock?

______

Math Focus

3. Calculate Determine the surface areas of the large cube and the eight smaller cubes. Write these values in the blank lines on the figure.

TAKE A LOOK

4. Apply Concepts Why do the edges and corners of the cubes weather faster than the faces?

______

Copyright © by Holt, Rinehart and Winston. All rights reserved.

How Does Climate Affect Weathering?

The rate of weathering in an area is affected by the climate of that area. *Climate* is the average weather conditions of an area over a long period of time. Some features of climate that affect weathering are temperature, moisture, elevation, slope, and living things.

TEMPERATURE

Temperature is a major factor in both chemical and mechanical weathering. Cycles of freezing and thawing increase the chance that ice wedging will take place. Areas in the world that have many freezes and thaws have faster mechanical weathering than other regions do. In addition, many chemical reactions happen faster at higher temperatures. These reactions can break down rock quickly in warm areas. ☑

READING CHECK

5. Identify What is one way that temperature affects mechanical weathering?

____

____

____

MOISTURE

Water can interact with rock as precipitation, as running water, or as water vapor in the air. Water can speed up many chemical reactions. For example, oxidation can happen faster when water is present.

Water is also important in many kinds of mechanical weathering. For example, ice wedging cannot happen without water. Abrasion is also faster when water is present.

ELEVATION, SLOPE, AND LIVING THINGS

Factor	How the factor affects weathering
Elevation	Rocks at high elevations are exposed to low temperatures and high winds. They can weather very quickly. Rocks at sea level can be weathered by ocean waves.
Slope	The steep sides of mountains and hills make water flow down them faster. Fast-moving water has more energy to break down rock than slow-moving water. Therefore, rocks on steep slopes can weather faster than rocks on level ground.
Living things	Organisms in the soil can produce acid, which causes chemical weathering. Plant roots and burrowing animals can move soil and rocks. Weathering may happen faster in places with many living things.

TAKE A LOOK

6. Explain Why do rocks on the sides of mountains weather faster than rocks on level ground?

____

____

____

____

Copyright © by Holt, Rinehart and Winston. All rights reserved.

Name ______________________ Class ______________ Date ______________

Section 2 Review

6.2.c

SECTION VOCABULARY

differential weathering the process by which softer, less weather resistant rocks wear away at a faster rate than harder, more weather resistant rocks do	

1. Identify What are five factors that affect how fast weathering happens?

____

____

____

____

2. Explain Why does it take less time for small rocks to wear away than it does for large rocks to wear away?

____

____

____

3. Describe Imagine a rock on a beach and a rock on the side of a mountain. How would the factors that control weathering be different for these two rocks?

____

____

____

____

4. Apply Concepts Two rivers run into the ocean. One river is very long. The other is very short. Which river probably drops the smallest rock pieces near the ocean? Explain your answer.

____

____

____

____

Copyright © by Holt, Rinehart and Winston. All rights reserved.

Name ______________________ Class ______________ Date ______________

CHAPTER 9 Weathering and Soil Formation

SECTION 3 From Bedrock to Soil

California Science Standards
6.5.e

BEFORE YOU READ

After you read this section, you should be able to answer these questions:

- What is soil?
- How do the features of soil affect the plants that grow in it?
- What is the effect of climate on soil?

STUDY TIP

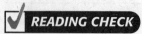

Summarize in Pairs Read this section quietly to yourself. Then, talk about the material with a partner. Together, try to figure out the parts that you didn't understand.

Where Does Soil Come From?

What do you think of when you think of soil? Most people think of dirt. However, soil is more than just dirt. **Soil** is a loose mixture of small mineral pieces, organic material, water, and air. All of these things help to make soil a good place for plants to grow.

Soils are made from weathered rocks. The rock that breaks down and forms a soil is called the soil's **parent rock**. Different parent rocks are made of different chemicals. Therefore, the soils that form from these rocks are also made of different chemicals. ☑

READING CHECK

1. Explain Why are different soils made of different chemicals?

____

____

Bedrock is the layer of rock beneath soil. In some cases, the bedrock is the same as the parent rock. In these cases, the soil remains in place above its parent rock. Soil that remains above its parent rock is called *residual soil.*

The material in soil is easily moved. Therefore, not very many soils are residual soils. Soil can be carried away from its parent rock by wind, water, ice, or gravity. Soil that has been moved away from its parent rock is called *transported soil.*

The soil is weathered from bedrock.	The soil is carried in from another place.
The bedrock is the same as the parent rock.	The bedrock is not the same as the parent rock.
______________	______________

TAKE A LOOK

2. Identify Fill in the blanks with the terms *residual soil* and *transported soil.*

Copyright © by Holt, Rinehart and Winston. All rights reserved.

What Are the Properties of Soil?

Some soils are great for growing plants. However, plants cannot grow in some other soils. Why is this? To better understand how plants can grow in soil, you must know about the properties of soil. These properties include soil composition, soil texture, and soil fertility. ☑

READING CHECK

3. Identify What are three properties of soil?

____

____

____

SOIL COMPOSITION

Soil is made up of mineral fragments, organic material, water, and air. Different soils have different amounts of these materials. For example, some soils contain more water or air than other soils do. *Soil composition* describes the kinds and amounts of different materials in the soil.

The composition of a soil can affect the kinds of plants that grow in it. For example, dark-colored soil is often rich in organic material. More kinds of plants can grow in dark-colored soil than in soil that has less organic matter.

CALIFORNIA STANDARDS CHECK

6.5.e Students know the number and type of organisms an ecosystem can support depends on the resources available and on abiotic factors, such as quantities of light and water, a range of temperatures, and soil composition.

Word Help: resource
anything that can be used to take care of a need

4. Explain How can soil composition affect the plants that can grow in soil?

____

____

____

____

____

SOIL TEXTURE

Soil is made of particles of different sizes. Some particles, such as sand, are fairly large. Other particles are so small that they cannot be seen without a microscope. *Soil texture* describes the amounts of soil particles of different sizes that a soil contains.

Soil texture affects the consistency of soil and how easily water can move into the soil. *Consistency* describes how easily a soil can be broken up for farming. For example, soil that contains a lot of clay can be hard, which makes breaking up the soil difficult. Most plants grow best in soils that can be broken up easily.

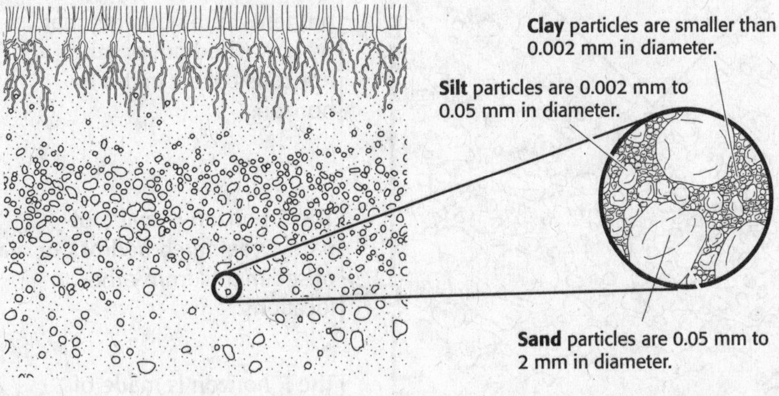
Clay particles are smaller than 0.002 mm in diameter.
Silt particles are 0.002 mm to 0.05 mm in diameter.
Sand particles are 0.05 mm to 2 mm in diameter.

Soil contains particles of many different sizes. However, all of the particles are smaller than 2 mm in diameter.

Math Focus

5. Calculate How many times larger is the biggest silt particle than the biggest clay particle?

____

Copyright © by Holt, Rinehart and Winston. All rights reserved.

SOIL FERTILITY

Plants need to get nutrients from soil in order to grow. Some soils are rich in nutrients. Other soils may have few nutrients or may be unable to give the nutrients to plants. The ability of soil to hold nutrients and to supply nutrients to plants is called *soil fertility*.

Many of the nutrients in a soil come from its parent rock. Other nutrients come from **humus**. Humus is the organic material that forms in soil from the remains of plants and animals. These remains are broken down into nutrients by decomposers, such as bacteria and fungi. It is humus that gives dark-colored soils their color. ☑

READING CHECK

6. Define What is humus?

______

What Are the Different Layers in Soil?

Most soil forms in layers. The layers are horizontal, so soil scientists call them *horizons*.

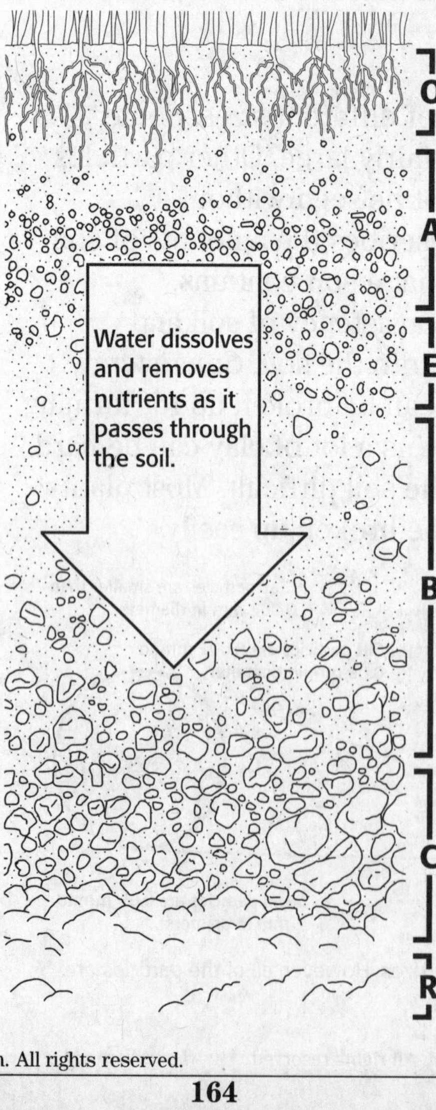
Water dissolves and removes nutrients as it passes through the soil.
O
A
E
B
C
R

Horizon name	Description
O	The O horizon is made of decaying material from dead organisms. It is found in some areas, such as forests, but not in others.
A	The A horizon is made of topsoil. Topsoil contains more humus than any other soil horizon does.
E	The E horizon is a layer of sediment with very few nutrients in it. The nutrients in the E horizon have been removed by water.
B	The B horizon is very rich in nutrients. The nutrients that were washed out of other horizons collect in the B horizon.
C	The C horizon is made of partly weathered bedrock or of sediments from other locations.
R	The R horizon is made of bedrock that has not been weathered very much.

TAKE A LOOK

7. Infer Name a kind of environment where an O horizon would probably not be found.

______

8. Identify Which three soil horizons probably have the most nutrients?

______

Copyright © by Holt, Rinehart and Winston. All rights reserved.

Why Is the pH of a Soil Important?

The *pH scale* is used to measure how acidic or basic something is. The scale ranges from 0 to 14. A pH of 7 is a *neutral* pH. Soil that has a pH below 7 is *acidic*. Soil that has a pH above 7 is *basic*.

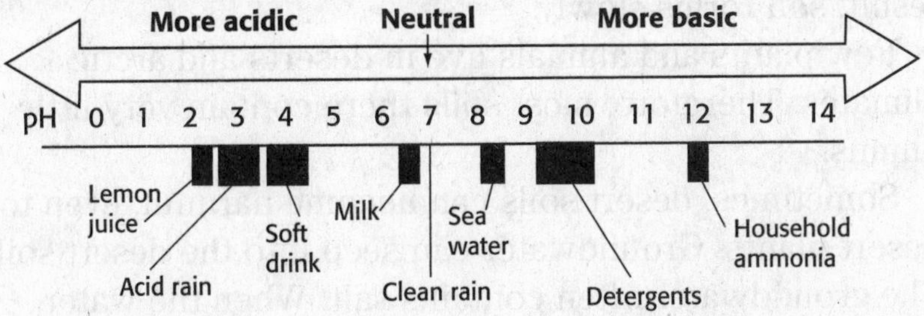
More acidic
Neutral
More basic
pH 0 1 2 3 4 5 6 7 8 9 10 11 12 13 14
Lemon juice
Acid rain
Soft drink
Milk
Clean rain
Sea water
Detergents
Household ammonia

TAKE A LOOK

9. Identify Which is more acidic, lemon juice or a soft drink?

____

The pH of a soil affects how nutrients dissolve in the soil. Many plants are unable to get certain nutrients from soils that are very acidic or basic. The pH of a soil therefore has a strong effect on soil fertility. Most plants grow best in soil with a pH of 5.5 to 7.0. A few plants grow best in soils with higher or lower pH.

Soil pH is determined partly by the soil's parent rock. Soil pH is also affected by the acidity of rainwater, the use of fertilizers, and the amount of chemical weathering. ☑

READING CHECK

10. List What are three things that affect soil pH?

____

____

____

How Does Climate Affect Soil?

Soil types vary from place to place. The kinds of soils that develop in an area depend on its climate. The different features of these soils affect the number and kinds of organisms that can survive in different areas.

TROPICAL CLIMATES

Tropical rain forests receive a lot of direct sunlight and rain. Because of these factors, plants grow year-round. The heat and moisture also cause dead organisms to decay easily. This decay produces a lot of rich humus in the soil.

Even though a lot of humus can be produced in tropical rain forests, their soils are often poor in nutrients. One reason for this is that tropical rain forests have heavy rains. The heavy rains in this climate zone can remove, or *leach*, nutrients from the topsoil. The rainwater carries the nutrients deep into the soil, where the plants can't reach them. In addition, the many plants that grow in tropical climates can use up the nutrients in the soil. ☑

READING CHECK

11. Explain Why are many tropical soils poor in nutrients?

____

____

Copyright © by Holt, Rinehart and Winston. All rights reserved.

DESERTS AND ARCTIC CLIMATES

Deserts and arctic climates receive little rainfall. Therefore, the nutrients in the soil are not leached by rainwater. However, the small amount of rain in these climates makes weathering happen more slowly. As a result, soil forms slowly.

Few plants and animals live in deserts and arctic climates. Therefore, most soils there contain very little humus.

Sometimes, desert soils can become harmful, even to desert plants. Groundwater can seep into the desert soil. The groundwater often contains salt. When the water evaporates, the salt is left in the soil. The salt can build up in the soil and harm plants.

Critical Thinking

12. Apply Concepts As in deserts, groundwater in arctic climates can contain salt. Salt does not build up in arctic soils as quickly as in desert soils. What do you think is the reason for this?

____

____

____

TEMPERATE FORESTS AND GRASSLANDS

Most of the continental United States has a temperate climate. Because the temperature changes often, mechanical weathering happens quickly in temperate climates. Thick layers of soil can build up.

Temperate areas get a medium amount of rain. The rain is enough to weather rock quickly, but it is not enough to remove nutrients from the soil.

Temperate soils are very productive. Many different kinds of plants can grow in them. Therefore, they contain a lot of humus. The large amount of humus makes the soils very rich in nutrients. The most fertile soils in the world are found in temperate climates. For example, the Midwestern part of the United States is often called the "breadbasket" because of the many crops that grow there.

Type of climate	Description of climate	Features of the soil in this climate
Tropical climates	warm temperatures a lot of rain many living things	
Deserts and arctic climates	very little rain few living things	
Temperate forests and grasslands	medium amount of rain temperature changes often	

TAKE A LOOK

13. Describe Fill in the chart to show the features of soils in different climates.

Copyright © by Holt, Rinehart and Winston. All rights reserved.

Name ______________________ Class ______________ Date ______________

Section 3 Review

6.5.e

SECTION VOCABULARY

bedrock the layer of rock beneath soil

humus dark, organic material formed in soil from the decayed remains of plants and animals

parent rock a rock formation that is the source of soil

soil a loose mixture of rock fragments, organic material, water, and air that can support the growth of vegetation

1. Summarize What are three properties of soil?

____

____

2. Compare What climate feature do arctic climates and desert climates share that makes their soils similar?

____

____

3. Analyze How can flowing water affect the fertility of soils?

____

____

____

4. Identify How does soil pH affect plant growth?

____

____

____

5. Explain What determines a soil's texture?

____

____

6. Apply If the soil in your yard is mostly clay, how can you change its properties to make it better for growing plants?

____

____

____

Copyright © by Holt, Rinehart and Winston. All rights reserved.

Name ______________________ Class ____________ Date ____________

CHAPTER 9 Weathering and Soil Formation

SECTION 4

Soil Conservation

California Science Standards

6.2.a, 6.5.e, 6.6.b

BEFORE YOU READ

After you read this section, you should be able to answer these questions:

- Why is soil important?
- How can farmers conserve soil?

STUDY TIP

Compare Create a chart that shows the similarities and differences between the ways that farmers can help conserve soil.

Why Is Soil Important?

You have probably heard about endangered plants and animals. Did you know that soil can be endangered, too? Soil can take many years to form. It is not easy to replace. Therefore, soil is considered a nonrenewable resource.

Soil is important for many reasons. Soil provides nutrients for plants. If the soil loses its nutrients, plants will not be able to grow. Soil also helps to support plant roots so the plants can grow well.

Animals get their energy from plants. The animals get energy either by eating plants or by eating animals that have eaten plants. If plants are unhealthy because the soil has few nutrients, then animals will be unhealthy, too.

Soil also provides a home, or *habitat*, for many living things. Bacteria, insects, mushrooms, and many other organisms live in soil. If the soil disappears, so does the habitat for these living things. ☑

READING CHECK

1. Explain Why is soil important for animals?

____

____

____

Soil is very important for storing water. Soil holds water that plants can use. Soil also helps to prevent floods. When rain falls, the soil can soak it up. The water is less likely to run over the land and cause floods.

What does soil provide?	Why is it important?
Nutrients	
Habitat	
Water storage	

TAKE A LOOK

2. Identify In the table, fill in the reasons that nutrients, habitat, and water storage are important.

If we do not take care of soils, we could lose them. In order to keep our soils from being lost, we need to conserve them. Soil conservation means protecting soils from erosion and nutrient loss. **Soil conservation** can help to keep soils fertile and healthy.

Copyright © by Holt, Rinehart and Winston. All rights reserved.

How Can Soil Be Lost?

Soil loss is a major problem around the world. One cause of soil loss is soil damage. Soil can be damaged if it is overused. Overused soil can lose its nutrients and become infertile. Plants can't grow in infertile soil.

Plants help to hold water in the soil. If plants can't grow somewhere because the soil is infertile, the area can become a desert. This process is known as *desertification.*

EROSION

Another cause of soil loss is erosion. **Erosion** happens when wind, water, or gravity transports soil and sediment from one place to another. If soil is not protected, it can be exposed to erosion.

Plant roots help to keep soil in place. They prevent water and wind from carrying the soil away. If there are no plants, soil can be lost through erosion.

CALIFORNIA STANDARDS CHECK

6.2.a Students know water running downhill is the dominant process in shaping the landscape, including California's landscape.

Word Help: dominant
having the greatest effect

Word Help: process
a set of steps, events, or changes

3. Describe How do plant roots prevent erosion?

____

____

____

____

____

How Can Farmers Help to Conserve Soil?

Farming can cause soil damage. However, farmers can prevent soil damage if they use certain methods when they plow, plant, and harvest their fields.

CONTOUR PLOWING

Water that runs straight down a hill can carry away a lot of soil. Farmers can plow their fields in special ways to help slow the water down. When the water moves more slowly down a hill, it carries away less soil. *Contour plowing* means plowing a field in rows that run across the slope of a hill.

Contour plowing helps water to run more slowly down hills. This reduces erosion because _______________

TAKE A LOOK

4. Identify Fill in the blank line in the figure to explain how contour plowing reduces erosion.

Copyright © by Holt, Rinehart and Winston. All rights reserved.

Critical Thinking

5. Infer What do you think is the reason farmers use terraces only on very steep hills?

____

____

____

____

TERRACES

On very steep hills, farmers can use terraces to prevent soil erosion. *Terraces* change one very steep field into many smaller, flatter fields.

Terraces keep water from running downhill very quickly.

NO-TILL FARMING

In no-till farming, farmers leave the stalks from old crops lying on the field while the newer crops grow. The old stalks protect the soil from rain and help reduce erosion.

The stalks left behind in no-till farming reduce erosion by protecting the soil from rain.

CALIFORNIA STANDARDS CHECK

6.5.e Students know the number and types of organisms an ecosystem can support depends on the resources available and on abiotic factors, such as quantities of light and water, a range of temperatures, and soil composition.

Word Help: resource
anything that can be used to take care of a need

6. Apply Concepts How can crop rotation affect the number of plants that soil can support?

____

____

____

____

____

COVER CROPS

Cover crops are crops that are planted between harvests of a main crop. Cover crops can help to replace nutrients in the soil. They can also prevent erosion by providing cover from wind and rain.

CROP ROTATION

If the same crop is grown year after year in the same field, the soil can lose certain nutrients. To slow this process, a farmer can plant different crops in the field every year. Different crops use different nutrients from the soil. Some crops used in crop rotation can replace soil nutrients.

Copyright © by Holt, Rinehart and Winston. All rights reserved.

Name _______________ Class _______________ Date _______________

Section 4 Review

6.2.a, 6.5.e, 6.6.b

SECTION VOCABULARY

erosion the process by which wind, water, ice, or gravity transports soil and sediment from one location to another	**soil conservation** a method to maintain the fertility of soil by protecting the soil from erosion and nutrient loss

1. Explain Why is soil a nonrenewable resource?

____

____

____

2. Describe How is weathering different from erosion?

____

____

____

3. Identify What are two causes of soil loss?

____

____

4. Explain How does no-till farming help to reduce erosion?

____

____

____

5. List What are five ways that farmers can help to conserve soil?

____

____

____

____

____

6. Apply Many gardeners put mulch, which is usually shredded bark or straw, around their plants in the fall. How do you think this helps conserve soil?

____

____

Copyright © by Holt, Rinehart and Winston. All rights reserved.

Name ______________________ Class ______________ Date ____________

CHAPTER 10 Agents of Erosion and Deposition

SECTION 1

Shoreline Erosion and Deposition

California Science Standards

6.2.a, 6.2.c, 6.3.a

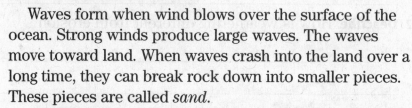

BEFORE YOU READ

After you read this section, you should be able to answer these questions:

- What is a shoreline?
- How do waves shape shorelines?

STUDY TIP

Summarize Read this section quietly to yourself. With a partner, talk about what you learned. Together, try to figure out the answers to any questions that you have.

How Do Waves Form?

Waves form when wind blows over the surface of the ocean. Strong winds produce large waves. The waves move toward land. When waves crash into the land over a long time, they can break rock down into smaller pieces. These pieces are called *sand.*

A **shoreline** is a place where the land and the water meet. Most shorelines contain sand. The motion of waves helps to shape shorelines. During *erosion*, waves remove sand from shorelines. During *deposition*, waves add sand to shorelines. ☑

READING CHECK

1. Compare How is wave erosion different from wave deposition?

____

____

____

____

WAVE TRAINS

Waves move in groups called *wave trains*. The waves in a wave train are separated by a period of time called the *wave period*. You can measure the wave period by counting the seconds between waves breaking on the shore. Most wave periods are 10 s to 20 s long.

When a wave reaches shallow water, the bottom of the wave drags against the sea floor. The bottom slows down. As the water gets shallower, the wave gets taller. The wave gets so tall, it can't support itself. It begins to curl, fall over, and break. Breaking waves are called *surf*.

Waves travel in groups called wave trains. The time between one wave and the next is the wave period.

TAKE A LOOK

2. Identify On the figure, label the wave train.

Copyright © by Holt, Rinehart and Winston. All rights reserved.

POUNDING SURF

The energy in waves is constantly breaking rock into smaller and smaller pieces. Crashing waves can break solid rock and throw the pieces back toward the shore. Breaking waves can enter cracks in the rock and break off large boulders. Waves also pick up fine grains of sand. The loose sand wears down other rocks on the shore. ☑

READING CHECK

3. Identify Give two ways that waves can break rock into smaller pieces.

______

______

______

______

What Are the Effects of Wave Erosion?

Wave erosion can produce many features along a shoreline. For example, *sea cliffs* form when waves erode rock to form steep slopes. As waves strike the bottom of the cliffs, the waves wear away soil and rock and make the cliffs steeper.

How fast sea cliffs erode depends on how hard the rock is and how strong the waves are. Cliffs made of hard rock, such as granite, erode slowly. Cliffs made of soft rock, such as shale, erode more quickly.

During storms, large, high-energy waves can erode the shore very quickly. These waves can break off large chunks of rock. Many of the features of shorelines are shaped by storm waves. The figures below and at the top of the next page show some of these features.

Critical Thinking

4. Identify Relationships When may a storm not produce high-energy waves?

______

______

______

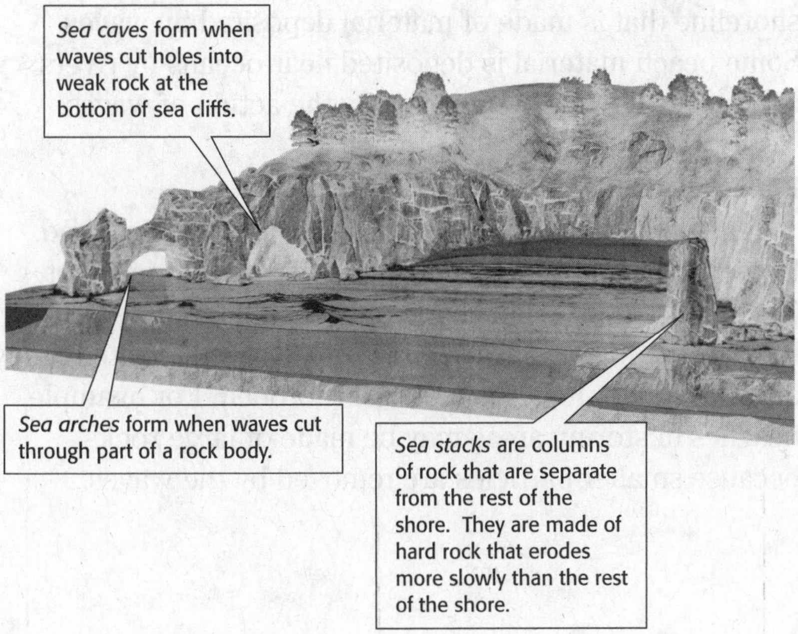
Sea caves form when waves cut holes into weak rock at the bottom of sea cliffs.
Sea arches form when waves cut through part of a rock body.
Sea stacks are columns of rock that are separate from the rest of the shore. They are made of hard rock that erodes more slowly than the rest of the shore.

TAKE A LOOK

5. Compare How is a sea stack different from a sea arch?

______

______

______

Copyright © by Holt, Rinehart and Winston. All rights reserved.

Headlands are finger-shaped bodies of rock that stick out into the sea. They are made of harder rock than the rest of the shore.
Wave-cut terraces form when sea cliffs are worn back from the shore. This produces a nearly flat platform beneath the water at the base of the cliff.

TAKE A LOOK

6. Define What is a headland?

____

____

What Are the Effects of Wave Deposition?

Waves carry many materials, such as sand, shells, and small rocks. When the waves deposit these materials on the shoreline, a beach forms. A **beach** is any area of shoreline that is made of material deposited by waves. Some beach material is deposited near oceans by rivers and moves down the shoreline by the action of waves.

CALIFORNIA STANDARDS CHECK

6.2.c Students know beaches are dynamic systems in which the sand is supplied by rivers and moved along the coast by the action of waves.

Word Help: dynamic
active

7. Define Write your own definition for *beach*.

____

____

____

____

____

BEACH MATERIALS

Many people think that all beaches are made of sand. However, beaches may be made of many materials, not just sand. The size and shape of beach material depend on how far the material traveled before it was deposited. They also depend on how the material is eroded. For example, beaches in stormy areas may be made of large rocks because smaller particles are removed by the waves.

Copyright © by Holt, Rinehart and Winston. All rights reserved.

BEACH COLOR

The color of a beach depends on the particles deposited there. Many beaches are light-colored because light-colored sand is the most common beach material. Most light-colored sand is made of the mineral quartz, which comes from eroded sandstone. Many Florida beaches are made of quartz sand. On many tropical beaches, the sand is white. It is made of finely ground white coral. ☑

Beaches can also be black or dark-colored. Black-sand beaches are found in Hawaii. Their sands are made of eroded lava from volcanoes. This lava is rich in dark-colored minerals, so the sand is also dark-colored. The figures below show some examples of beaches from across the United States.

READING CHECK

8. Identify What mineral is most light-colored sand made of?

____

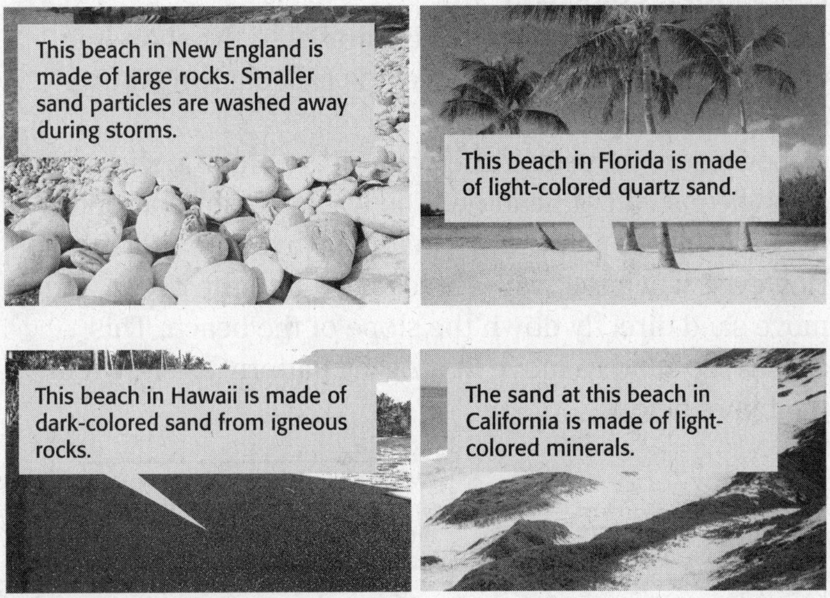
This beach in New England is made of large rocks. Smaller sand particles are washed away during storms.
This beach in Florida is made of light-colored quartz sand.
This beach in Hawaii is made of dark-colored sand from igneous rocks.
The sand at this beach in California is made of light-colored minerals.

TAKE A LOOK

9. Explain Why are some beaches made mostly of larger rock pieces, instead of sand?

____

____

CALIFORNIA BEACHES

California has both rocky and sandy beaches. Many rocky beaches are found where mountains or cliffs meet the ocean. Many sandy beaches form on the edges of gently sloping land. Most sandy beaches get some of their sand from rivers.

The materials that make up California beaches also vary. Materials from different sources have different colors and compositions. For example, the sand at Shelter Cove in Humboldt County is grey because it is made of eroded shale. The white sand at Carmel is made of light-colored quartz and feldspar. The dark sand near Santa Cruz contains dark grains of magnetite from igneous rock.

Copyright © by Holt, Rinehart and Winston. All rights reserved.

BEACH SIZE

The amount of sand on a beach can change with the seasons. In California, beaches can become narrower in winter. Large waves from winter storms erode sand away from the beaches. The eroded sand can be trapped in areas away from the shore. Small waves return the sand to the beaches during the summer. ☑

READING CHECK

10. Explain Why do some beaches in California get smaller in the winter?

____

____

____

SHORE CURRENTS

After waves crash on the beach, the water glides back to the ocean. It flows underneath the incoming waves. This kind of water movement is called **undertow**. Undertow carries sand and pieces of rock away from the shore.

LONGSHORE CURRENTS

In some cases, water travels parallel to the shoreline very near the shore. This current is called a **longshore current**.

Longshore currents form when waves hit the shore at an angle instead of head-on. The waves wash sand onto the shore at the same angle that the waves are moving. However, when the waves wash back into the ocean, they move sand directly down the slope of the beach. This causes the sand to move in a zigzag pattern, as shown in the figure below.

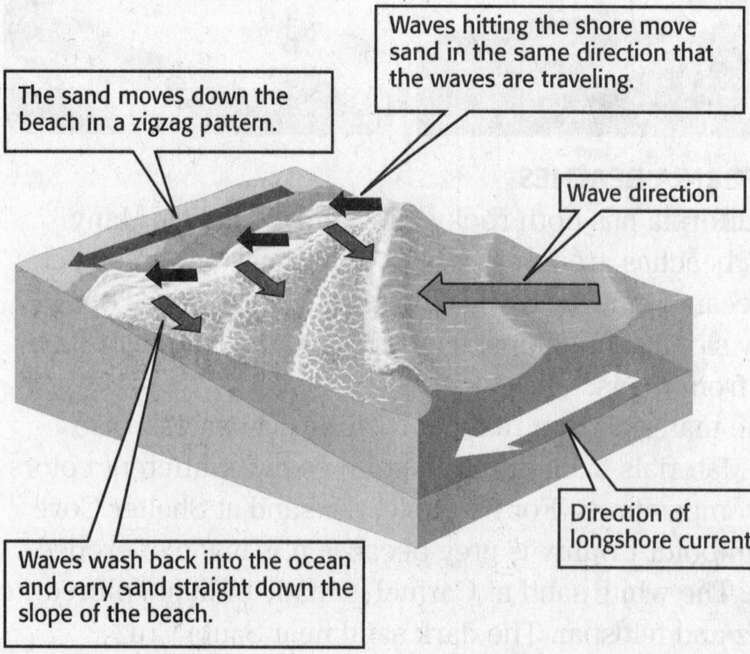
Waves hitting the shore move sand in the same direction that the waves are traveling.
The sand moves down the beach in a zigzag pattern.
Wave direction
Direction of longshore current
Waves wash back into the ocean and carry sand straight down the slope of the beach.

TAKE A LOOK

11. Infer Why don't longshore currents form in places where waves hit the shore head-on?

____

____

____

____

Copyright © by Holt, Rinehart and Winston. All rights reserved.

OFFSHORE DEPOSITS

Longshore currents can carry beach material offshore. This process can produce landforms in open water. These landforms include sandbars, barrier spits, and barrier islands.

A *sandbar* is a ridge of sand, gravel, or broken shells that is found in open water. Sandbars may be completely underwater or they may stick above the water. ☑

A *barrier spit* is a sandbar that sticks above the water and is connected to the shoreline. A *barrier island* is a long, narrow island that forms parallel to the shoreline. Most barrier islands are made of sand.

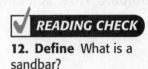

READING CHECK

12. Define What is a sandbar?

Cape Cod, Massachusetts, is an example of a barrier spit. Barrier spits form when sandbars are connected to the shoreline.

TAKE A LOOK

13. Identify What is a barrier spit?

CALIFORNIA ISLANDS

California is different from many other states because it has offshore islands. For example, the Channel Islands are found off the southern California coast. They formed millions of years ago. They were once part of the mainland. Now they are separated from the mainland by the Santa Monica and San Pedro Channels. The Channel Islands protect the coastline. They reduce the erosion of coastal cliffs by storm waves.

Copyright © by Holt, Rinehart and Winston. All rights reserved.

Name ______________________ Class ______________ Date ______________

Section 1 Review

6.2.a, 6.2.c, 6.3.a

SECTION VOCABULARY

beach an area of the shoreline that is made up of deposited sediment **longshore current** a water current that travels near and parallel to the shoreline	**shoreline** the boundary between land and a body of water **undertow** a subsurface current that is near shore and that pulls objects out to sea

1. **Compare** How is a longshore current different from undertow?

____

____

2. **Explain** Where does the energy to change the shoreline come from? Explain your answer.

____

____

____

3. **Identify** Give two examples of different-colored beach sand and explain why each kind is a certain color.

____

____

____

____

4. **Explain** How do longshore currents move sand?

____

____

____

5. **List** Give five landforms that are produced by wave erosion.

____

____

____

____

____

Copyright © by Holt, Rinehart and Winston. All rights reserved.

Name ________________ Class ________________ Date ________________

SECTION 2

Wind Erosion and Deposition

BEFORE YOU READ

After you read this section, you should be able to answer these questions:

- How can wind erosion shape the landscape?
- How can wind deposition shape the landscape?

California Science Standards

6.2.a, 6.7.e

How Can Wind Erosion Affect Rocks?

Wind can move soil, sand, and small pieces of rock. Therefore, wind can cause erosion. However, some areas are more likely to have wind erosion than other areas. For example, plant roots help to hold soil and rock in place. Therefore, areas with few plants, such as deserts and coastlines, are more likely to be eroded by wind. These areas also may be made of fine, loose rock particles. Wind can move these particles easily. ☑

Wind can shape rock pieces in three ways: saltation, deflation, and abrasion.

STUDY TIP

Learn New Words As you read this section, underline words you don't understand. When you figure out what they mean, write the words and their definitions in your notebook.

READING CHECK

1. Explain How do plant roots help to prevent wind erosion?

____

____

SALTATION

Wind moves large grains of soil, sand, and rock by saltation. **Saltation** happens when sand-sized particles skip and bounce along in the direction that the wind is moving. When moving sand grains hit one another, some of the grains bounce up into the air. These grains fall back to the ground and bump other grains. These other grains can then move forward.

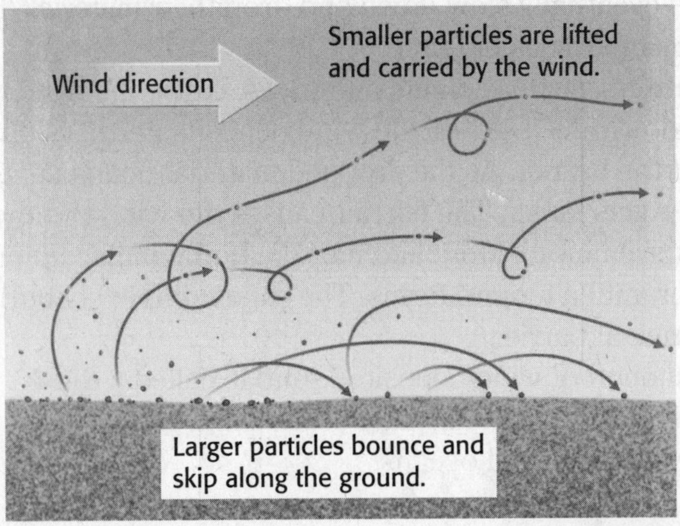
Wind direction
Smaller particles are lifted and carried by the wind.
Larger particles bounce and skip along the ground.

TAKE A LOOK

2. Apply Concepts Why can't the wind lift and carry large particles?

____

____

____

____

Copyright © by Holt, Rinehart and Winston. All rights reserved.

DEFLATION

Wind can blow tiny particles away from larger rock pieces during deflation. **Deflation** happens when wind removes the top layers of fine sediment or soil and leaves behind larger rock pieces. ☑

READING CHECK

3. Define What is deflation?

Deflation can form certain land features. It can produce *desert pavement*, which is a surface made of pebbles and small, broken rocks. In some places, the wind can scoop out small, bowl-shaped areas from fine sediment. These areas are called *deflation hollows*.

ABRASION

Wind can grind and wear down rocks by abrasion. **Abrasion** happens when rock or sand wears down larger pieces of rock. Abrasion happens in areas where there are strong winds, loose sand, and soft rocks. The wind blows the loose sand against the rocks. The sand acts like sandpaper to erode, smooth, and polish the rocks.

TAKE A LOOK

4. Complete Fill in the blank spaces in the table.

Process	Description
	Large particles bounce and skip along the ground.
Deflation	
Abrasion	

What Landforms Are Produced by Wind Deposition?

Wind can carry material over long distances. The wind can carry different amounts and sizes of particles depending on its speed. Fast winds can carry large particles and may move a lot of material. However, all winds eventually slow down and drop their material. The heaviest particles fall first.

Critical Thinking

5. Infer What do you think is the reason that fast winds can carry larger particles than slower winds?

Barriers, such as plants and rocks, cause the wind to slow down. As it slows, the wind deposits particles on top of the barrier. As the dropped material builds up, the barrier gets larger. The barrier causes the wind to slow down even more. More and more material builds up on the barrier until a mound forms. The sand completely buries the original barrier.

A mound of wind-deposited sand is called a **dune**. Dunes are common in sandy deserts and along sandy shores of lakes and oceans.

Copyright © by Holt, Rinehart and Winston. All rights reserved.

CALIFORNIA DUNES

There are many formations of dunes in California. They are found in coastal and desert areas. People use these areas for activities such as camping.

One major coastal dune system is the Monterey Bay Dunes, which covers 40 square miles. The largest desert sand-dune system in the state is the Algodones. The Algodones are in southeast California. They cover 1,000 square miles.

Math Focus

6. Calculate How many times larger is the Algodones sand-dune system than the Monterey Bay Dunes?

____

THE MOVEMENT OF DUNES

Wind conditions affect a dune's shape and size. In general, dunes move in the same direction the wind is blowing. A dune has one gently sloped side and one steep side. The gently sloped side faces the wind. It is called the *windward slope*. The wind constantly moves sand up this side. As sand moves over the top of the dune, the sand slides down the steep side. The steep side is called the *slip face*. ☑

READING CHECK

7. Identify In what direction do dunes generally move?

____

____

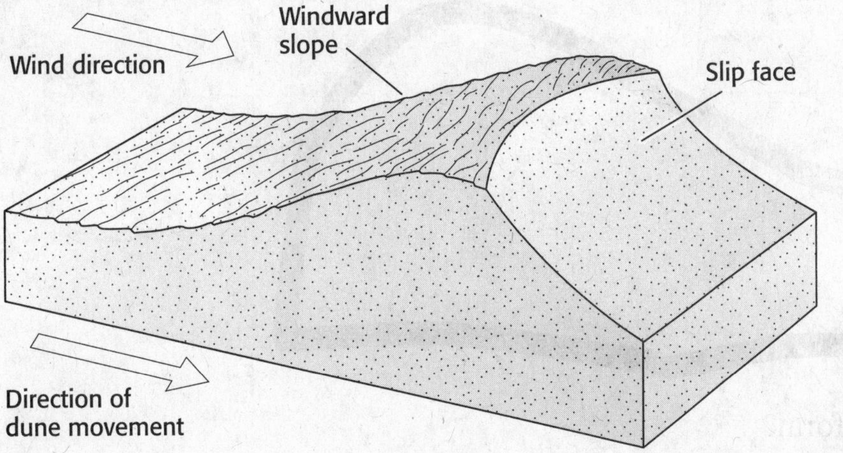
Wind direction
Windward slope
Slip face
Direction of dune movement

The wind blows sand up the windward slope of the dune. The sand moves over the top of the dune and falls down the steep slip face. In this way, dunes move across the land in the direction that the wind blows.

TAKE A LOOK

8. Compare How is the windward slope of a dune different from the slip face?

____

____

Copyright © by Holt, Rinehart and Winston. All rights reserved.

Name ______________________ Class ______________ Date ____________

Section 2 Review

6.2.a, 6.7.e

SECTION VOCABULARY

abrasion the grinding and wearing away of rock surfaces through the mechanical action of other rock or sand particles **deflation** a form of wind erosion in which fine, dry soil particles are blown away	**dune** a mound of wind-deposited sand that moves as a result of the action of wind **saltation** the movement of sand or other sediments by short jumps and bounces that is caused by wind or water

1. **Identify** Give two land features that can form because of deflation.

____

____

2. **Describe** What areas are most likely to be affected by wind erosion? Give two examples.

____

____

3. **Label** The figure shows a drawing of a sand dune. Label the windward slope and the slip face. Add an arrow to show the direction of the wind.

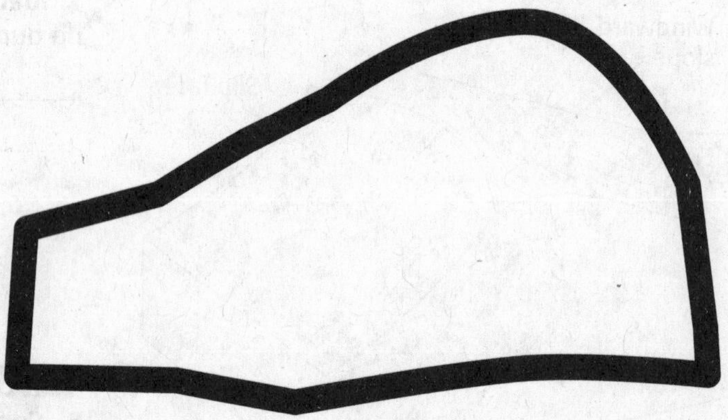

4. **Explain** How do dunes form?

____

____

____

5. **Apply Concepts** Wind can transport particles of many different sizes. What sized particles are probably carried the farthest by the wind?

____

Copyright © by Holt, Rinehart and Winston. All rights reserved.

Name ______________________ Class ______________ Date ______________

CHAPTER 10 Agents of Erosion and Deposition

SECTION 3

Erosion and Deposition by Ice

BEFORE YOU READ

After you read this section, you should be able to answer these questions:

- What are glaciers?
- How do glaciers affect the landscape?

California Science Standards

6.2.a

What Are Glaciers?

A **glacier** is a huge piece of moving ice. The ice in glaciers contains most of the fresh water on Earth. Glaciers are found on every continent except Australia.

There are two kinds of glaciers: continental and alpine. *Continental glaciers* are ice sheets that can spread across entire continents. *Alpine glaciers* are found on the tops of mountains. Both continental and alpine glaciers can greatly affect the landscape. ☑

Glaciers form in areas that are so cold that snow stays on the ground all year round. For example, glaciers are common in polar areas and on top of high mountains. In these areas, layers of snow build up year after year. Over time, the weight of the top layers pushes down on the lower layers. The lower layers change from snow to ice.

STUDY TIP

Compare As you read, make a table comparing the landforms that glaciers can produce.

READING CHECK

1. Identify What are the two kinds of glaciers?

____

____

HOW GLACIERS MOVE

Glaciers can move in two ways: by sliding and by flowing. As more ice builds up on a slope, the glacier becomes heavier. The glacier can start to slide downhill, the way a skier slides downhill. Glaciers can also move by flowing. The solid ice in glaciers can move slowly, like soft putty or chewing gum.

Thick glaciers move faster than thin glaciers. Glaciers on steep slopes move faster than those on gentler slopes.

This is McBride Glacier in Alaska.

TAKE A LOOK

2. Define What is a glacier?

____

____

Copyright © by Holt, Rinehart and Winston. All rights reserved.

Critical Thinking

3. Identify Relationships How is erosion by glaciers an example of water shaping the landscape?

____

____

How Do Glaciers Affect the Landscape?

Glaciers can produce many different features as they move over Earth's surface. As a glacier moves, it can pick up and carry the rocks in its path. Glaciers can carry rocks of many different sizes, from dust all the way up to boulders. These rocks can scrape grooves into the land below the glacier as the glacier moves.

Continental glaciers tend to flatten the land that they pass over. However, alpine glaciers can produce sharp, rugged landscapes. The figure below shows some of the features that alpine glaciers can form.

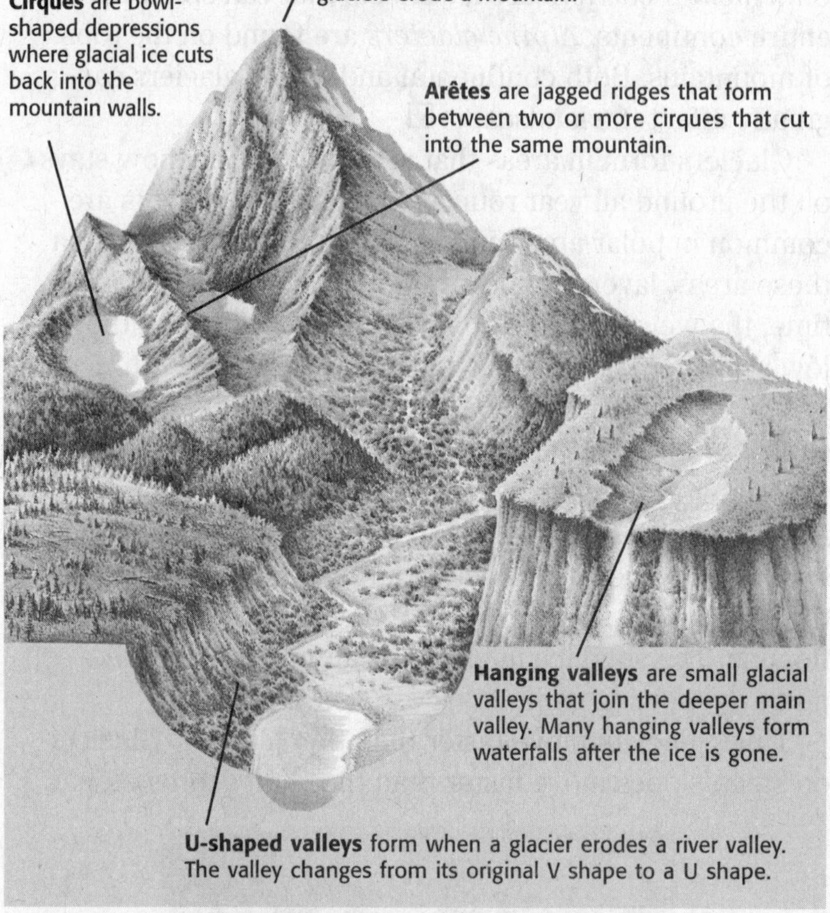
Horns are sharp, pyramid-shaped peaks that form when three or more cirque glaciers erode a mountain.
Cirques are bowl-shaped depressions where glacial ice cuts back into the mountain walls.
Arêtes are jagged ridges that form between two or more cirques that cut into the same mountain.
Hanging valleys are small glacial valleys that join the deeper main valley. Many hanging valleys form waterfalls after the ice is gone.
U-shaped valleys form when a glacier erodes a river valley. The valley changes from its original V shape to a U shape.

TAKE A LOOK

4. Explain How are horns, cirques, and arêtes related?

____

____

____

____

GLACIAL DEPOSITS

As a glacier melts, it drops all of the material that it is carrying. The material that is carried and deposited by glaciers is called **glacial drift**. There are two kinds of glacial drift: till and stratified drift.

Copyright © by Holt, Rinehart and Winston. All rights reserved.

TILL DEPOSITS

Till is unsorted rock material that is deposited by melting glacial ice. It is called "unsorted" because the rocks are of all different sizes. Till contains fine sediment as well as large boulders. When the ice melts, it deposits this material onto the ground. ☑

The most common till deposits are *moraines*. Moraines form ridges along the edges of glaciers. There are many types of moraines. They are shown in the figure below.

READING CHECK

5. Explain Why is till considered to be unsorted?

____

____

Lateral moraines form along each side of a glacier.

Medial moraines form when valley glaciers that have lateral moraines meet.

Ground moraines form from unsorted materials left beneath a glacier.

Terminal moraines form when sediment is dropped at the front of the glacier.

Say It

Learn New Words Look up the words *lateral*, *medial*, and *terminal* in a dictionary. In a group, talk about why these words are used to describe different kinds of moraines.

STRATIFIED DRIFT

When a glacier melts, the water forms streams that carry rock material away from the glacier. The streams deposit the rocks in different places depending on their size. The rocks form a sorted deposit called **stratified drift**. The large area where the stratified drift is deposited is called an *outwash plain*. ☑

In some cases, a block of ice is left in the outwash plain as the glacier melts. As the ice melts, sediment builds up around it. The sediment forms a bowl-shaped feature called a *kettle*. Kettles can fill with water and become ponds or lakes.

READING CHECK

6. Define Write your own definition for *stratified drift*.

____

____

____

____

Copyright © by Holt, Rinehart and Winston. All rights reserved.

Name ______________________ Class ______________ Date ______________

Section 3 Review

6.2.a

SECTION VOCABULARY

glacial drift the rock material carried and deposited by glaciers **glacier** a large mass of moving ice	**stratified drift** a glacial deposit that has been sorted and layered by the action of streams or meltwater **till** unsorted rock material that is deposited directly by a melting glacier

1. List Give two kinds of glacial drift.

____

____

2. Identify What are four kinds of moraines?

____

____

____

____

3. Compare How are continental glaciers different from alpine glaciers?

____

____

4. Explain How do glaciers form?

____

____

____

5. Describe How does a kettle form?

____

____

____

6. Infer How can a glacier deposit both unsorted and sorted material?

____

____

____

____

Copyright © by Holt, Rinehart and Winston. All rights reserved.

Name ______________________ Class ______________ Date ____________

SECTION 4

Erosion and Deposition by Mass Movement

BEFORE YOU READ

After you read this section, you should be able to answer these questions:

- What is mass movement?
- How does mass movement shape Earth's surface?
- How can mass movement affect living things?

California Science Standards

6.2.d

What Is Mass Movement?

Gravity can cause erosion and deposition. Gravity makes water and ice move. It also causes rock, soil, snow, or other material to move downhill in a process called **mass movement**.

STUDY TIP

Ask Questions As you read this section, write down any questions you have. Talk about your questions in a small group.

ANGLE OF REPOSE

Particles in a sand pile move downhill. They stop when the slope of the pile becomes stable. The *angle of repose* is the steepest angle, or slope, at which the loose material no longer moves downhill. If the slope of a pile of material is larger than the angle of repose, mass movement happens. ☑

READING CHECK

1. Define What is the angle of repose?

35°

The slope of this pile of sand is equal to the sand's angle of repose. The sand pile is stable. The sand particles are not moving.

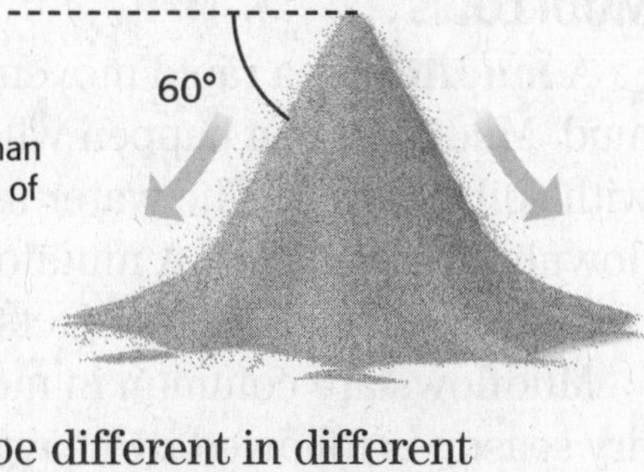
60°

The slope of this pile of sand is larger than the angle of repose. Therefore, particles of sand move down the slope of the pile.

TAKE A LOOK

2. Explain Why are sand particles moving downhill in the bottom picture?

The angle of repose can be different in different situations. The composition, size, weight, and shape of the particles in a material affect its angle of repose. The amount of water in a material can also change the material's angle of repose.

Copyright © by Holt, Rinehart and Winston. All rights reserved.

What Are the Kinds of Mass Movement?

Mass movement can happen suddenly and quickly. Rapid mass movement can be very dangerous. It can destroy or bury everything in its path.

CALIFORNIA STANDARDS CHECK

6.2.d Students know earthquakes, volcanic eruptions, landslides, and floods change human and wildlife habitats.

3. Describe How can landslides affect wildlife habitats?

____

____

____

LANDSLIDES

A **landslide** happens when a large amount of material moves suddenly and rapidly downhill. Landslides can carry away or bury plants and animals and destroy their habitats. Several factors can make landslides more likely.

- Heavy rains can make soil wet and heavy, which makes the soil more likely to move downhill.
- Tree roots help to keep land from moving. Therefore, *deforestation*, or cutting down trees, can make landslides more likely.
- Earthquakes can cause rock and soil to start moving.
- People may build houses and other buildings on unstable hillsides. The extra weight of the buildings can cause a landslide. ☑

READING CHECK

4. Identify Give three factors that can make landslides more likely.

____

____

____

____

____

The most common kind of landslide is a *slump*. Slumps happen when a block of material moves downhill along a curved surface.

ROCK FALLS

A **rock fall** happens when loose rocks fall down a steep slope. Many such slopes are found on the sides of roads that run through mountains. Gravity can cause the loose and broken rocks above the road to fall. The rocks in a rock fall may be many different sizes.

MUDFLOWS

A **mudflow** is a rapid movement of a large amount of mud. Mudflows can happen when a lot of water mixes with soil and rock. The water makes the slippery mud flow downhill very quickly. A mudflow can carry away cars, trees, houses, and other objects that are in its path. ☑

READING CHECK

5. Define What is a mudflow?

____

____

Mudflows are common in mountain regions when a long dry season is followed by heavy rain. Mudflows may also happen when trees and other plants are cut down. Without plant roots to hold soil in place and help water drain away, large amounts of mud can quickly form.

Copyright © by Holt, Rinehart and Winston. All rights reserved.

CREEP

Not all mass movement is fast. In fact, very slow mass movement is happening on almost all slopes. **Creep** is the name given to this very slow movement of material downhill. Even though creep happens very slowly, it can move large amounts of material over a long period of time. ☑

Many factors can affect creep. Water can loosen soil and rock so that they move more easily. Plant roots can cause rocks to crack and can push soil particles apart. Burrowing animals, such as moles and gophers, can loosen rock and soil particles. All of these factors may make creep more likely.

READING CHECK

6. Compare How is creep different from the other kinds of mass movement that are mentioned in this section?

______

______

Type of mass movement	Description
Landslide	Material moves suddenly and rapidly down a slope.
Rock fall	
Mudflow	
	Material moves downhill very slowly.

TAKE A LOOK

7. Describe Fill in the blank spaces in the table.

How Does Mass Movement Affect Land Use?

Mass movement can cause serious property damage, injury, and even death. Therefore, people need to consider mass movement when they decide how to use land. For example, when people build roads through mountains, rock falls may happen on the sides of the roads. People may put metal fences on the sides of the road to prevent rocks from damaging cars on the road.

Many of the short-term effects of sudden, large mass movements are harmful to people and to other organisms. However, the long-term effects of mass movements may be helpful. For example, minerals and rocks may be exposed at the surface after a landslide. The minerals and rocks can be used by people or can break down to form new, rich soil. ☑

READING CHECK

8. Explain How can mass movements be helpful?

______

______

______

Copyright © by Holt, Rinehart and Winston. All rights reserved.

Name ____________ Class ____________ Date ____________

Section 4 Review

6.2.d

SECTION VOCABULARY

creep the slow downhill movement of weathered rock material **landslide** the sudden movement of rock and soil down a slope	**mass movement** the movement of a large mass of sediment or a section of land down a slope **mudflow** the flow of a mass of mud or rock and soil mixed with a large amount of water **rock fall** the rapid mass movement of rock down a steep slope or cliff

1. List What are four kinds of mass movement?

2. Explain Why is it important for people to think about mass movement when they decide how to use land?

3. Identify Relationships How is mass movement related to the angle of repose?

4. Identify What force causes mass movements?

5. Compare How are landslides different from mudflows?

6. List Give four things that can affect a material's angle of repose.

Copyright © by Holt, Rinehart and Winston. All rights reserved.

Name ______________________ Class ____________ Date ____________

CHAPTER 11 Rivers and Groundwater

SECTION 1

The Active River

BEFORE YOU READ

After you read this section, you should be able to answer these questions:

- How does moving water change the surface of Earth?
- What is the water cycle?
- Which factors affect the rate of stream erosion?

California Science Standards

6.2.a, 6.2.b, 6.4.a

What Is Erosion?

Six million years ago, the Colorado River began carving through rock to form the Grand Canyon. Today, the river has carved through 1.6 km (about 1 mi) of rock!

Before the Grand Canyon was formed, the land was flat. Then the rock in the area began to lift upward because of plate tectonics. As Earth's crust lifted upward, water began running downhill. The moving water cut into the rock and started forming the Grand Canyon.

Over millions of years, water cut into rock through the process of erosion. During **erosion**, wind, water, ice, and gravity move soil and rock from one place to another. Water is the main force in forming the Grand Canyon and in changing the Earth's landscape. ☑

STUDY TIP

Make A List As you read, make a list of the different ways in which water can change the landscape of Earth.

READING CHECK

1. Identify What formed the Grand Canyon?

____

____

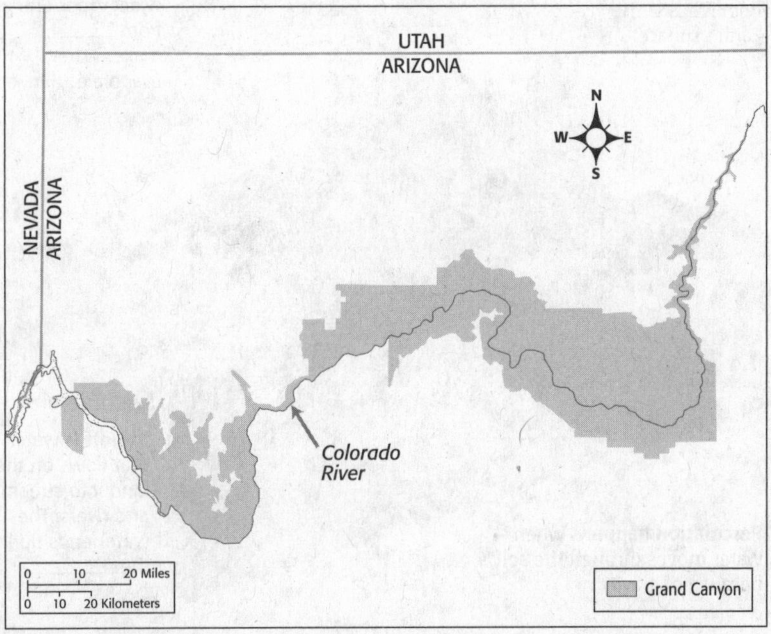

UTAH
ARIZONA
NEVADA
ARIZONA
N
W
E
S
Colorado River
0 10 20 Miles
0 10 20 Kilometers
Grand Canyon

Six million years ago, the Colorado River started flowing through northern Arizona. Today, it has carved the Grand Canyon, which is about 1.6 km deep and 446 km long.

Math Focus

2. Calculate How long is the Grand Canyon in miles? Show how you got your answer.

____

____

____

Copyright © by Holt, Rinehart and Winston. All rights reserved.

CALIFORNIA STANDARDS CHECK

6.4.a Students know the sun is the major source of energy for phenomena on Earth's surface; it powers winds, ocean currents, and the water cycle.

Word Help: major
of great importance or large scale

Word Help: energy
what makes things happen

Word Help: phenomena
any facts or events that can be sensed or described scientifically

3. Explain How does the sun help the water cycle work?

____

____

____

____

How Does the Water Cycle Work?

Have you ever wondered where the water in rivers comes from? It is part of the water cycle. The **water cycle** is the nonstop movement of water between the air, the land, and the oceans. The major source of energy that drives the water cycle is the sun.

In the water cycle, water comes to Earth's surface from the clouds in the form of rain, snow, sleet, or hail. The water moves downward through the soil or flows over the land. Water that flows over the land collects in streams and rivers and flows to the oceans.

Energy from the sun changes the water on Earth's surface into a gas that rises up to form clouds. The gas is called *water vapor*. The water vapor moves through the atmosphere until it falls to Earth's surface again.

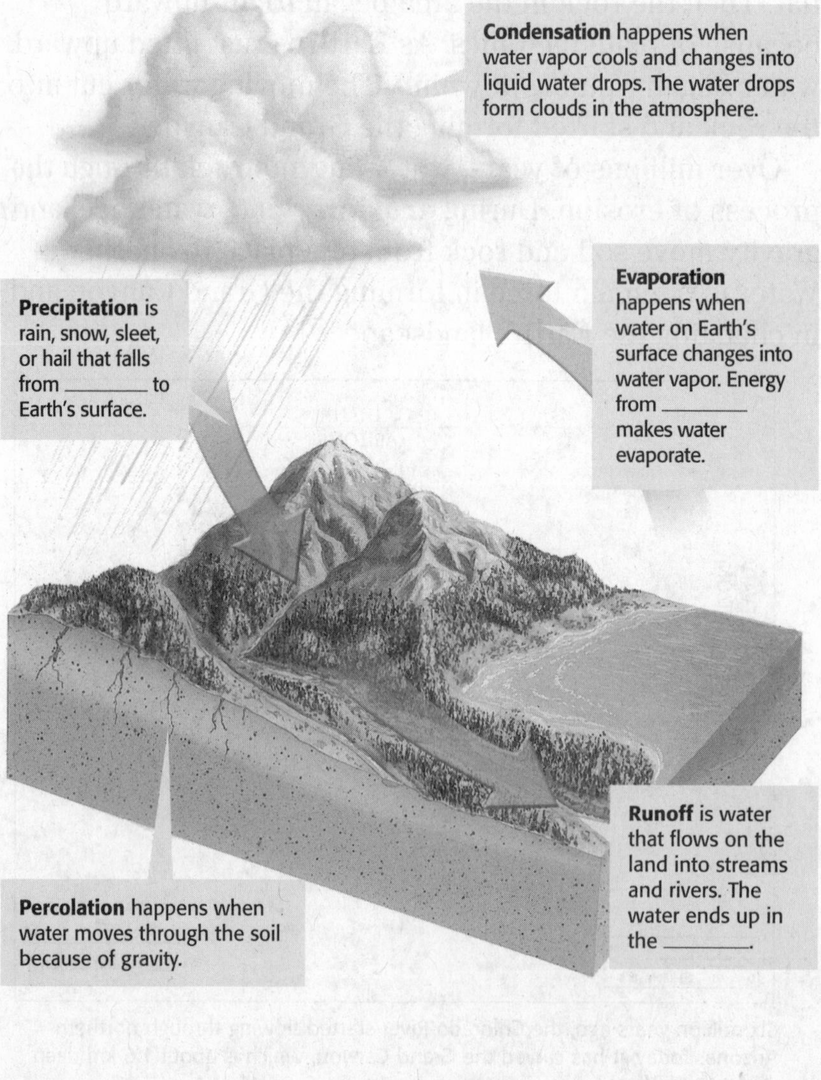
Condensation happens when water vapor cools and changes into liquid water drops. The water drops form clouds in the atmosphere.
Precipitation is rain, snow, sleet, or hail that falls from ________ to Earth's surface.
Evaporation happens when water on Earth's surface changes into water vapor. Energy from ________ makes water evaporate.
Runoff is water that flows on the land into streams and rivers. The water ends up in the ________.
Percolation happens when water moves through the soil because of gravity.

TAKE A LOOK

4. Identify In the figure, fill in the blank lines with the correct words.

Copyright © by Holt, Rinehart and Winston. All rights reserved.

What Is a River System?

What happens when you turn on the shower in your family's bathroom? When water hits the shower floor, the individual drops of water join together to form small streams. The small streams join together to form larger streams. The larger streams carry the water down the drain.

The water in your shower is like the water in a river system. A *river system* is a group of streams and rivers that drain an area of land. A *river* is a stream that has many tributaries. A **tributary** is a stream that flows into a lake or a larger stream. ☑

READING CHECK

5. Define What is a river system?

____

____

____

How Do River Systems Work?

River systems are divided into areas called watersheds. A **watershed** is the land that is drained by a river system. Many tributaries join together to form the rivers in a watershed.

The largest watershed in the United States is the Mississippi River watershed. It covers over one third of the United States. It has hundreds of tributaries. The Mississippi River watershed drains into the Gulf of Mexico.

Watersheds are separated from each other by an area of higher ground called a **divide**. All of the rivers on one side of a divide flow away from it in one direction. All of the rivers on the other side of the divide flow away from it in the opposite direction. The Continental Divide separates the Mississippi River watershed from the watersheds in the western United States.

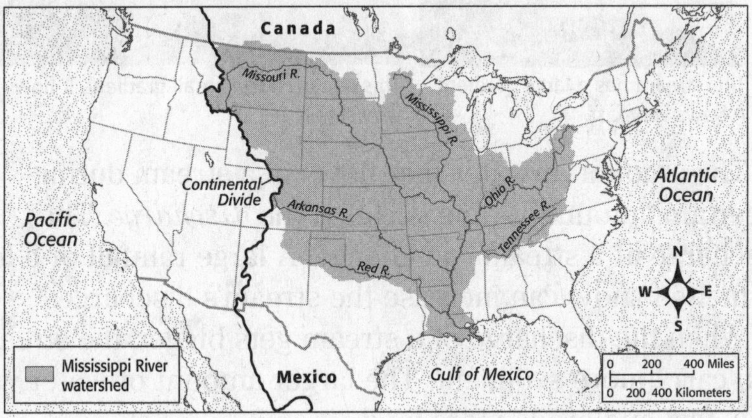
Canada
Missouri R.
Mississippi R.
Continental Divide
Arkansas R.
Ohio R.
Tennessee R.
Red R.
Pacific Ocean
Atlantic Ocean
N
W
E
S
Mississippi River watershed
Mexico
Gulf of Mexico
0 200 400 Miles
0 200 400 Kilometers

TAKE A LOOK

6. List Name three rivers that are tributaries to the Mississippi River.

____

____

____

Copyright © by Holt, Rinehart and Winston. All rights reserved.

What Is a Stream Channel?

A stream forms as water wears away soil and rock to make a channel. A **channel** is the path that a stream follows. As more soil and rock are washed away, the channel gets wider and deeper.

Over time, tributaries flow into the main channel of a river. The main channel has more water in it than the tributaries. The larger amount of water makes the main channel longer and wider. ☑

READING CHECK

7. Explain Why is the main channel of a river longer and wider than the channels of its tributaries?

____

____

____

What Causes Stream Erosion?

Gradient is a measure of the change in height over a certain distance. Gradient can be used to measure how steep a stream is. The left-hand picture below shows a stream with a high gradient. The water in this stream is moving very fast. The fast-moving water easily washes away rock and soil.

A river that is flat, as in the right-hand picture, has a low gradient and flows slowly. The slow water washes away less rock and soil.

TAKE A LOOK

8. Explain How does the gradient of a stream affect how much erosion it causes?

____

____

____

____

This stream has a large gradient. It flows very fast.

This stream has a small gradient. It flows very slowly.

The amount of water that flows in a stream during a certain amount of time is called the *discharge*. The discharge of a stream can change. A large rainfall or a lot of melted snow can increase the stream's discharge.

When the discharge of a stream gets bigger, the stream can carry more sediment. The larger amount of water will flow fast and erode more land.

Copyright © by Holt, Rinehart and Winston. All rights reserved.

How Does a Stream Carry Sediment?

A stream's **load** is the material carried in the stream's water. A fast-moving stream can carry large rocks. The large rocks speed up the stream's erosion by knocking away more rock and soil.

A slow-moving stream carries smaller rocks in its load. The smaller particles erode less rock and soil. The stream also carries material that is dissolved in the water.

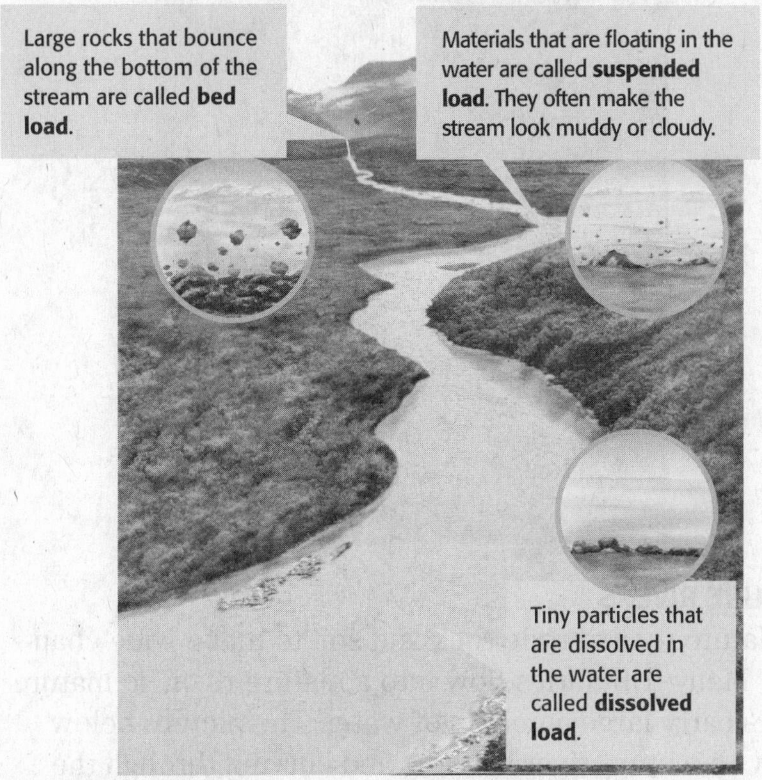
Large rocks that bounce along the bottom of the stream are called **bed load**.
Materials that are floating in the water are called **suspended load**. They often make the stream look muddy or cloudy.
Tiny particles that are dissolved in the water are called **dissolved load**.

How Do Scientists Describe Rivers?

All rivers have different features. These features can change with time. Many factors, such as weather, surroundings, gradient, and load, control the changes in a river. Scientists use special terms to describe rivers with certain features.

CALIFORNIA STANDARDS CHECK

6.2.b. Students know that rivers and streams are dynamic systems that erode, transport sediment, change course, and flood their banks in natural and recurring patterns.

Word Help: dynamic
active; tending toward change

Word Help: erode
to wear away

Word Help: transport
to carry from one place to another

9. Identify How does a stream carry material from one place to another? Give three ways.

____

____

____

Copyright © by Holt, Rinehart and Winston. All rights reserved.

YOUTHFUL RIVERS

Youthful rivers are fast-flowing rivers with steep gradients. They often flow over rapids and waterfalls. Youthful rivers make narrow, deep channels for the water to flow in. The picture below shows a youthful river.

TAKE A LOOK
10. Describe What features of this river tell you that it is a youthful river?

____

____

MATURE RIVERS

Mature rivers erode rock and soil to make wide channels. Many tributaries flow into a mature river, so mature rivers carry large amounts of water. The picture below shows a mature river bending and curving through the land. The curves and bends are called *meanders*. ☑

READING CHECK
11. Explain Why do mature rivers carry a lot of water?

____

____

TAKE A LOOK
12. Identify Label the meanders on this picture of a mature river.

Copyright © by Holt, Rinehart and Winston. All rights reserved.

REJUVENATED RIVERS

Rejuvenated rivers form where land has been raised up by plate tectonics. This gives a river a steep gradient. Therefore, rejuvenated rivers flow fast and have deep channels. As shown in the picture below, steplike gradients called *terraces* usually form along the sides of rejuvenated rivers.

TAKE A LOOK

13. Identify Label the terraces on this picture of a rejuvenated river.

OLD RIVERS

Old rivers have very low gradients. Instead of widening and deepening its channel, an old river deposits soil and rock along its channel. Since very few tributaries flow into an old river, the river does not quickly erode land. Old rivers have wide, flat floodplains and many meanders. In the picture below, a bend in an old river's channel has eroded into a lake. This is called an *oxbow lake*.

Old river
Oxbow lake

Critical Thinking

14. Infer Very few tributaries flow into an old river. Do you think it will have a large or a small discharge?

____

Copyright © by Holt, Rinehart and Winston. All rights reserved.

Name ______________________ Class ______________ Date ______________

Section 1 Review

6.2.a, 6.2.b, 6.4.a

SECTION VOCABULARY

channel the path that a stream follows

divide the boundary between drainage areas that have streams that flow in opposite directions

erosion the process by which wind, water, ice, or gravity transports soil and sediment from one location to another

load the materials carried in a stream

tributary a stream that flows into a lake or into a larger stream

water cycle the continuous movement of water between the atmosphere, the land, and the oceans

watershed the area of land that is drained by a river system

1. Explain Why do most rivers have wider channels than most streams?

____

____

2. Show a Sequence Fill in the Process Chart to show what happens in the water cycle.

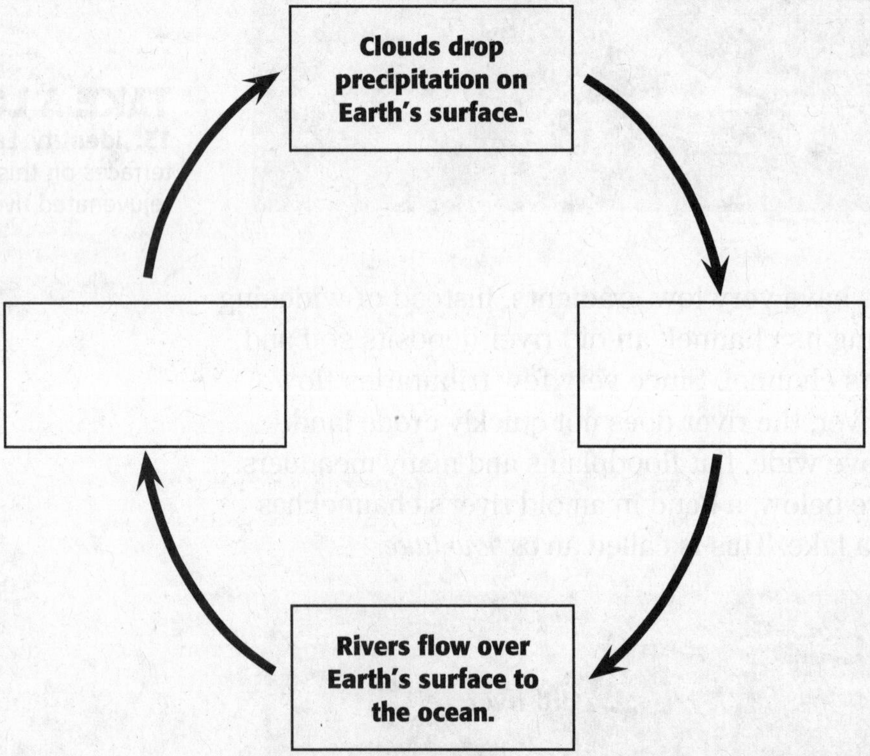

Clouds drop precipitation on Earth's surface.
Rivers flow over Earth's surface to the ocean.

3. Identify What is the main source of energy for the water cycle?

____

4. Describe How do rivers change Earth's surface?

____

____

____

Copyright © by Holt, Rinehart and Winston. All rights reserved.

CHAPTER 11 Rivers and Groundwater

SECTION 2

Stream and River Deposits

BEFORE YOU READ

After you read this section, you should be able to answer these questions:

- What types of deposits are caused by streams?
- Why do floods happen?
- How do floods from rivers and streams affect humans?

California Science Standards

6.2.a, 6.2.b, 6.2.d

How Do Rivers and Streams Rebuild Land?

Earlier, you learned that rivers and streams erode rock and soil. However, rivers also carry this loose rock and soil to new places. Rivers can help to form new land through deposition. **Deposition** is the process in which material is laid down, or dropped.

The rock and soil that a river erodes from the land move downstream in the river's load. When the flow of water slows down, the river deposits, or drops, some of its load. Material, such as rock and soil, that is deposited by rivers and streams is called *sediment.* ☑

In a river, erosion happens on the outside of a bend, where the water flows quickly. Deposition happens on the inside of a bend where the water flows more slowly.

STUDY TIP

Graphic Organizer In your notebook, create a table to compare the different landforms created by stream deposits.

READING CHECK

1. Define What is sediment?

____

____

____

Erosion happens along the outside of a bend, where water is flowing ______________.
Deposition happens along the inside of a bend, where water is flowing ______________.

TAKE A LOOK

2. Identify Fill in the blanks in the figure with the correct words.

Copyright © by Holt, Rinehart and Winston. All rights reserved.

What Are Deltas and Alluvial Fans?

Rivers deposit their loads of rocks and soil when their flow of water slows down. When a river enters the ocean, it flows much more slowly. Therefore, it deposits its load into the ocean.

When the river enters the ocean, it deposits its load in a fan-shaped pattern called a **delta**. The river deposits form new land and build new coastline.

Rivers and streams can also deposit their loads on dry land. When a fast-moving stream flows from a mountain onto flat land, the stream slows down quickly. As it slows down, the stream deposits its rocks and soil in a fan-shaped pattern known as an **alluvial fan**.

Say It

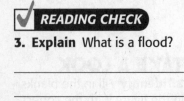

Investigate Learn about a place in the world where people live on a delta. Give a short talk to your class about the place you studied.

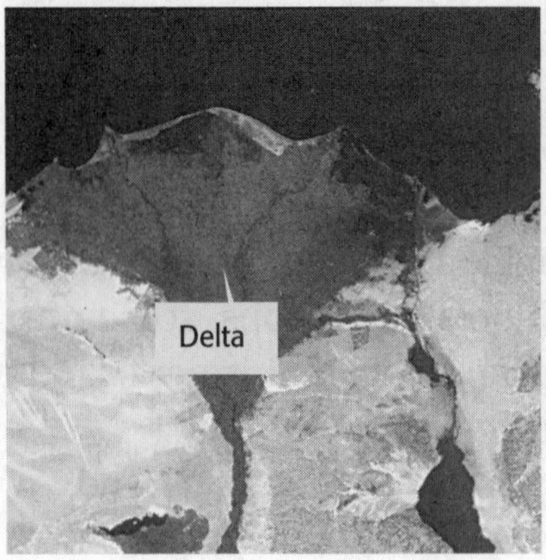
Delta

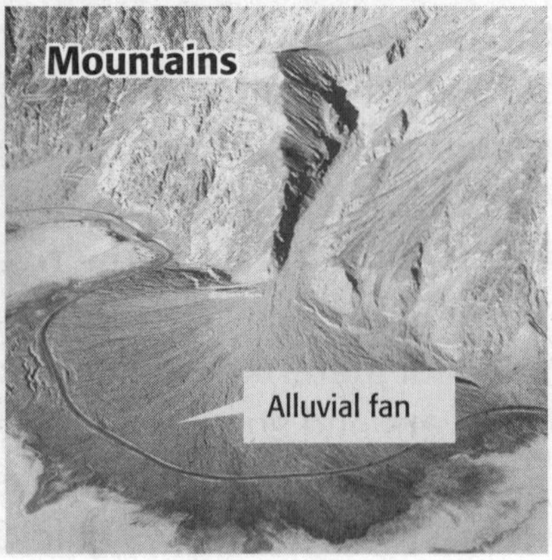
Mountains
Alluvial fan

Why Do Rivers and Streams Flood?

Rivers and streams are always changing. Rivers and streams have different amounts of water in them in different seasons. If there is a lot of rain or a lot of snow melting, a stream will have a lot of water in it.

Floods are natural events that happen with the change of seasons. A *flood* happens when there is too much water for the channel to hold. The stream flows over the sides of its channel. ☑

During a flood, the land along the sides of the stream is covered in water. The stream drops its sediment on this land. This area is called a **floodplain**. Floodplains are good areas for farming because flooding brings new soil to the land.

READING CHECK

3. Explain What is a flood?

____

____

____

____

____

Copyright © by Holt, Rinehart and Winston. All rights reserved.

What Are the Effects of Floods?

Floods are powerful and can cause a lot of damage. They can ruin homes and buildings. They can also wash away land where animals and people live.

In 1993, when the Mississippi River flooded, farms and towns in nine states were damaged. Floods can cover roads and carry cars and people downstream. They can also drown people and animals. ☑

Floods sometimes happen very fast. After a very heavy rainstorm, water can rush over the land and cause a *flash flood*. Flash floods can be hard to predict and are dangerous.

READING CHECK

4. List List three ways that floods can be harmful to people and animals.

____

____

____

Floodwater can flow strongly enough to move cars. These people were trapped in their car in a flash flood.

TAKE A LOOK

5. Explain Why can a flash flood be dangerous to people driving in cars?

____

____

____

____

Humans try to control flooding by building barriers. One kind of barrier is called a dam. A *dam* is a barrier that can guide floodwater to a reservoir.

Another barrier that people build is called a levee. A *levee* is a wall of sediment on the side of a river. The barrier helps keep the river from flooding the nearby land. ☑

READING CHECK

6. Identify What do people do to try to protect themselves from a flood?

____

____

____

____

Copyright © by Holt, Rinehart and Winston. All rights reserved.

Name ______________________ Class ______________ Date ______________

Section 2 Review

6.2.a, 6.2.b, 6.2.d

SECTION VOCABULARY

alluvial fan a fan-shaped mass of material deposited by a stream when the slope of the land decreases sharply **delta** a fan-shaped mass of material deposited at the mouth of a stream	**deposition** the process in which material is laid down **floodplain** an area along a river that forms from sediments deposited when the river overflows its banks

1. Explain Why do floods happen?

____

2. Compare Complete the table to describe the features of different kinds of stream deposits.

Type of Deposit	How is it formed?	Where is it formed?	When is it formed?
Alluvial fan			
Delta			
Floodplain			

3. Explain How can a flood be both helpful and harmful to people?

____

____

____

____

4. Compare How are dams and levees different?

____

____

5. Explain What causes floods?

____

____

____

Copyright © by Holt, Rinehart and Winston. All rights reserved.

CHAPTER 11 Rivers and Groundwater

SECTION 3

Using Water Wisely

BEFORE YOU READ

After you read this section, you should be able to answer these questions:

- How does California get its water supply?
- Why should people in California conserve water?
- How can people protect their water supply?

California Science Standards

6.6.b

Where Is Fresh Water Found?

Although the Earth is covered with oceans, lakes, and rivers, only 3% of the Earth's water is fresh water. Most of that 3% is frozen in the polar icecaps. Therefore, people must take care of and protect their water resources.

Some of the Earth's fresh water is found in streams and lakes. However, a large amount of water is also found underground. Rainwater and water from streams move through the soil and into the spaces between rocks underground. The water found inside the underground rocks is called *groundwater*. The rock layer that stores groundwater is called an **aquifer**. ☑

Groundwater is found beneath the Earth's surface. The upper surface of groundwater is called the **water table**. The water table is the top of an aquifer. When there is less water in the aquifer, the water table will be lower.

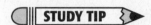

STUDY TIP

Summarize As you read, underline the important ideas in this section. When you are finished reading, write a one- or two-paragraph summary of the section, using the underlined ideas.

READING CHECK

1. Define What is groundwater?

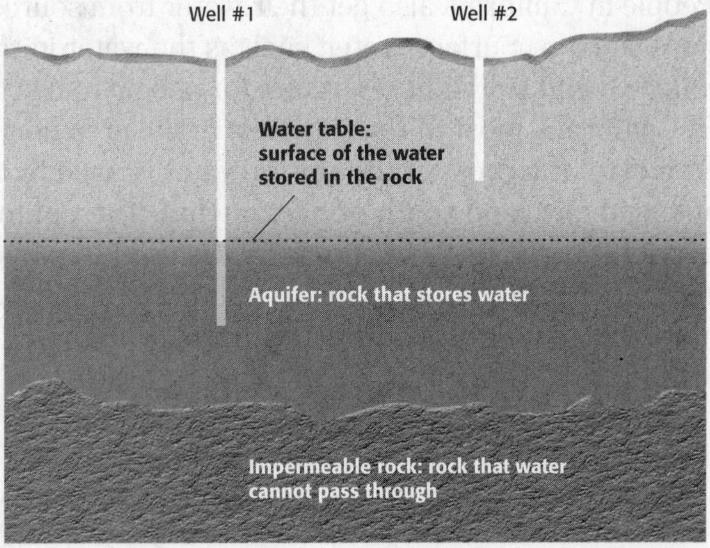

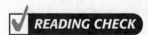

Well #1
Well #2
Water table: surface of the water stored in the rock
Aquifer: rock that stores water
Impermeable rock: rock that water cannot pass through

TAKE A LOOK

2. Identify Two people drilled wells to try to get water out of the ground. The white bars in the figure show where the two people drilled their wells. Which of the wells will probably produce water? Explain your answer.

Copyright © by Holt, Rinehart and Winston. All rights reserved.

Where Does Water in California Come From?

People in California get some of their water from aquifers. The shaded areas on the map below show the state's aquifers. However, the aquifers in California can't supply enough water for all the people who live there. Many of these aquifers are polluted. In others, *overdraft* occurs. That means more water is being taken out than is being replaced by rain. ☑

READING CHECK

3. Explain Why can't all of the water in California's aquifers be used for drinking?

____

____

TAKE A LOOK

4. Identify On the map, draw an arrow showing where most of the rain in California falls. Draw another arrow showing where most of the water in California is used.

Aquifers in California

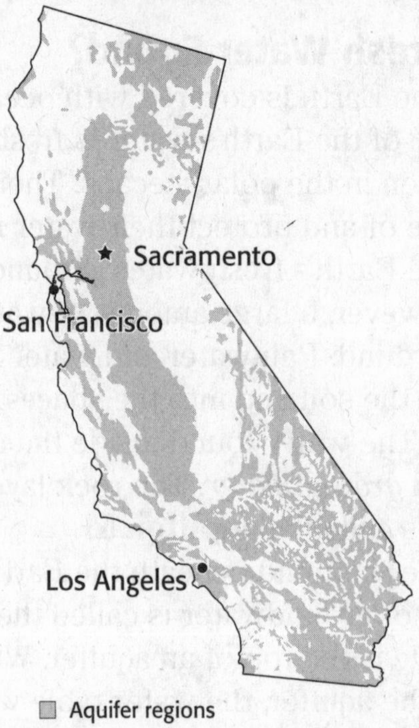
Sacramento
San Francisco
Los Angeles
Aquifer regions

People in California also get their water from sources besides aquifers. Surface water, such as the water in rivers, is collected and stored in reservoirs for people to use.

In California, most of the rain falls in the area north of Sacramento. But most of the state's water is used by the people and farmland south of Sacramento. The California Aqueduct is a system of channels that is used to move water from Northern California to Southern California. Some water in California also comes from Oregon, Colorado, and Mexico. ☑

READING CHECK

5. Explain Why does California need to get some of its water from other states and countries?

____

____

Copyright © by Holt, Rinehart and Winston. All rights reserved.

How Is Water in California Used?

California is the largest producer of crops in the United States. Up to 50% of the state's water supply is used to support the farming industry. The water supply is also used to generate power and energy for use by the state. Businesses and households use the remaining water supply.

What Is Water Pollution?

Water pollution is any waste material or harmful substance that is found in water. Water pollution can make water unsafe for organisms to live in, drink, or touch.

Most water pollution comes from the waste of cities, factories, and farms. Surface water and groundwater can become so polluted that they can kill organisms. ☑

READING CHECK

6. Describe Why is water pollution harmful?

____

____

Where Does Water Pollution Come From?

Water pollution can come from just one source or from many sources. Most of these sources are created by humans.

Pollution from a single place can be traced back to its source, and the problem can be corrected. However, when pollution comes from many sources, people must work together to figure out how to correct the problem.

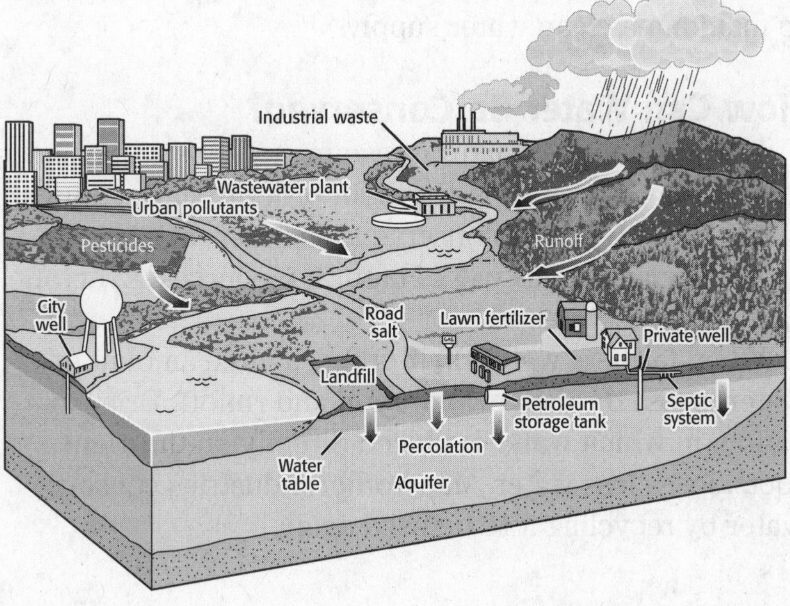
Industrial waste
Wastewater plant
Urban pollutants
Pesticides
Runoff
City well
Road salt
Lawn fertilizer
Private well
Landfill
Petroleum storage tank
Septic system
Percolation
Water table
Aquifer

TAKE A LOOK

7. Identify List four sources of groundwater pollution.

____

____

____

____

Copyright © by Holt, Rinehart and Winston. All rights reserved.

How Do Laws Help Keep Water Clean?

In 1969, the Cuyahoga River in Cleveland, Ohio, was so polluted with oil that it caught on fire. To try to prevent such shocking events from happening again, Congress passed the Clean Water Act of 1972.

The goal of the Clean Water Act of 1972 was to make all surface water clean enough for swimming and fishing by 1983. Although this goal was not reached, many states have passed tougher water-quality standards. The number of clean lakes and rivers has also increased.

The table below shows some of the laws that have been passed to help protect our water supply.

Name of the act	What the act did
Clean Water Act of 1972	
Marine Protection, Research, and Sanctuaries Act of 1972	strengthened laws against dumping chemicals and waste into the oceans
Drinking Water Act of 1975	helped to protect groundwater and surface water from pollution
Oil Pollution Act of 1990	required that all ships that carry oil near the United States have two layers of metal in their hulls

Although some of their goals have not been met, these and many other laws have improved the quality of water in the United States. However, people, businesses, and governments will have to continue to cooperate in order to maintain a clean water supply.

Say It

Investigate Learn more about one of the laws that help to protect our water supply. Share what you find out with your class or a small group.

TAKE A LOOK

8. Identify Fill in the blank box in the table to describe the Clean Water Act of 1972.

How Can Water Be Conserved?

Water is a limited natural resource. There is only a certain amount of water on Earth. In order to have enough clean water, we must conserve it. **Conservation** is the protection and smart use of natural resources. Everyone can conserve water. ☑

When farmers water their crops, a large amount of water is lost through evaporation and runoff. Drip irrigation, in which water is placed directly on the plant's roots, loses less water. Many other industries conserve water by recycling it to be used again.

READING CHECK

9. Define What is conservation?

____

____

Copyright © by Holt, Rinehart and Winston. All rights reserved.

What Can You Do?

Individual families can conserve water also. Toilets and shower heads that work with less water are good choices for conservation. If you have to water your lawn, water it at night and use a drip watering system.

Some people use an idea called xeriscaping. *Xeriscaping* is landscaping in a way that conserves water. This method uses plants that are local and used to the weather in an area, so they easily survive without extra water.

Each person can do his or her part to conserve water. Simple choices, such as taking shorter showers and turning off the water when you brush your teeth, are helpful. If everyone tries to conserve water, we can make a big difference.

Things You Can Do to Conserve Water
•
•
•
•
•

CALIFORNIA STANDARDS CHECK

6.6.b Students know different natural energy and material resources, including air, soil, rocks, minerals, petroleum, fresh water, wildlife, and forests, and know how to classify them as renewable or nonrenewable.

Word Help: energy
what makes things happen

Word Help: resource
anything that can be used to take care of a need

10. Draw Conclusions Why is it important for people to try to conserve water?

____

____

____

____

____

TAKE A LOOK

11. Brainstorm With your class, talk about ways that you can conserve water. Write down five of those ways in the table.

Copyright © by Holt, Rinehart and Winston. All rights reserved.

Name ______________________ Class ______________ Date ____________

Section 3 Review

6.6.b

SECTION VOCABULARY

aquifer a body of rock or sediment that stores groundwater and allows the flow of groundwater Wordwise the root *aqui-* means "water;" the suffix *–fer* means "to carry" **conservation** the preservation and wise use of natural resources	**water pollution** waste matter or other material that is introduced into water and that is harmful to organisms that live in, drink, or are exposed to the water **water table** the upper surface of underground water

1. Explain Why is it important to protect our water from pollution?

____

____

____

2. Analyze Why does Northern California transport water to Southern California?

____

____

3. Describe How can farmers and other industries conserve water?

____

____

4. Infer How can watering your lawn at night help to conserve water?

____

____

5. List What are four laws that help to protect our water supply?

____

____

____

____

6. Identify Name four sources of water pollution.

____

____

Copyright © by Holt, Rinehart and Winston. All rights reserved.

Name ______________________ Class ______________ Date ______________

CHAPTER 12 Exploring the Oceans

SECTION 1

Earth's Oceans

California Science Standards

6.3.c, 6.4.d

BEFORE YOU READ

After you read this section, you should be able to answer these questions:

- What affects the salinity of ocean water?
- What affects the temperature of ocean water?
- How does density affect ocean currents?

What Is the Global Ocean?

Earth has more liquid water on its surface than any other planet in the solar system. In fact, 71% of Earth's surface is covered by liquid water. Most of Earth's water is found in its oceans. There are five main oceans on Earth. However, the oceans are all connected to each other. Therefore, scientists often refer to all the oceans on Earth as the *global ocean*.

The continents divide the global ocean into the five main oceans. The largest ocean is the *Pacific Ocean*. The *Atlantic Ocean* is the second largest ocean. It has half the volume of the Pacific Ocean. The *Indian Ocean* is the third largest ocean. The *Southern Ocean* extends from the coast of Antarctica to 60°S latitude. The *Arctic Ocean* is the smallest ocean. Much of its surface is covered by ice. The figure below shows where these oceans are found. ☑

STUDY TIP

Ask Questions As you read this section, write down any questions you have. After you read, discuss your questions in a small group.

READING CHECK

1. Describe What divides the global ocean into five parts?

____

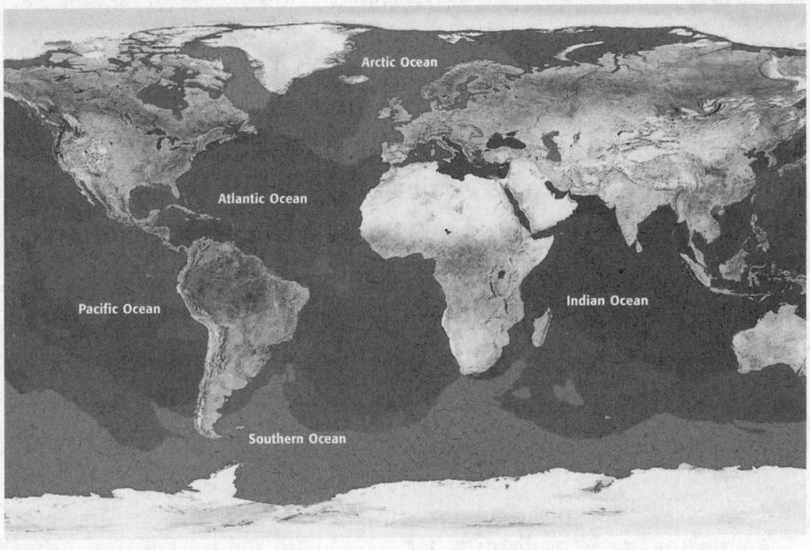

Arctic Ocean
Atlantic Ocean
Pacific Ocean
Indian Ocean
Southern Ocean

TAKE A LOOK

2. Identify What are the five main oceans?

____

____

____

____

____

Copyright © by Holt, Rinehart and Winston. All rights reserved.

Why Is Ocean Water Salty?

Ocean water is different from the water that we drink. People cannot use ocean water for drinking because it is salty.

Most of the salt in the ocean is the same kind of salt we use on our food. This type of salt is called sodium chloride. It is a compound made from the elements sodium, Na, and chlorine, Cl. Ocean water also contains other dissolved solids, including magnesium and calcium, which make the water taste salty.

Math Focus

3. Read a Graph Which two elements make up most of the dissolved solids in sea water?

______

______

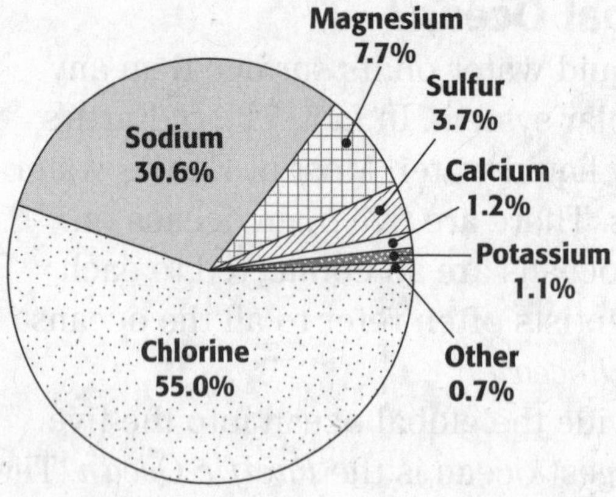
Magnesium
7.7%
Sulfur
3.7%
Sodium
30.6%
Calcium
1.2%
Potassium
1.1%
Chlorine
55.0%
Other
0.7%

This graph shows the amounts of different kinds of solids in ocean water.

Salts have collected in the ocean for billions of years. As rivers and streams flow toward the ocean, they dissolve minerals from rocks and soil. The rivers carry the dissolved minerals to the ocean. At the same time, liquid water in the oceans evaporates to form water vapor. As the water evaporates, it leaves behind the salts that were dissolved in it. ☑

READING CHECK

4. Identify Where does most of the salt in the ocean come from?

______

______

What Is Salinity?

Salinity is a measure of the amount of solid material that is dissolved in a certain amount of liquid. It is usually measured as grams of dissolved solids per kilogram. On average, ocean water has 35 g/kg of dissolved solids in it. This means that 1 kg of ocean water has about 35 g of solids dissolved in it. If you evaporated 1 kg of ocean water, 35 g of solids would remain.

Copyright © by Holt, Rinehart and Winston. All rights reserved.

EFFECTS OF LOCATION ON SALINITY

Some parts of the ocean are saltier than others. Most oceans in hot, dry climates have high salinities. In these areas, the hot weather causes water to evaporate quickly. Salt is left behind. For example, the Red Sea in the Middle East is very salty. The climate there is very hot and dry. ☑

Some parts of the ocean are less salty than others. Along the coastlines, fresh water from streams and rivers runs into the ocean. As fresh water mixes with ocean water, the salinity of the ocean water decreases. For example, the salinity of the ocean waters near the Amazon River is much lower than the salinity in other parts of the ocean.

READING CHECK

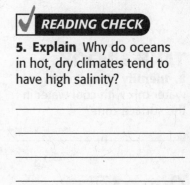

5. Explain Why do oceans in hot, dry climates tend to have high salinity?

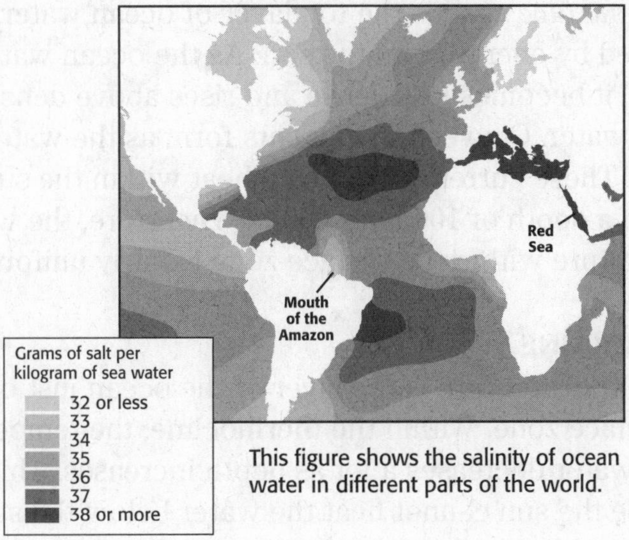
Red Sea
Mouth of the Amazon
Grams of salt per kilogram of sea water
32 or less
33
34
35
36
37
38 or more

This figure shows the salinity of ocean water in different parts of the world.

TAKE A LOOK

6. Infer The Gulf of Mexico is located between Mexico and Florida. Why is the ocean water in the Gulf of Mexico less salty than in other places?

EFFECTS OF WATER MOVEMENT ON SALINITY

The movement of water also affects salinity. Slow-moving ocean water tends to have higher salinity than fast-moving water. Parts of the ocean with slow-moving water, such as gulfs, bays, and seas, often have high salinities. Parts of the ocean without currents are likely to have higher salinities as well. ☑

READING CHECK

7. Describe How does water movement affect salinity?

What Affects Ocean Temperatures?

The temperature of ocean water decreases as depth increases. However, the temperature change is not uniform. Scientists can divide the water in the ocean into three layers based on how temperature changes. These three layers are the surface zone, the thermocline, and the deep zone.

Copyright © by Holt, Rinehart and Winston. All rights reserved.

TAKE A LOOK

8. Identify How does warm water mix with cool water in the surface zone?

____

Critical Thinking

9. Predict Consequences What would happen to the water temperature in the surface zone if convection currents did not form there? Explain your answer.

____

____

____

____

____

Ocean Temperature and Depth

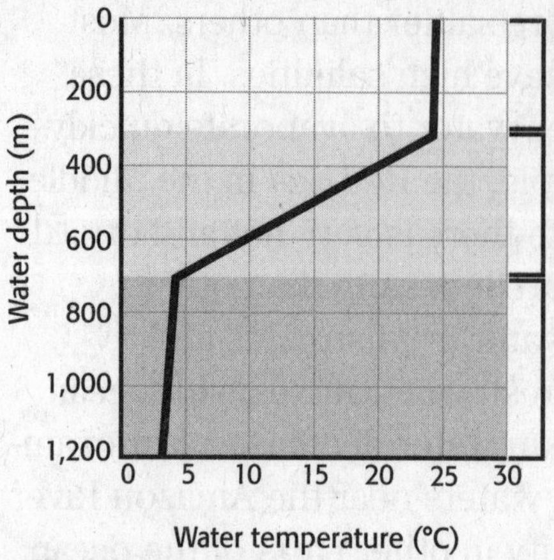
Water depth (m)
0
200
400
600
800
1,000
1,200
0 5 10 15 20 25 30
Water temperature (°C)
Surface zone The surface zone extends to about 300 m below the surface. Sunlight heats the top 100 m. Convection currents mix the heated water with the cooler water below.
Thermocline The thermocline extends to between 700 m and 1,000 m below the surface. Temperature decreases quickly as depth increases.
Deep zone The deep zone is the deepest layer of the ocean. The temperature in the deep zone is only about 2°C.

SURFACE ZONE

The *surface zone* is the top layer of ocean water. It is heated by energy from the sun. As the ocean water is heated, it becomes less dense and rises above denser, cooler water. Convection currents form as the water moves. These currents can move heat within the surface zone to a depth of 100 m to 300 m. Therefore, the water temperature within the surface zone is fairly uniform.

THERMOCLINE

The **thermocline** is the layer of the ocean just beneath the surface zone. Within the thermocline, the temperature of the water decreases a lot as depth increases. This is because the sun cannot heat the water below the surface zone. In addition, the warm water of the surface zone cannot mix easily with the water below.

The depth of the thermocline is different in different places. It can extend from 100 m to almost 1,000 m below the surface of the ocean.

DEEP ZONE

The *deep zone* is the layer below the thermocline. In the deep zone, the temperature of the water is about 2°C. This very cold water is very dense. It moves slowly across the ocean floor and forms the deep ocean currents.

Copyright © by Holt, Rinehart and Winston. All rights reserved.

CHANGES IN SURFACE TEMPERATURE

The temperature of water at the surface of the ocean is different in different places. Surface water along the equator is warmer than water at the poles. This is because more sunlight reaches the equator than the poles. Water near the surface at the equator can be up to 30°C. In the polar oceans, water at the surface can be as cold as −1.9°C. ☑

The temperature of water at the surface of the ocean can also change during different times of year. Many areas receive more sunlight in the summer than in the winter. In these areas, the surface water in the oceans is warmer in the summer.

READING CHECK

10. Explain Why is surface water warmer at the equator than at the poles?

____

____

Why Does Ocean Water Move?

Salinity and temperature both affect the density of ocean water. The large amount of solids dissolved in ocean water make it more dense than fresh water. The higher the salinity of ocean water, the higher its density.

Temperature can have a large effect on the density of ocean water. In fact, water temperature has a larger effect on the density of ocean water than salinity does. Cold water is denser than warm water. Therefore, cold water will sink below warm water in the oceans.

The differences in density in the ocean cause water to circulate in convection currents. These currents distribute heat, water, and dissolved materials throughout the oceans.

CALIFORNIA STANDARDS CHECK

6.4.d Students know convection currents <u>distribute</u> heat in the atmosphere and oceans.

Word Help: <u>distribute</u>
to spread out over an area

11. Describe How does heat move in Earth's oceans?

____

____

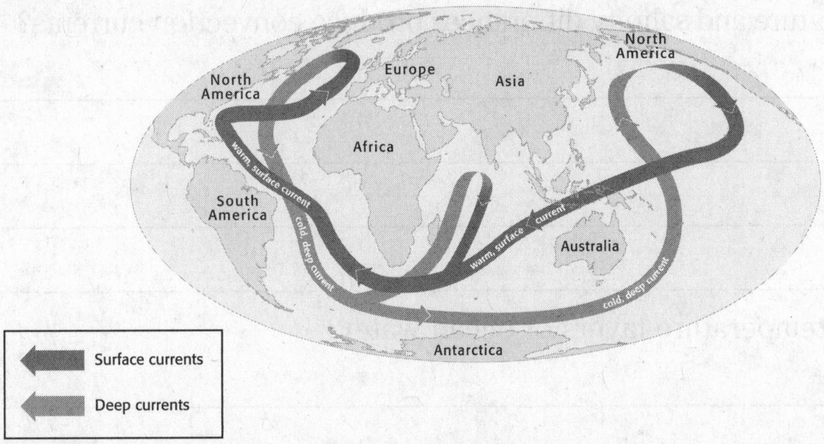
North America
Europe
Asia
North America
Africa
South America
Australia
Antarctica
warm, surface current
cold, deep current
warm, surface current
cold, deep current
Surface currents
Deep currents

The density of ocean water varies because of differences in temperature and salinity. These differences cause the water to move in currents. Cold, dense water moves from the poles toward the equator. Warm, less dense water moves from the equator to the poles. The currents move heat throughout the oceans.

Copyright © by Holt, Rinehart and Winston. All rights reserved.

Name ______________________ Class ______________ Date ____________

Section 1 Review

6.3.c, 6.4.d

SECTION VOCABULARY

salinity a measure of the amount of dissolved salts in a given amount of liquid	**thermocline** a layer in a body of water in which water temperature drops with increased depth faster than it does in other layers **Wordwise** The root *therm* means "heat." The root *clino* means "slope." Other examples are *incline, decline,* and *thermometer.*

1. Explain Why do scientists call the ocean water on Earth the global ocean?

____

____

2. Identify What are two factors that can affect salinity?

____

____

3. Identify Relationships How does salinity affect the density of ocean water?

____

____

4. Explain Why does the temperature in the thermocline decrease quickly with depth?

____

____

5. Describe How do temperature and salinity differences produce convection currents?

____

____

____

____

6. List What are the three temperature layers of ocean water?

____

____

____

Copyright © by Holt, Rinehart and Winston. All rights reserved.

Name ______________________ Class ______________ Date ______________

CHAPTER 12 Exploring the Oceans

SECTION 2

The Ocean Floor

BEFORE YOU READ

After you read this section, you should be able to answer these questions:

- What are the features of the ocean floor?
- How do the features of the ocean floor form?

California Science Standards

6.1.a, 6.1.d, 6.1.e

How Do Scientists Study the Ocean Floor?

The ocean floor can be as far as 11,000 m below the surface of the ocean. People cannot survive at such great depths without a lot of special equipment. Scientists sometimes use such equipment to travel to the ocean floor. However, scientists can also study the ocean floor from the surface of the ocean. They can even study the ocean floor from outer space!

STUDY TIP

Compare As you read, make a chart showing the similarities and differences between the different ways that scientists study the ocean floor.

STUDYING THE OCEAN FLOOR WITH SONAR

Sonar stands for ***sound navigation and ranging***. Sonar instruments on a ship send pulses of sound down into the ocean. The sound moves through the water, bounces off the ocean floor, and returns to the ship. The deeper the water, the longer it takes for the sound to return to the ship. ☑

Scientists use sonar to determine how deep the ocean floor is. They know how fast sound travels in ocean water (about 1,500 m/s). They measure how long it takes for the sound to return to the ship. Then, they use this information to figure out how deep the ocean floor is.

READING CHECK

1. Identify What is sonar?

____

____

____

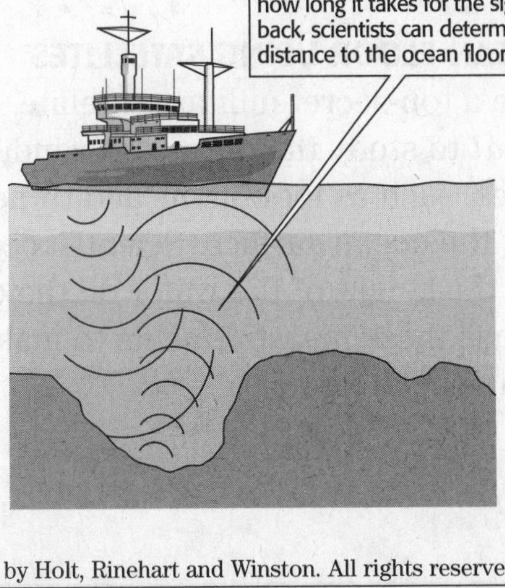

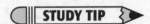

Sonar equipment is carried on a ship. The equipment sends out a pulse of sound. The sound bounces off of the ocean floor and travels back to the ship. By timing how long it takes for the signal to bounce back, scientists can determine the distance to the ocean floor.

Math Focus

2. Calculate The ship sends out a sonar signal. It picks up the return signal 4 s later. About how deep is the ocean floor? The speed of sound in ocean water is about 1,500 m/s. Show your work.

____

____

____

Copyright © by Holt, Rinehart and Winston. All rights reserved.

READING CHECK

3. Explain Why did scientists have to build a special ship to explore the deep zone?

____

____

____

Critical Thinking

4. Infer The *Deep Flight* and other deep-ocean vessels have very thick windows and walls. Why do you think this is?

____

____

____

READING CHECK

5. Describe How do scientists in the IODP study the ocean floor?

____

____

____

STUDYING THE OCEAN FLOOR WITH SHIPS

People cannot survive the high pressures on the ocean floor. Therefore, in order to explore the deep zone, scientists had to build an underwater vessel that can survive under high pressures. This vessel is called *Deep Flight.* Future models of *Deep Flight* may be able to carry scientists to 11,000 m below the ocean's surface. ☑

Scientists can also use Remotely Operated Vehicles, or ROVs, to explore the ocean floor directly. These ROVs can explore parts of the ocean floor that are too dangerous for people to travel to.

DEEP FLIGHT

The *Deep Flight* is a vessel that can carry scientists deep beneath the ocean surface.

STUDYING THE OCEAN FLOOR BY DRILLING

Scientists involved in the Integrated Ocean Drilling Program (IODP) use drilling to study the ocean floor. Scientists use special equipment to cut cores, or long tubes of rock and sediment, from the ocean floor. The layers of rock and sediment in the cores can tell scientists about the history of Earth. ☑

STUDYING THE OCEAN FLOOR USING SATELLITES

Geosat was once a top-secret military satellite. Today, scientists use *Geosat* to study the ocean floor indirectly. Underwater features, such as mountains and trenches, affect the height of the ocean surface. Scientists can use *Geosat* to measure the height of the water on the ocean surface. They can use these measurements to make detailed maps of the ocean floor.

Copyright © by Holt, Rinehart and Winston. All rights reserved.

What Are the Features of the Ocean Floor?

Because most of the ocean floor is covered by kilometers of water, it has not been thoroughly explored. However, scientists do know that many of the features of the ocean floor are caused by plate tectonics.

THE MAJOR REGIONS OF THE OCEAN FLOOR

Scientists divide the ocean floor into two major regions: the continental margin and the deep-ocean basin. The *continental margin* is the edge of a continent that is covered by ocean water. The *deep-ocean basin* begins at the edge of the continental margin and extends under the deepest parts of the ocean. ☑

There are three main parts of the continental margin: the continental shelf, the continental slope, and the continental rise. The figure below shows where these parts are found.

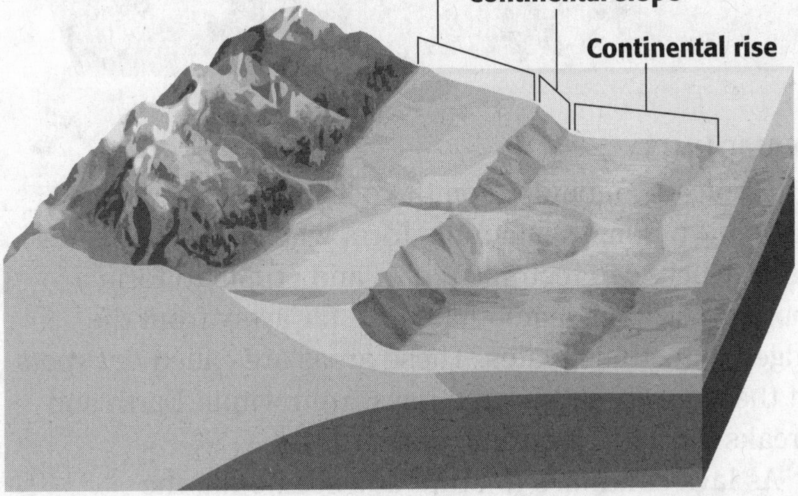
Continental shelf
Continental slope
Continental rise

The deep-ocean basin includes the abyssal plain, mid-ocean ridges, and ocean trenches. The **abyssal plain** is the flat, almost level part of the ocean basin. The abyssal plain is covered by layers of sediment. Some of the sediment comes from land. Some of it is made of the remains of dead sea creatures. These remains settle to the ocean floor when the creatures die.

READING CHECK

6. Identify What are the two major regions of the ocean floor?

____

____

TAKE A LOOK

7. Arrange Write the names of the three parts of the continental margin in order from shallowest to deepest.

____

____

____

Copyright © by Holt, Rinehart and Winston. All rights reserved.

PLATE TECTONICS AND OCEAN FLOOR FEATURES

Convection within Earth causes tectonic plates to move. When the plates touch, they can slide past each other, collide with each other, or move away from each other. Remember that the movement of tectonic plates creates features, such as mountains, on land. Tectonic movements can also form features on the ocean floor. These features include seamounts, mid-ocean ridges, and ocean trenches.

CALIFORNIA STANDARDS CHECK

6.1.e Students know major geologic events, such as earthquakes, volcanic eruptions, and mountain building, result from plate motions.

Word Help: major
of great importance or large scale

8. Infer What would the ocean floor look like if plate tectonics did not happen?

____

____

____

The Deep-Ocean Basin

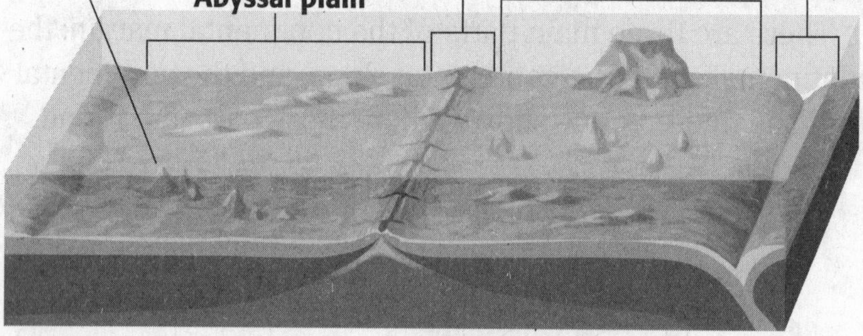
Seamount
Abyssal plain
Mid-ocean ridge
Abyssal plain
Ocean trench

SEAMOUNTS

A volcanic mountain on the ocean floor is called a *seamount*. Some seamounts form when magma pushes its way between tectonic plates and erupts to form a mountain. Other seamounts form far away from the edges of tectonic plates. These areas are called *hot spots*. At these locations, magma rises from within Earth and breaks through a tectonic plate. ☑

As lava continues to erupt at a seamount, the mountain gets taller. If it gets tall enough to rise above the ocean's surface, it is called a volcanic island. The islands of the state of Hawaii are volcanic islands.

READING CHECK

9. Define What is a seamount?

____

____

Copyright © by Holt, Rinehart and Winston. All rights reserved.

MID-OCEAN RIDGES

A long mountain chain that forms on the floor of the ocean is a **mid-ocean ridge**. Mid-ocean ridges form where tectonic plates move apart. This motion produces a crack in the ocean floor called a *rift*. Magma rises through the rift and cools to form new rock. The ridge is built up from this new rock. ☑

Most mid-ocean ridges are far below the ocean surface. However, Iceland is one place on Earth where the mid-ocean ridge has risen above the ocean's surface.

READING CHECK

10. Identify Where do mid-ocean ridges form?

OCEAN TRENCHES

Long, narrow valleys in the deep-ocean basins are called **ocean trenches**. Trenches form where tectonic plates are moving together. As the plates move toward each other, one plate sinks under the other plate in a process called *subduction*. ☑

Subduction causes pressure to increase on the plate that is sinking. Water and other fluids are squeezed out of the rock and sediments on this plate. These fluids cause mantle rock to melt and form magma. The magma rises to the surface and erupts to form a chain of volcanoes.

Ocean trenches are some of the deepest places on Earth. For example, the Marianas Trench in the Pacific Ocean is nearly 11,000 m deep.

READING CHECK

11. Define Write your own definition for *ocean trench*.

Feature	What it is	Where it is found
Seamount		
	flat part of the ocean basin, covered in sediment	in the deep-ocean basins
		where plates move apart
Ocean trench	long, narrow valley on the ocean floor	

TAKE A LOOK

12. Describe Fill in the table to compare the different features of the ocean floor.

Copyright © by Holt, Rinehart and Winston. All rights reserved.

Name ____________ Class ____________ Date ____________

Section 2 Review

6.1.a, 6.1.d, 6.1.e

SECTION VOCABULARY

abyssal plain a large, flat, almost level area of the deep-ocean basin

mid-ocean ridge a long, undersea mountain chain that forms along the floor of the major oceans

ocean trench a long, narrow, and steep depression on the ocean floor that forms when one tectonic plate subducts beneath another plate; trenches run parallel to volcanic island chains or to the coastlines of continents; also called a trench or a deep-ocean trench

1. List Name four ways that scientists study the ocean floor.

____

____

____

____

2. Describe How do plate movements form ocean trenches?

____

____

____

3. Identify Where does the sediment on the abyssal plain come from? Give two sources.

____

____

4. Explain How do mid-ocean ridges form?

____

____

____

____

5. Explain How can scientists map the ocean floor using the satellite *Geosat*?

____

____

____

____

Copyright © by Holt, Rinehart and Winston. All rights reserved.

SECTION 3

Resources from the Ocean

BEFORE YOU READ

After you read this section, you should be able to answer these questions:

- What are the living resources from the ocean?
- What are the nonliving resources from the ocean?

California Science Standards

6.6.a, 6.6.b, 6.6.c

What Are the Living Resources of the Ocean?

People have been harvesting plants and animals from the ocean for thousands of years. Today, harvesting food from the ocean is a multi-billion-dollar industry. As the population of humans on Earth has grown, the demand for these resources has increased. However, the availability of these resources has not increased as much.

Harvesting fish from the ocean can harm the environment. Usually, fish can reproduce faster than people can catch them. However, new technology, such as drift nets, has allowed people to catch more fish in less time. This may allow people to take fish faster than they can reproduce. This could cause the population of fish in the oceans to decrease. Also, other animals, such as dolphins and turtles, can be caught in fishing nets.

Recently, laws have been passed that regulate fishing more strictly. These laws are designed to help reduce the amount of damage fishing causes to the environment. As a result of these laws, people have begun to raise fish and other types of seafood, such as shellfish and seaweed, in farms near the shore.

New technology, such as these drift nets, allows people to catch more fish in less time. However, other animals, such as dolphins and turtles, can sometimes get caught in the nets.

STUDY TIP

Summarize As you read, make a chart describing the resources that people use from the ocean.

CALIFORNIA STANDARDS CHECK

6.6.b Students know different natural energy and material resources, including air, soil, rocks, minerals, petroleum, fresh water, wildlife, and forests, and know how to classify them as renewable or nonrenewable.

Word Help: resource
anything that can be used to take care of a need

1. Explain Why can fish be considered to be a renewable resource? When would fish not be considered a renewable resource?

____

____

____

____

____

____

____

Copyright © by Holt, Rinehart and Winston. All rights reserved.

What Are the Nonliving Resources of the Ocean?

Fish and other seafood are important resources that people take from the oceans. However, people also take many nonliving resources from the oceans. These resources include energy resources and material resources.

FRESH WATER

Fresh water is often considered a renewable resource. However, in parts of the world where the climate is dry, fresh water is limited. In these parts of the world, ocean water is desalinated to provide fresh drinking water. **Desalination** is the process of removing salt from sea water. ☑

Most desalination plants heat ocean water to cause the water to evaporate. The water vapor, which is not salty, is collected and condensed into liquid fresh water. Another method of desalination involves passing the ocean water through a membrane to leave the salts behind. However, no matter what process is used, desalination is expensive and can be slow.

READING CHECK

2. Define What is desalination?

____

____

TIDAL ENERGY

The ocean is constantly moving as tides come in and go out. People can use the motion of the water to generate electricity. Energy that is generated from the movement of tides is called *tidal energy*. Tidal energy is clean, inexpensive, and renewable. However, it can only be used in certain parts of the world.

Critical Thinking

3. Infer Why can't tidal energy be used everywhere?

____

____

____

High tide
Gate closes
Gate closed
Low tide
Gate opens

❶ As the tide rises, water enters a bay behind a dam. The gate closes when high tide reaches its peak.

❷ The gate remains closed as the tide falls.

❸ At low tide, the gate opens, and water rushes through the dam. The water moves turbines, which generate electricity.

TAKE A LOOK

4. Explain Why is tidal energy considered to be renewable?

____

____

Copyright © by Holt, Rinehart and Winston. All rights reserved.

OIL AND NATURAL GAS

Oil and natural gas are considered the most valuable resources in the ocean. Oil and natural gas form from the remains of tiny plants and animals. These remains take millions of years to turn into oil and natural gas. Therefore, oil and gas are nonrenewable resources.

Many deposits of oil and natural gas are found in rock near the continental margins. In order to obtain these resources, engineers must drill wells through the rock. About one-fourth of the world's oil is now obtained from wells in rock beneath the oceans.

Oil is refined by manufacturers to make gasoline. Gasoline powers vehicles and generators that make electricity. Oil is also used to make plastic and other products. ☑

Say It

Discuss In a small group, talk about different ways that you use resources from the ocean every day.

READING CHECK

5. List Give two ways that people use oil or natural gas.

____

____

MINERALS

Many different kinds of minerals can be found on the ocean floor. These minerals are commonly in the form of nodules. *Nodules* are potato-shaped lumps of minerals that form from chemicals dissolved in ocean water.

Nodules can be made of many different kinds of minerals. Most nodules contain the element manganese. Manganese can be used to make certain kinds of steel. Some nodules contain the valuable metals iron, copper, nickel, or cobalt. Some contain phosphorus, which can be used in fertilizer.

Nodules can be very large. They may contain a large amount of valuable minerals. However, they form in the very deep parts of the ocean. For this reason, they are difficult to mine. ☑

READING CHECK

6. Explain Why are nodules hard to mine?

____

____

Minerals can be found on the ocean floor in the form of nodules. These nodules are difficult to mine because they are found in very deep water.

Copyright © by Holt, Rinehart and Winston. All rights reserved.

Name ____________________ Class ____________ Date ____________

Section 3 Review

6.6.a, 6.6.b, 6.6.c

SECTION VOCABULARY

desalination a process of removing salt from ocean water	

1. List Name two living resources from the ocean.

2. Define Write your own definition for desalination.

3. Describe How can the tides can be used to generate electricity?

4. Identify Give five minerals that may be found in nodules.

5. List Give four nonliving ocean resources.

6. Infer Why are people starting to farm the oceans instead of harvesting wild organisms?

Copyright © by Holt, Rinehart and Winston. All rights reserved.

Name ________________ Class ____________ Date ____________

CHAPTER 12 Exploring the Oceans

SECTION 4

Ocean Pollution

California Science Standards

6.6.a

BEFORE YOU READ

After you read this section, you should be able to answer these questions:

- What are the different types of pollution in the ocean?
- How can we preserve ocean resources?

What Pollutes the Ocean?

Many human activities produce pollution that can harm the oceans. Some of this pollution comes from a specific source. Pollution that can be traced to one source is called **point-source pollution**. However, some pollution comes from many sources. Pollution that cannot be traced to a single source is called **nonpoint-source pollution**. ☑

STUDY TIP

Summarize As you read, underline the main ideas in each paragraph. When you finish reading, write a short summary of the section using the ideas you underlined.

READING CHECK

1. Define Write your own definition for nonpoint-source pollution.

____

____

____

TRASH DUMPING

People dump trash in many places, including the ocean. In the 1980s, scientists became alarmed by the kinds of trash that were washing up on beaches. Bandages, vials of blood, and other medical wastes were found among the trash.

The Environmental Protection Agency (EPA) found that hospitals in the United States were dumping medical wastes into the oceans. Much of this waste is now buried in sanitary landfills. However, dumping trash in the ocean is still common in many countries.

Trash thrown into the ocean can affect the organisms that live there. It also affects the organisms, such as people, that depend on the ocean for food. For example, most plastic material that is thrown into the ocean does not break down for thousands of years. Animals can mistake plastic material for food and choke on it.

Marine animals, such as this bird, can choke on plastic trash that is thrown into the oceans.

TAKE A LOOK

2. Describe How can trash harm the organisms that live in the oceans? Give two ways.

____

____

____

____

Copyright © by Holt, Rinehart and Winston. All rights reserved.

SLUDGE DUMPING

Raw sewage is all of the liquid and solid wastes that are flushed down toilets and poured down drains. In most places, raw sewage is collected and sent to a treatment plant. The treatment removes solid waste and cleans the raw sewage. The solid waste that remains is called *sludge.* ☑

In many places, people dump sludge into the ocean. Currents can stir up the sludge and move it closer to shore. The sludge can pollute beaches and kill ocean life. Many countries have banned sludge dumping. However, it still happens in many parts of the world.

READING CHECK

3. Compare How is raw sewage different from sludge?

OIL SPILLS

Oil is used by most of the world as an energy source. However, oil is only found in certain places around the world. Therefore, large tankers must transport billions of barrels of oil across the oceans. Sometimes, the tankers break open and the oil spills out of them.

Oil spills can cause many problems for the environment. Oil is poisonous to plants and animals. It is also very hard to clean up oil spills, so their effects can last for a long time. ☑

In 1990, the Oil Pollution Act was passed. This law states that all oil tankers that travel in United States waters must have two hulls. If the outer hull of a ship is damaged, the inner hull can keep oil from spilling into the ocean.

The Oil Pollution Act may help to prevent large oil spills. However, as the figure below shows, big spills only cause about 5% of the oil pollution in the ocean. Most of the oil in the ocean comes from nonpoint-source pollution on land.

READING CHECK

4. Explain Why can the effects of an oil spill last a long time?

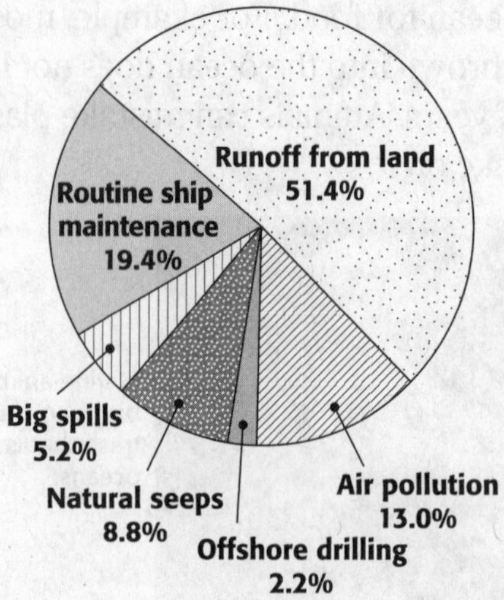
Runoff from land
51.4%
Routine ship maintenance
19.4%
Big spills
5.2%
Natural seeps
8.8%
Offshore drilling
2.2%
Air pollution
13.0%

Math Focus

5. Read a Graph What are the three largest sources of oil pollution in the oceans?

Copyright © by Holt, Rinehart and Winston. All rights reserved.

NONPOINT-SOURCE POLLUTION

Nonpoint-source pollution is pollution that comes from many sources instead of a single place. Most ocean pollution is nonpoint-source pollution. Things that people do on land can pollute rivers. The rivers can carry the pollution into the oceans. ☑

Nonpoint-source pollution is hard to control because it enters the water in many different ways. However, there are things that people can do to help reduce nonpoint-source pollution. For example, we can throw away chemicals, such as used motor oil, properly instead of washing them down the drain.

READING CHECK

6. Explain How can human activities on land cause ocean pollution?

______

______

______

Oil and gasoline can leak out of cars and trucks and onto the streets. Rain can wash the oil and gasoline into rivers, which carry them to the oceans.

Boats and other watercraft can leak oil, gasoline, and other chemicals into the water.

People use chemicals to help their lawns grow. Rain can wash the chemicals into rivers, which carry them to the oceans.

TAKE A LOOK

7. Identify Give three examples of nonpoint-source pollution.

______

______

______

How Can We Protect Our Ocean Resources?

People have begun to take steps to save and protect our ocean resources. From international treaties to volunteer cleanups, efforts to conserve and protect ocean resources are making a difference around the world.

Copyright © by Holt, Rinehart and Winston. All rights reserved.

NATIONS TAKE NOTICE

In the 1970s and 1980s, ocean pollution was very bad. Many countries realized that they would need to work together to reduce ocean pollution. In 1989, 64 countries signed a treaty that bans the dumping of many harmful materials into the ocean.

Many other treaties and laws have also been passed to help protect the oceans. For example, Congress passed the Clean Water Act in 1972. This law gave the Environmental Protection Agency more control over the trash that is dumped into the ocean.

Another law, the U.S. Marine Protection, Research, and Sanctuaries Act, was also passed in 1972. This law forbids people from dumping harmful materials into the oceans. These laws have helped to reduce the pollution entering the oceans. However, waste dumping and oil spills still happen.

Critical Thinking

8. Infer Why do countries need to work together to reduce ocean pollution?

____

____

____

____

____

____

____

CITIZENS TAKE CHARGE

Citizens of many different countries have demanded that their governments do more to prevent ocean pollution. They have also begun to take the matter into their own hands. For example, people began to organize beach cleanups. Millions of tons of trash have been gathered from beaches. Also, people are helping to spread the word about the problems with dumping wastes into the oceans. ☑

READING CHECK

9. Explain How can individual people help to reduce ocean pollution?

____

____

____

During a beach cleanup, many people work together to remove trash from a beach. This helps make the beach safer for people, other animals, and plants.

Copyright © by Holt, Rinehart and Winston. All rights reserved.

Name ______________________ Class ______________ Date ______________

Section 4 Review

6.6.a

SECTION VOCABULARY

nonpoint-source pollution pollution that comes from many sources rather than from a single, specific site	**point-source pollution** pollution that comes from a specific site

1. **Compare** How is point-source pollution different from nonpoint-source pollution?

____

____

2. **Describe** Fill in the table below to describe different sources of ocean pollution.

Type of pollution	Description
Trash dumping	
Oil spills	
Chemical leaks	

3. **Infer** Most of the trash in the United States is buried instead of being dumped into the oceans. However, there is still trash in the oceans. Why is this?

____

____

____

4. **Identify** What type of pollution is most ocean pollution?

____

____

5. **Describe** Why is nonpoint-source pollution hard to control?

____

____

____

6. **List** Give two United States laws that protect ocean resources.

____

____

Copyright © by Holt, Rinehart and Winston. All rights reserved.

Name ______________________ Class ______________ Date ______________

CHAPTER 13 The Movement of Ocean Water

SECTION 1

Currents

California Science Standards

6.3.c, 6.4.a, 6.4.d

BEFORE YOU READ

After you read this section, you should be able to answer these questions:

- What factors affect ocean currents?
- Why are ocean currents important?

What Are Surface Currents?

STUDY TIP

Summarize As you read, make a diagram showing the types of ocean currents and the factors that affect them.

Imagine that you are stranded on an island. You write a note and put it into a bottle. You throw the bottle into the ocean to communicate with the outside world. Can you predict where the bottle will end up? If you understand ocean currents, you can! The oceans contain many streamlike movements of water called *ocean currents*. There are two main kinds of ocean currents: surface currents and deep currents.

Surface currents are horizontal, streamlike movements of water that are found at or near the surface of the ocean. Surface currents can be up to several hundred meters deep. They can be as long as several thousand kilometers. Three factors affect surface currents: global winds, the Coriolis effect, and continental deflections. ☑

READING CHECK

1. Define Write your own definition for *surface current*.

____

____

____

____

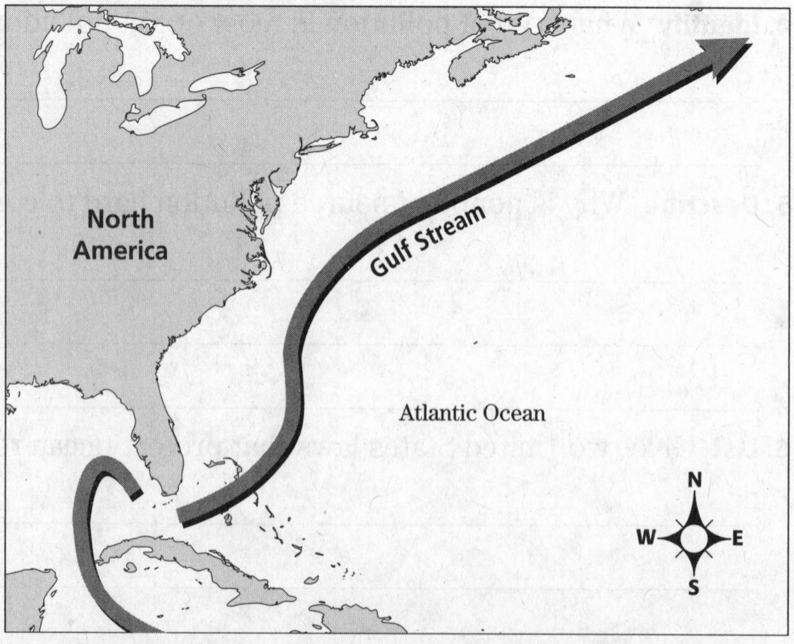
North America
Gulf Stream
Atlantic Ocean
N
W
E
S

The Gulf Stream is one of the largest surface currents in the world. Every year, it transports at least 25 times as much water as all of the rivers on Earth combined!

TAKE A LOOK

2. Read a Map In what direction does the Gulf Stream flow?

____

Copyright © by Holt, Rinehart and Winston. All rights reserved.

GLOBAL WINDS

Have you ever blown across a bowl of hot soup? You may have noticed that your breath pushes the soup across the surface of the bowl. In much the same way, winds that blow across the surface of Earth's oceans push water across Earth's surface. This process causes surface currents in the ocean. ☑

Many winds blow across Earth's surface, but they do not all blow in the same direction. Near the equator, the winds blow mostly east to west. Between 30° and 60° north and south latitudes, the winds blow mostly west to east.

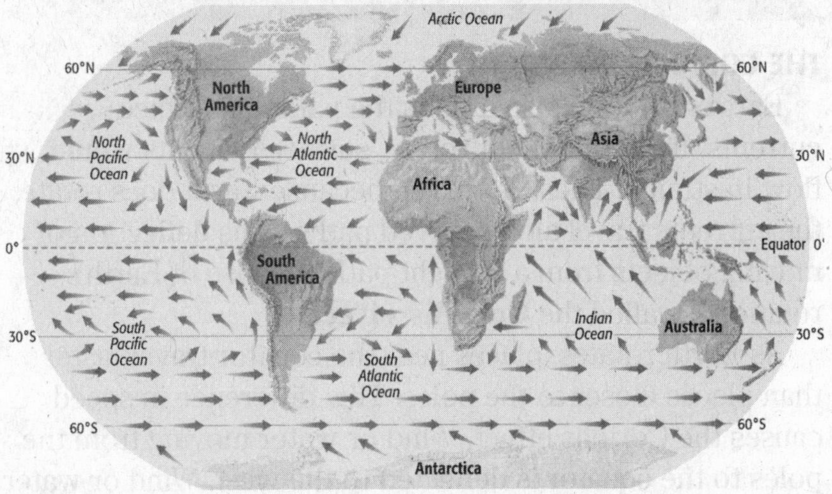
Arctic Ocean
60°N
60°N
North America
Europe
North Pacific Ocean
North Atlantic Ocean
Asia
30°N
30°N
Africa
0°
Equator 0'
South America
Indian Ocean
Australia
30°S
South Pacific Ocean
30°S
South Atlantic Ocean
60°S
60°S
Antarctica

Winds are important in producing surface currents. The winds near Earth's surface do not all blow in the same direction.

The sun causes winds to blow, and winds cause surface currents to form. Therefore, the major source of the energy that powers surface currents is the sun. The sun heats air near the equator more than it heats air at other latitudes. As a result, the air near the equator is warmer and less dense than the surrounding air. The warm air rises and produces an area of low pressure. Cooler air rushes into this area of low pressure, producing wind.

CONTINENTAL DEFLECTIONS

If Earth's surface were covered only with water, surface currents would travel across the oceans in a uniform pattern. However, water does not cover all of Earth's surface. Continents cover about one-third of Earth's surface. When surface currents meet continents, the currents *deflect*, or change direction. The figure on the top of the next page shows this process.

READING CHECK

3. Explain How do winds cause surface currents?

____

____

____

____

CALIFORNIA STANDARDS CHECK

6.4.a Students know the sun is the major source of energy for phenomena on Earth's surface; it powers winds, ocean currents, and the water cycle.

Word Help: major
of great importance or large scale

Word Help: source
the thing from which something else comes

Word Help: energy
what makes things happen

Word Help: phenomenon
(plural, *phenomena*) any facts or events that can be sensed or described scientifically

4. Explain How does the sun power surface currents?

____

____

____

____

Copyright © by Holt, Rinehart and Winston. All rights reserved.

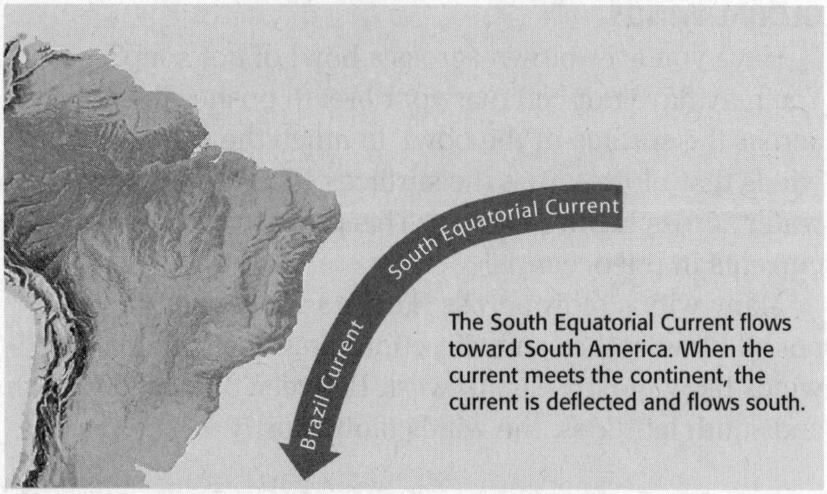
South Equatorial Current
Brazil Current
The South Equatorial Current flows toward South America. When the current meets the continent, the current is deflected and flows south.

TAKE A LOOK

5. Predict Consequences What would probably happen to the South Equatorial Current if South America were not there?

____

____

____

THE CORIOLIS EFFECT

Earth's rotation also affects the paths of surface currents. If Earth did not rotate, surface currents would flow in straight lines. However, because Earth does rotate, the currents travel along curved paths. This deflection of moving objects from a straight path because of Earth's rotation is called the **Coriolis effect**. ☑

READING CHECK

6. Describe How does Earth's rotation affect the paths of surface currents?

____

____

____

As Earth rotates, places near the equator travel faster than places closer to the poles. This difference in speed causes the Coriolis effect. Wind or water moving from the poles to the equator is deflected to the west. Wind or water moving from the equator to the poles is deflected east. The figure below shows examples of these paths.

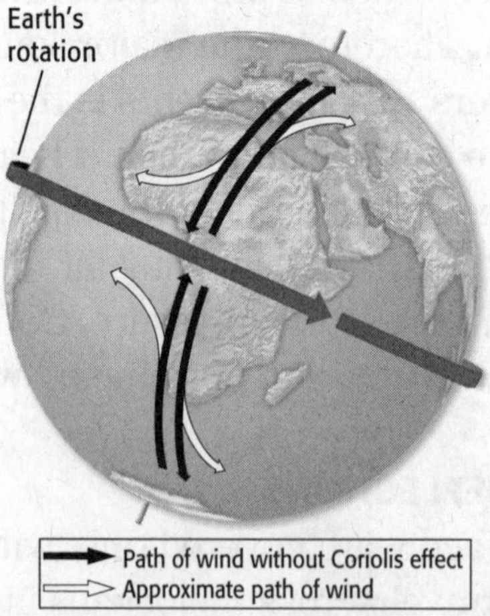
Earth's rotation
Path of wind without Coriolis effect
Approximate path of wind

The Coriolis effect causes wind and water to move along curved paths.

TAKE A LOOK

7. Apply Concepts A surface current starts at the equator near the west coast of Africa and begins moving north. In which direction will the current end up moving?

____

The Coriolis effect is most noticeable for objects that travel very fast or travel over long distances. Over short distances or with slow moving objects, the rotation of the Earth does not make much of a difference.

Copyright © by Holt, Rinehart and Winston. All rights reserved.

HOW SURFACE CURRENTS DISTRIBUTE HEAT

Surface currents help to move heat from one part of Earth to another. Water near the equator absorbs heat energy from the sun. Then, warm-water currents carry the heat from the equator to other parts of the ocean. The heat from the warm-water currents moves into colder water or into the atmosphere. This transfer of energy through the movement of matter is called *convection*.

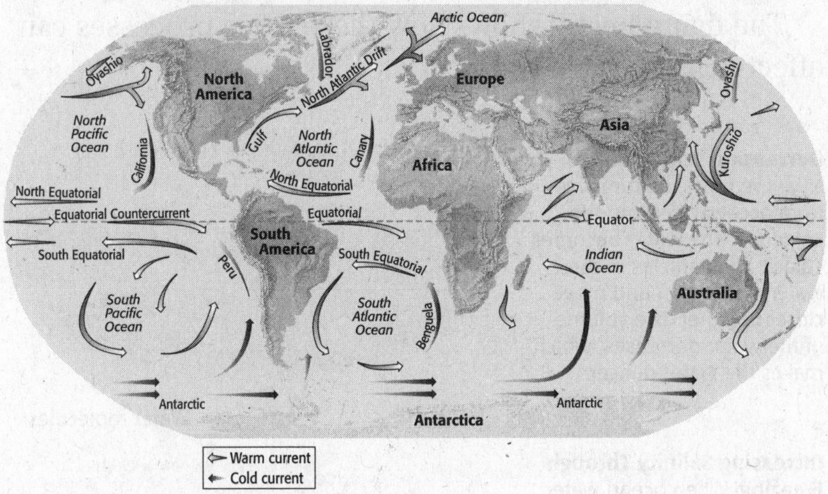
Arctic Ocean
Labrador
North Atlantic Drift
Europe
Oyashio
North America
Oyashio
North Pacific Ocean
Gulf
North Atlantic Ocean
Canary
Asia
California
Africa
Kuroshio
North Equatorial
North Equatorial
Equatorial Countercurrent
Equatorial
Equator
South America
South Equatorial
Peru
South Equatorial
Indian Ocean
Australia
South Pacific Ocean
South Atlantic Ocean
Benguela
Antarctic
Antarctica
Antarctic
Warm current
Cold current

This map shows Earth's major surface currents. Surface currents help to distribute heat across Earth's surface.

TAKE A LOOK

8. Identify Which surface current carries warm water along the equator toward the west coast of South America?

____

What Are Deep Currents?

Not all ocean currents are found at the surface. Movements of ocean water far below the surface are called **deep currents**. Unlike surface currents, deep currents are not controlled by wind. The movements of deep currents are controlled by differences in water density. ☑

Density is the amount of matter in a given space or volume. The density of ocean water is affected by salinity and temperature. *Salinity* is a measure of the amount of salts or solids dissolved in a liquid. Water with a high salinity is denser than water with a low salinity. Cold water is denser than warm water.

READING CHECK

9. Compare How are deep currents different from surface currents? Give two ways.

____

____

____

____

Copyright © by Holt, Rinehart and Winston. All rights reserved.

HOW DEEP CURRENTS FORM

Deep currents form when the density of ocean water increases, causing it to sink toward the bottom of the ocean. There are three main ways that the density of ocean water can increase.

- Temperature can decrease when heat moves from warm water to cold air or water.
- Salinity can increase when ocean water freezes.
- Salinity can increase when ocean water evaporates.

The figure below shows how these three processes can affect density.

Critical Thinking

10. Infer Why do most deep currents form near the poles?

____

____

____

____

Decreasing Temperature Near the poles, heat moves from ocean water into the colder air. The water becomes colder. The particles in the water slow down and move closer together. The volume of the water decreases, which makes the water denser.

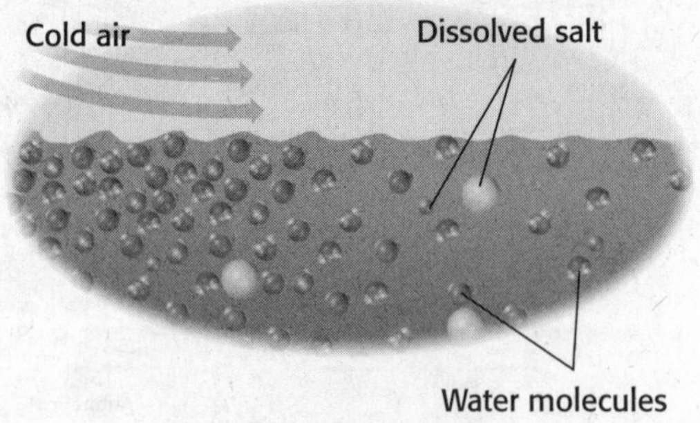
Cold air
Dissolved salt
Water molecules

TAKE A LOOK

11. Explain How does freezing cause ocean water to become denser?

____

____

____

____

____

Increasing Salinity Through Freezing When ocean water freezes, the salt in the ocean water does not become part of the ice. The salt remains in the water that has not frozen. This process increases the salinity of the water, and the water becomes denser.

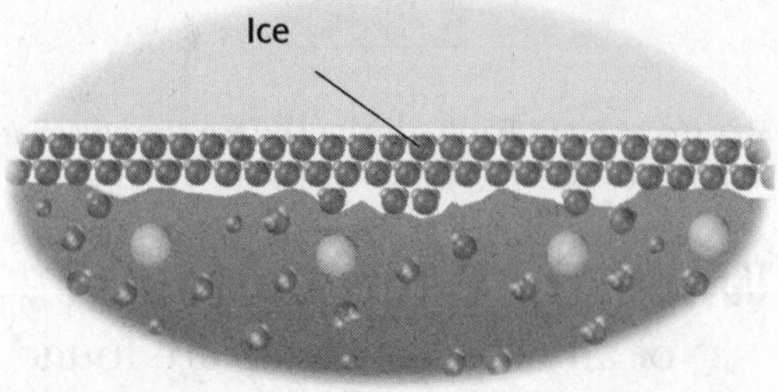
Ice

Increasing Salinity Through Evaporation When ocean water evaporates, the salt in the water remains in the liquid. This process increases the salinity of the water, and the water becomes denser.

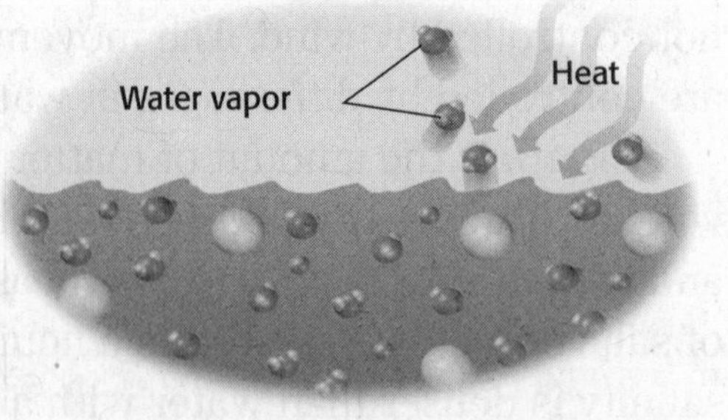
Water vapor
Heat

There are several main deep currents in the ocean. The deepest and densest water in the ocean is the Antarctic Bottom Water, which forms near Antarctica. North Atlantic Deep Water is less dense and forms in the North Atlantic Ocean. Water that is less dense stays above denser water. Therefore, North Atlantic Deep Water stays on top of Antarctic Bottom Water when the two meet.

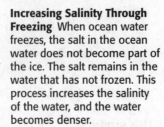

Copyright © by Holt, Rinehart and Winston. All rights reserved.

How Do Convection Currents Affect Earth?

Remember that convection is the transfer of heat as matter moves. Both surface currents and deep currents transfer energy. Therefore, they are also called **convection currents**. Convection currents carry heat from the equator toward the poles. The figure below shows a model of how convection currents work.

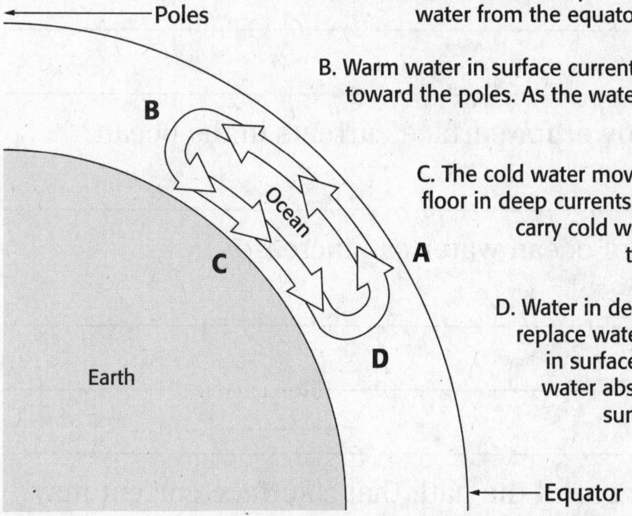
A. Water near the ocean's surface absorbs heat from the sun. Surface currents carry the warmer, less-dense water from the equator toward the poles.
B. Warm water in surface currents cools as it moves toward the poles. As the water cools, it becomes denser and sinks.
C. The cold water moves along the ocean floor in deep currents. The deep currents carry cold water from the poles toward the equator.
D. Water in deep currents rises to replace water that moves away in surface currents. The cold water absorbs heat from the sun as it gets near the surface, and the cycle continues.
Poles
B
Ocean
A
C
D
Earth
Equator

CALIFORNIA STANDARDS CHECK

6.4.d Students know convection currents distribute heat in the atmosphere and oceans.

Word Help: distribute
to spread out over an area

12. Explain Why are surface and deep currents considered convection currents?

______

______

______

Why Are Ocean Currents Important?

Deep currents and surface currents are like highways that molecules of water travel on. However, currents carry more than just water around Earth. Materials in the water also move within currents.

Remember that convection currents carry heat throughout the oceans. Ocean currents also carry oxygen and nutrients. Deep currents transport oxygen from the surface to the deep ocean. As water rises to the surface, nutrients from deep in the ocean are brought to the surface along with the water.

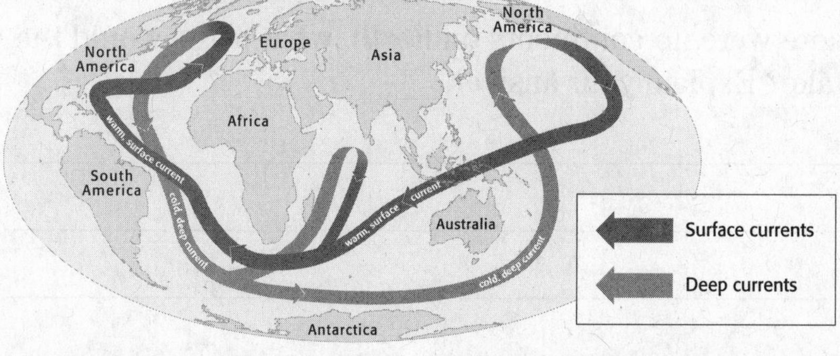
North America
Europe
Asia
North America
Africa
South America
warm, surface current
cold, deep current
warm, surface current
Australia
cold, deep current
Antarctica
Surface currents
Deep currents

This map shows the main paths of global ocean circulation.

TAKE A LOOK

13. Identify Name three things that ocean currents transport.

______

______

______

Copyright © by Holt, Rinehart and Winston. All rights reserved.

Name ______________________ Class ____________ Date ____________

Section 1 Review

6.3.c, 6.4.a, 6.4.d

SECTION VOCABULARY

convection current any movement of matter that results from differences in density; may be vertical, circular, or cyclical

Coriolis effect the curving of the path of a moving object from an otherwise straight path due to the Earth's rotation

deep current a streamlike movement of ocean water far below the surface

surface current a horizontal movement of ocean water that is caused by wind and that occurs at or near the ocean's surface

1. Define Write your own definition for *convection current.*

____

____

2. Identify What is the main source of power for surface currents in the ocean?

____

3. List Give three ways that the density of ocean water can increase.

____

____

____

4. Describe What are three factors that control the path that a surface current may take?

____

____

____

5. Infer Deep currents carry oxygen from the surface to the deep ocean. What is the most likely reason that there is more oxygen in water near the surface?

____

____

____

6. Predict Consequences If there were no continents on Earth, what paths would the ocean's surface currents take? Explain your answer.

____

____

____

____

Copyright © by Holt, Rinehart and Winston. All rights reserved.

SECTION 2

Currents and Climate

California Science Standards

6.4.e

BEFORE YOU READ

After you read this section, you should be able to answer these questions:

- How do surface currents affect climate?
- How do changes in surface currents affect climate?

How Do Surface Currents Affect Earth?

Surface currents can have a large impact on climate. The temperature of the water at the surface of the ocean affects the air above it. Warm water can heat air and produce warmer air temperatures. Cold water can absorb heat and produce cooler air temperatures.

STUDY TIP

Summarize in Pairs Read this section quietly to yourself. Then, talk about the section with a partner. Together, try to answer any questions that you have.

WARM-WATER CURRENTS AND CLIMATE

Surface currents can make coastal areas warmer than inland areas at the same latitude. For example, Great Britain and Newfoundland, Canada, are located at about the same latitude. However, the Gulf Stream flows close to Great Britain. The warm water of the Gulf Stream warms the air around Great Britain. As a result, Great Britain has a milder climate than Newfoundland.

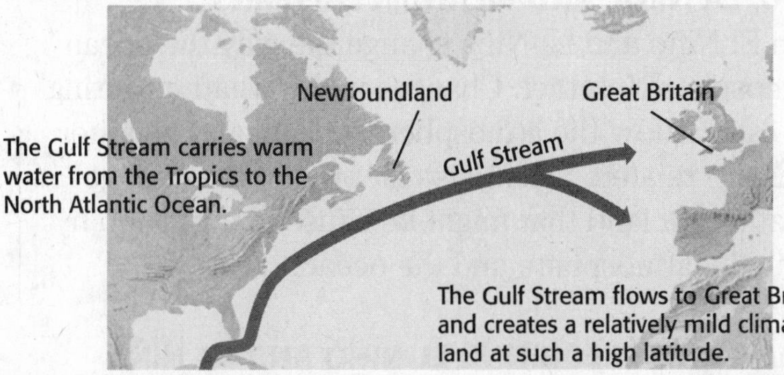
Newfoundland
Great Britain
Gulf Stream
The Gulf Stream carries warm water from the Tropics to the North Atlantic Ocean.
The Gulf Stream flows to Great Britain and creates a relatively mild climate for land at such a high latitude.

TAKE A LOOK

1. Explain Why is Great Britain's climate milder than Newfoundland's?

______

______

______

______

COLD-WATER CURRENTS AND CLIMATE

Cold-water currents also affect coastal areas. Coastal areas near cold currents tend to have cooler climates than inland areas at the same latitude. For example, the California Current is a cold-water current that flows near the West Coast of the United States. As a result, the climate along the West Coast is usually cooler than the climate of areas further inland. The figure on the top of the next page shows the location of the California Current.

Copyright © by Holt, Rinehart and Winston. All rights reserved.

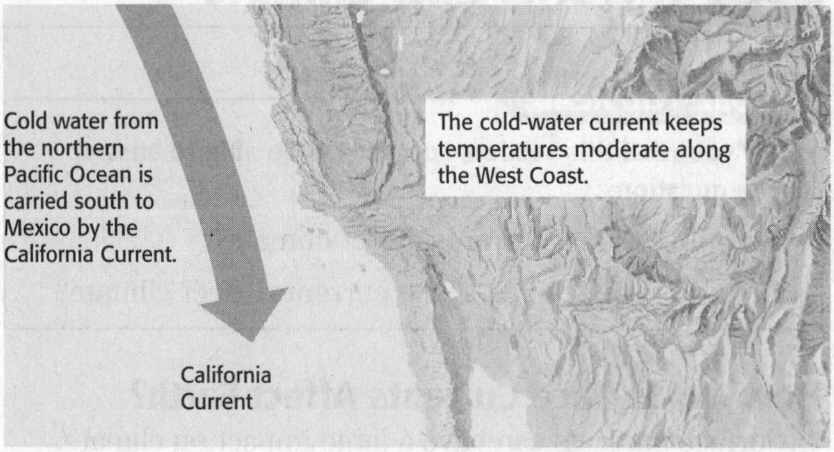
Cold water from the northern Pacific Ocean is carried south to Mexico by the California Current.
The cold-water current keeps temperatures moderate along the West Coast.
California Current

TAKE A LOOK

2. Identify In which direction does the California Current flow?

____

How Do El Niño and La Niña Affect Climate?

Every 2 to 12 years, the South Pacific trade winds move less warm water to the western Pacific than usual. As a result, surface-water temperatures along the west coast of South America rise. Over time, this warming spreads westward. This periodic change in the location of warm and cool surface waters is called **El Niño**. El Niño events can last a year or longer.

Sometimes, El Niño is followed by La Niña. **La Niña** happens when surface-water temperatures in the eastern Pacific become unusually cool.

EL NIÑO, LA NIÑA, AND WEATHER PATTERNS

Both El Niño and La Niña change the way the ocean and atmosphere interact. Changes in the weather during El Niño show how the atmosphere, ocean, and weather patterns are related. Scientists can predict the changes in the weather on land that might be caused by El Niño by studying the atmosphere and the ocean.

Critical Thinking

3. Apply Concepts El Niño happens when there is warmer water near the west coast of South America. Why do scientists collect information about air temperatures in order to help them predict El Niño and La Niña?

____

____

____

STUDYING AND PREDICTING EL NIÑO AND LA NIÑA

To study El Niño and La Niña, scientists collect data with buoys anchored to the ocean floor along the equator. The buoys record data about water temperature, air temperature, currents, and winds. The data sometimes show that the South Pacific trade winds are weaker than usual. The data may also show that the surface-water temperatures in the oceans have increased. Either of these changes can tell scientists that El Niño is likely to happen.

Copyright © by Holt, Rinehart and Winston. All rights reserved.

EFFECTS OF EL NIÑO

El Niño can have a major effect on weather patterns. Flash floods and mudslides may happen in areas of the world that usually receive little rain, such as Peru and the southern United States. Other areas of the world, such as Indonesia and Australia, may receive less rain than usual.

Say It

Discuss You may have heard news reports about the effects of El Niño. In a small group, talk about some of the effects of El Niño that you heard about on the news.

How Does Upwelling Affect Climate?

Ocean **upwelling** happens when warm surface water is replaced by cold, nutrient-rich water from the deep ocean. Upwelling is caused by local winds. These winds blow toward the equator along the northwest coast of South America and west coast of North America. The winds cause the local surface currents to move away from the shore. The warm surface water is then replaced by cold deep water. ☑

Upwelling is important for ocean life. Nutrients support the growth of plankton, which is the base of the food chain in the ocean. On the California coast, upwelling is usually strongest from March to September. El Niño can interrupt the process of upwelling and reduce the diversity of organisms near the ocean's surface.

READING CHECK

4. Define What is upwelling?

____

____

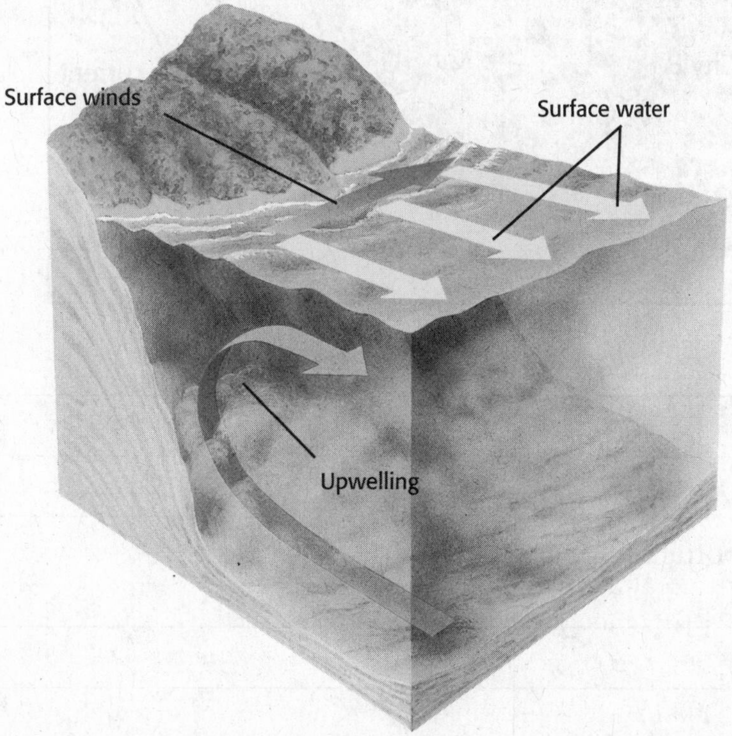
Surface winds
Surface water
Upwelling

Upwelling can happen along coastlines when the wind conditions are right. Winds can blow surface water away from the shore. Then, deep, nutrient-rich water can rise toward the surface.

TAKE A LOOK

5. Explain How do winds cause upwelling?

____

____

____

____

Copyright © by Holt, Rinehart and Winston. All rights reserved.

Name _______________ Class _______________ Date _______________

Section 2 Review

6.4.e

SECTION VOCABULARY

El Niño a change in the surface water temperature in the Pacific Ocean that produces a warm current **La Niña** a change in the eastern Pacific Ocean in which the surface water temperature becomes unusually cool	**upwelling** the movement of deep, cold, and nutrient-rich water to the surface

1. Explain Why do surface-water temperatures on the west coast of South America rise during El Niño?

____

____

____

2. Apply Concepts City A and city B are the same height above sea level. Based on the figure below, make a prediction about the average temperature in city A compared to city B. Explain your answer.

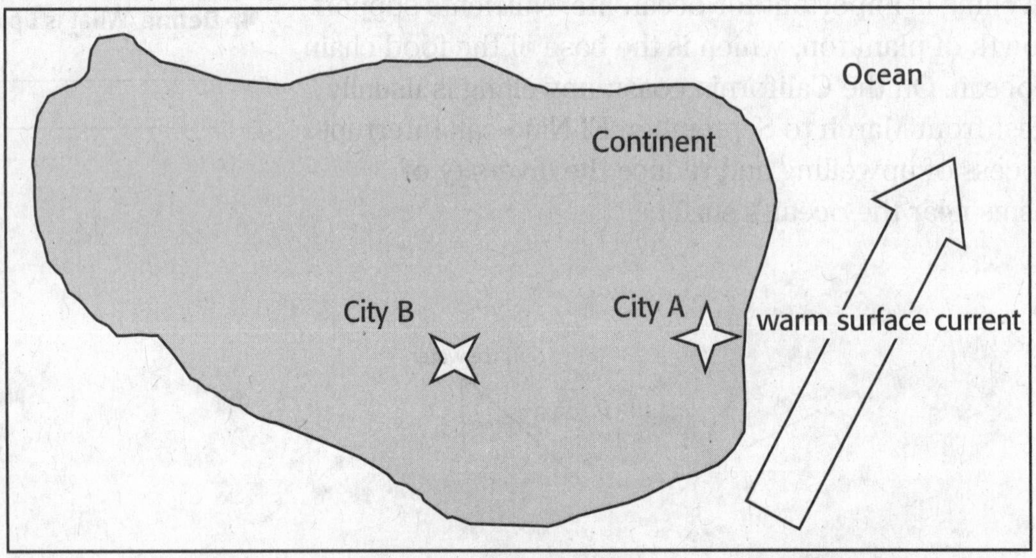
Ocean
Continent
City B
City A
warm surface current

____

____

____

3. Explain Why is upwelling important for marine life?

____

____

____

Copyright © by Holt, Rinehart and Winston. All rights reserved.

Name ______________________ Class ______________ Date ______________

CHAPTER 13 The Movement of Ocean Water

SECTION 3

Waves and Tides

BEFORE YOU READ

After you read this section, you should be able to answer these questions:

- How do waves form?
- What are the parts of a wave?
- What causes tides?

California Science Standards

6.3.a

How Does a Wave Form?

A **wave** is any disturbance that carries energy through matter or empty space. Waves in the ocean carry energy through water.

STUDY TIP

Compare As you read, make a chart describing the causes and features of high tides, low tides, spring tides, and neap tides.

WAVE FORMATION

Ocean waves form when energy is transferred from a source to the ocean water. Most ocean waves form as wind blows across the water's surface and transfers energy to the water. However, some waves form when energy from an earthquake or meteorite impact is transferred to the ocean water. ☑

Ocean waves can travel at different speeds. They can be very small or extremely large. The size and speed of a wave depend on the amount of energy the wave carries.

READING CHECK

1. Identify Give two sources of energy that can cause ocean waves.

____

____

PARTS OF A WAVE

Waves are made up of two main parts: crests and troughs. A *crest* is the highest point of a wave. A *trough* is the lowest point of a wave. The distance between one crest and the next, or one trough and the next, is the *wavelength* of the wave. The distance in height between the crest and the trough is called the *wave height*.

Parts of a Wave

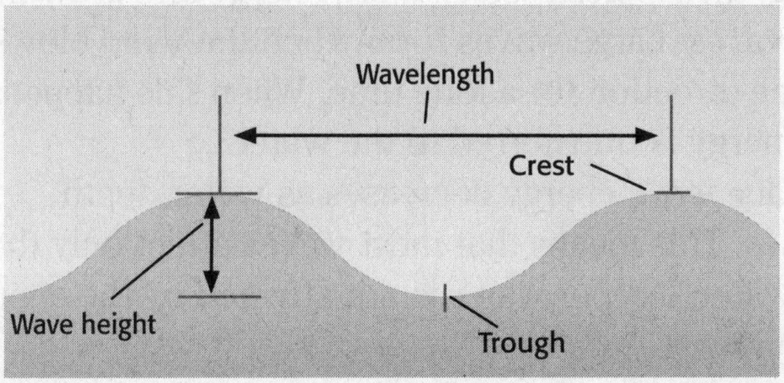

Wavelength
Crest
Wave height
Trough

TAKE A LOOK

2. Define Write your own definition for *wavelength*.

____

____

____

Copyright © by Holt, Rinehart and Winston. All rights reserved.

WAVE MOVEMENT

If you have ever watched ocean waves, you may have noticed that water seems to move across the ocean's surface. However, this movement is only an illusion. The energy in the wave causes the water to rise and fall in circular movements. The water does not move horizontally very much. The figure below shows how waves can move energy without moving water horizontally.

Critical Thinking

3. Predict Consequences People who own boats often leave the boats anchored a short distance away from the shore. The boats stay in about the same place over many days. What would happen to these boats if waves caused water to move horizontally?

____

____

____

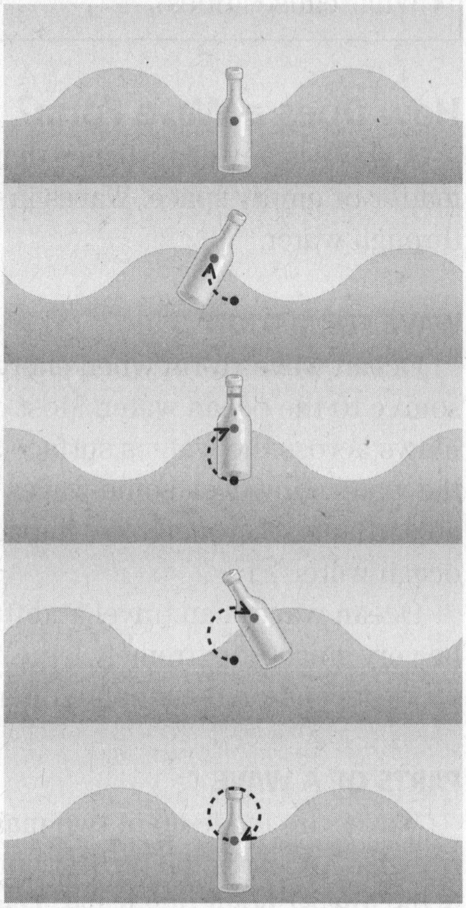

The bottle shows the circular motion of matter when a wave moves in the ocean. The energy in the wave makes matter near the surface move in circular motions. The matter does not move horizontally.

TAKE A LOOK

4. Describe What is the shape of the path that the bottle takes as the wave passes by it?

____

WAVE ENERGY

The size of a wave is related to the amount of energy it is carrying. When the wind begins to blow over the water, small waves called *ripples* form. If the wind keeps blowing, the ripples receive more energy and grow into larger waves. Large waves form when the wind blows in the same direction for a long time. When this happens, a lot of energy is transferred to the water.

Surface wave energy decreases as water depth increases. This means that most waves affect only the top of the water. Deeper water is not affected by the energy of surface waves.

Copyright © by Holt, Rinehart and Winston. All rights reserved.

WAVE SPEED

Waves travel at different speeds. To calculate wave speed, scientists must know the wavelength and the wave period. *Wave period* is the time between the passage of two wave crests or troughs at a fixed point. Dividing wavelength by wave period gives you wave speed, as shown below.

$$\frac{\textit{Wavelength}\ (\text{m})}{\textit{Wave period}\ (\text{s})} = \textit{Wave speed}\ (\text{m/s})$$

Increasing the wave period, the time between wave crests, will decrease the wave speed. Decreasing the wave period will increase the wave speed. The figure below shows how the period of a wave can be measured.

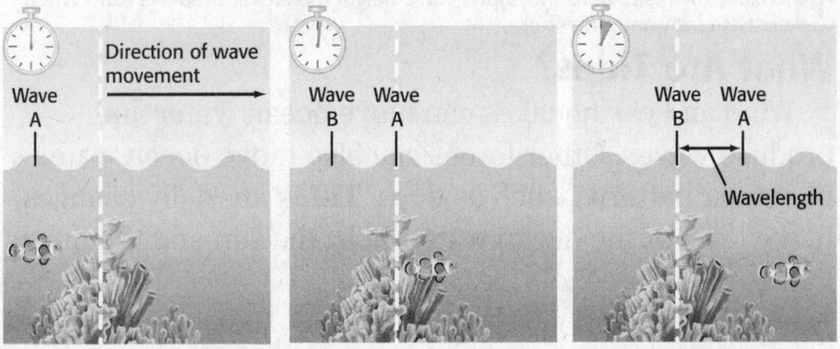
Direction of wave movement
Wave A
Wave B
Wave A
Wave B
Wave A
Wavelength

1. The waves are moving from left to right. The reef is a fixed point because it is not moving. The dotted line marks the center of the reef.

2. The timer begins running as the crest of wave A passes the center of the reef.

3. The timer stops when the crest of wave B passes the center of the reef. The time that the timer recorded, 5 s, is the wave period.

WAVES REACHING THE SHORE

When waves reach water shallower than one-half their wavelength, they begin to interact with the ocean floor. As waves begin to touch the ocean floor, the waves transfer energy from the wave to the ocean floor. As a result, the water at the bottom of the waves slows down. However, the water at the top of wave continues to travel at the original speed. Eventually, the wave crest crashes onto the shore as a *breaker*. ☑

The energy of the wave is transferred to the beach environment at the shore. The energy of the wave and the angle at which the wave hits the shore determine how much energy is transferred to the shore.

Math Focus

5. Calculate A water wave has a speed of 5 m/s. If its wavelength is 50 m, what is its wave period? Show your work.

____

____

____

____

TAKE A LOOK

6. Identify What part of wave B is passing the reef when the timer is stopped?

____

READING CHECK

7. Describe What happens to a wave's energy as the wave moves close to the shore?

____

____

Copyright © by Holt, Rinehart and Winston. All rights reserved.

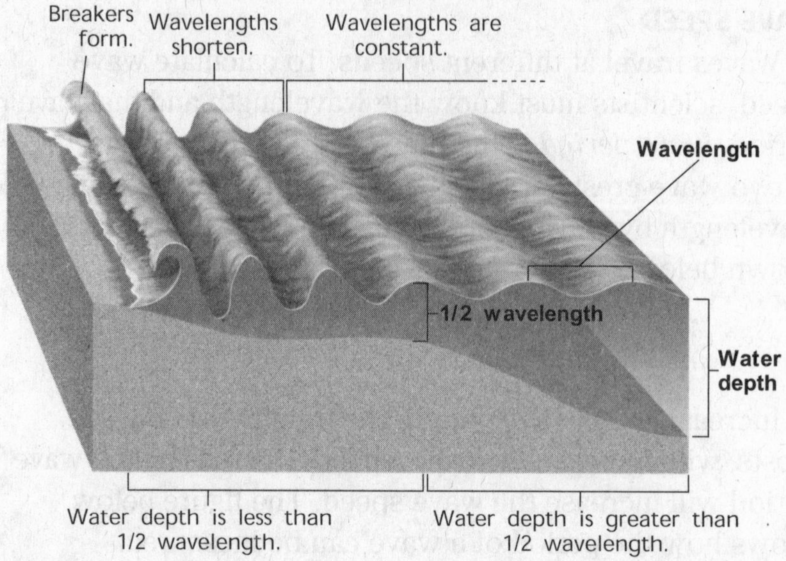
Breakers form.
Wavelengths shorten.
Wavelengths are constant.
Wavelength
1/2 wavelength
Water depth
Water depth is less than 1/2 wavelength.
Water depth is greater than 1/2 wavelength.

As the wave moves toward the shore, wave height increases. Breakers form when wave crests crash onto the shore.

TAKE A LOOK

8. Describe What are breakers?

____

____

What Are Tides?

Wind and earthquakes can move ocean water and produce waves. Other forces can also move ocean water in regular patterns, such as tides. **Tides** are daily changes in the level of the ocean water. Both the sun and the moon influence the level of tides. ☑

READING CHECK

9. Define Write your own definition for *tides*.

____

____

____

WHY TIDES HAPPEN

The moon's gravity pulls on every particle on Earth. However, the moon's gravity doesn't pull on every particle with the same strength. The moon's gravitational pull on Earth decreases with distance from the moon. Therefore, the pull on some parts of Earth is stronger than on other parts.

The difference in the moon's pull is more noticeable in liquids than in solids because liquids can move more easily. Therefore, the effects of the moon's pull on the oceans is more noticeable than on the land.

WHERE TIDES HAPPEN

The part of Earth that faces the moon is pulled toward the moon with the greatest force. Therefore, the water on the side of Earth that faces the moon bulges toward the moon. The water on Earth's opposite side is pulled toward the moon the least. Therefore, it bulges away from the moon. The figure at the top of the next page shows these bulges.

Copyright © by Holt, Rinehart and Winston. All rights reserved.

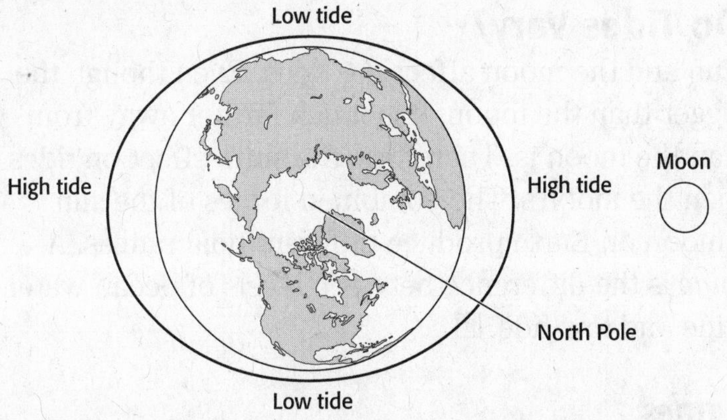
Low tide
High tide
High tide
Moon
North Pole
Low tide

Water bulges toward the moon on the side of Earth that faces the moon. Water bulges away from the moon on Earth's far side. As a result, these two sides of Earth experience high tide. In this image, the sizes and locations of Earth, the oceans, and the moon are not drawn to scale.

TAKE A LOOK

10. Explain Why does the water on the side of Earth that is farthest from the moon bulge outward?

____

HIGH TIDES AND LOW TIDES

The bulges that form in the oceans because of the moon's pull are called *high tides*. In high-tide areas, the water level is higher than average sea level. In areas between high tides, *low tides* form. In low-tide areas, the water level is lower than average sea level because the water is pulled toward high-tide areas.

Remember that Earth rotates on its axis. As a result, high tides happen in different places on Earth at different times of day. However, because Earth's rotation is predictable, the tides are also predictable. Many places on Earth experience two high tides and two low tides every day. ☑

READING CHECK

11. Describe Why are the tides predictable?

____

TIMING THE TIDES

The moon revolves around Earth much more slowly than Earth rotates. As the figure below shows, a place that is facing the moon takes 24 h and 50 min to rotate to face the moon again. Therefore, high and low tides at that place happen 50 min later each day.

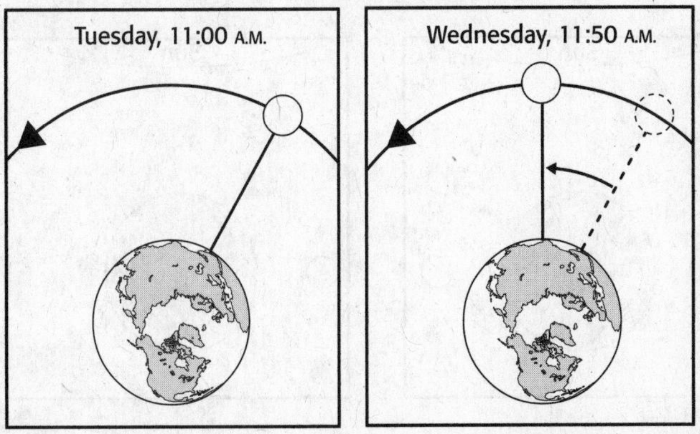
Tuesday, 11:00 A.M.
Wednesday, 11:50 A.M.

High and low tides happen about 50 min later each day at a given place. This happens because Earth rotates faster than the moon orbits Earth. If Earth rotated at the same speed as the moon orbits Earth, tides would not alternate between high and low.

Copyright © by Holt, Rinehart and Winston. All rights reserved.

How Do Tides Vary?

The sun and the moon affect the tides. Even though the sun is bigger than the moon, it is much farther away from Earth than the moon is. Therefore, the sun's effect on tides is less than the moon's. The combined forces of the sun and the moon on Earth produce different tidal ranges. A *tidal range* is the difference between levels of ocean water at high tide and low tide. ☑

READING CHECK

12. Explain Why does the sun affect the tides less than the moon does?

____

____

____

SPRING TIDES

Tides that have the largest daily tidal range are **spring tides**. Spring tides happen when the sun, Earth, and the moon are aligned, as shown in the figures below. Spring tides happen during the new-moon and full-moon phases, or every 14 days. During these times, the pull of the sun and moon produces one pair of very large tidal bulges.

TAKE A LOOK

13. Describe Draw an oval around Earth in each picture to show where the tides are highest and where they are lowest during spring tides.

Spring tides happen when the sun, the moon, and Earth are aligned. This can happen in two ways. One way is when the moon is between Earth and the sun, as shown in the left-hand figure. The other way is when Earth is between the moon and the sun, as shown in the right-hand figure.

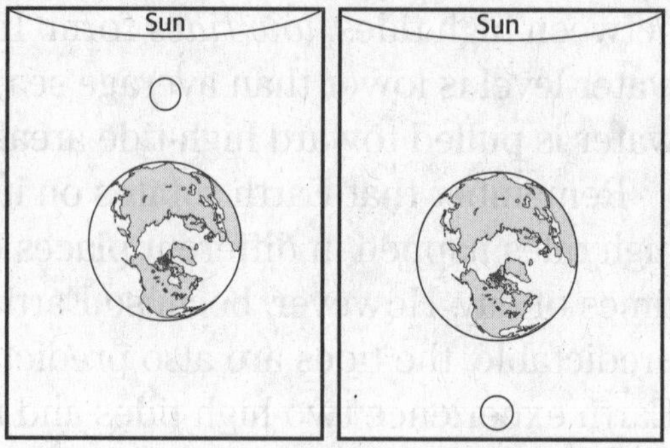
Sun
Sun

NEAP TIDES

Tides that have the smallest daily tidal range are called **neap tides**. Neap tides happen when the sun, Earth, and the moon form a 90° angle, as shown in the figures below. They happen halfway between the spring tides, during the first-quarter and third-quarter phases of the moon. During these times, the pull of the sun and moon produces smaller tidal bulges.

Neap tides happen when the sun, the moon, and Earth form a 90° angle.

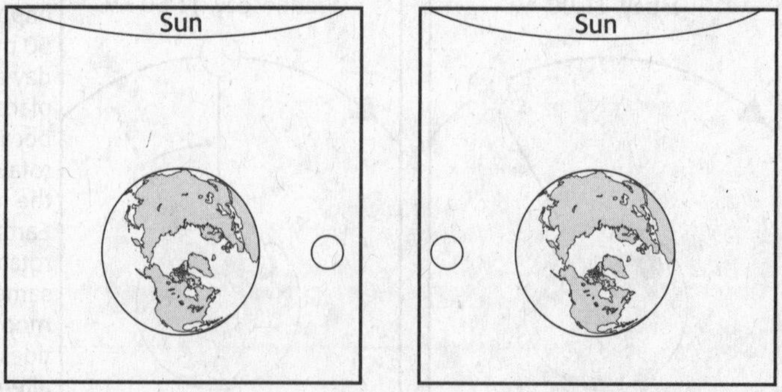
Sun
Sun

Copyright © by Holt, Rinehart and Winston. All rights reserved.

Name ______________________ Class ______________ Date ______________

Section 3 Review

6.3.a

SECTION VOCABULARY

neap tide a tide of minimum range that occurs during the first and third quarters of the moon

spring tide a tide of increased range that occurs two times a month, at the new and full moons

tide the periodic rise and fall of the water level in the oceans and other large bodies of water

wave a periodic disturbance in a solid, liquid, or gas as energy is transmitted through a medium

1. Describe How do ocean waves form?

____

____

____

2. Identify Relationships How does the length of time that wind blows over water affect the amount of energy in a wave? Explain your answer.

____

____

____

3. Calculate A wave has a wave period of 20 s and a wavelength of 100 m. What is its speed? How would the speed change if the wave period increased?

____

____

4. Summarize What causes tides?

____

____

____

5. Explain Why is the moon's pull more noticeable in liquids than in solids?

____

____

6. Apply Ideas If you have a calendar that shows only the phases of the moon, can you predict when spring tides and neap tides will happen? Explain your answer.

____

____

____

____

Copyright © by Holt, Rinehart and Winston. All rights reserved.

Name ______________________ Class ______________ Date __________

CHAPTER 14 The Atmosphere

SECTION 1

Characteristics of the Atmosphere

California Science Standards

6.4.b, 6.4.e

BEFORE YOU READ

After you read this section, you should be able to answer these questions:

- What is Earth's atmosphere made of?
- How do air pressure and temperature change as you move away from Earth's surface?
- What are the layers of the atmosphere?

STUDY TIP

Define When you come across a word you don't know, circle it. When you figure out what it means, write the definition in your notebook

What Is Earth's Atmosphere Made Of?

An **atmosphere** is a layer of gases that surrounds a planet or moon. On Earth, the atmosphere is often called just "the air." When you take a breath of air, you are breathing in atmosphere.

The air you breathe is made of many different things. Almost 80% of it is nitrogen gas. The rest is mostly oxygen, the gas we need to live. There is also water in the atmosphere. Some of it is invisible, in the form of a gas called *water vapor*. ☑

READING CHECK

1. List Which two gases make up most of Earth's atmosphere?

____

____

Water is also found in the atmosphere as water droplets and ice crystals, like those that make up clouds. The atmosphere contains tiny solid bits, or particles, too. These particles are things like dust and dirt from continents, salt from oceans, and ash from volcanoes.

Gases in Earth's Atmosphere

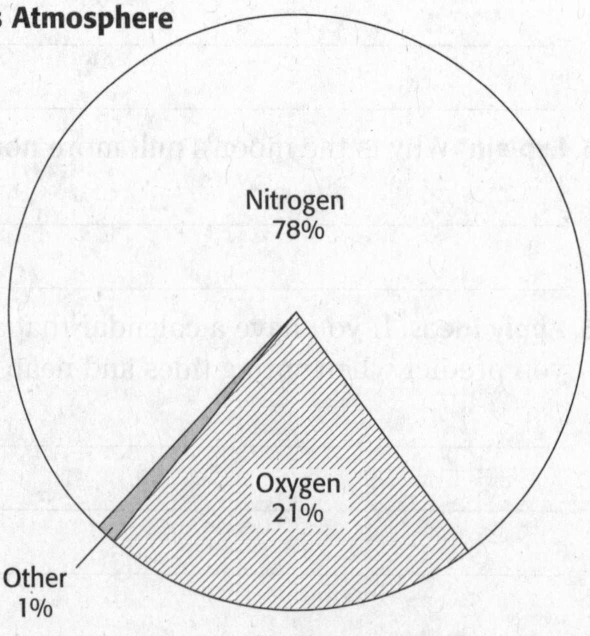

Nitrogen
78%
Oxygen
21%
Other
1%

Math Focus

2. Analyze Data About how many times more nitrogen than oxygen is there in Earth's atmosphere?

____

Copyright © by Holt, Rinehart and Winston. All rights reserved.

Why Does Air Pressure Change with Height?

Air pressure is how much the air above you weighs. It is a measure of how hard air molecules push on a surface. We don't normally notice air pressure, because our bodies are used to it.

As you move up from the ground and out toward space, there are fewer gas molecules pressing down from above. Therefore, the air pressure drops. The higher you go, the lower the air pressure gets. ☑

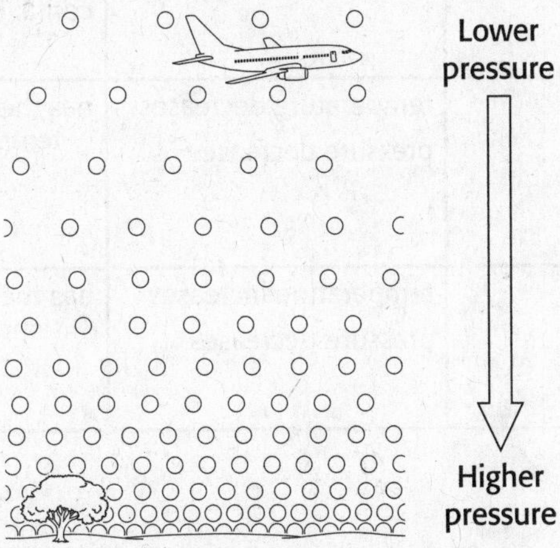
Lower pressure
Higher pressure

READING CHECK

3. Explain Why does air pressure drop as you move away from Earth's surface?

____

____

TAKE A LOOK

4. Compare How is the air pressure around the tree different from the air pressure around the plane?

____

____

Why Does Air Temperature Change with Height?

Like air pressure, air temperature changes as you move higher in the atmosphere. Air pressure always gets lower as you move higher, but air temperature can get higher or lower. The air can get hotter or colder. ☑

Air temperature depends on a number of things, including the gases that are present in the atmosphere. There are different layers of the atmosphere. Each layer is made of a different combination of gases. Some gases absorb energy from the sun better than others. When a gas absorbs energy from the sun, the air temperature goes up.

READING CHECK

5. Summarize What is the main idea of this paragraph?

____

____

____

What Are the Four Layers of the Atmosphere?

There are four layers of the atmosphere: **troposphere**, **stratosphere**, **mesosphere**, and **thermosphere**. You cannot actually see these different layers. The atmosphere is divided into layers based on how each layer's temperature changes. The table on the next page shows the features of the different layers of the atmosphere.

Copyright © by Holt, Rinehart and Winston. All rights reserved.

Critical Thinking

6. Explain Events How do you think scientists found out that there are four layers of the atmosphere?

Say It

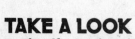

Make Up a Memory Trick In groups of two or three, make up a sentence to help you remember the order of the layers of the atmosphere. The words in the sentence should start with T, S, M, and T. For example, "Tacos Sound Mighty Tasty." A sentence like this is called a *mnemonic.*

Layer	How temperature and pressure change as you move higher	Important features
Troposhere	temperature decreases ↓ pressure decreases ↓	densest layer where most of the atmosphere is where weather happens where clouds are
Stratosphere	temperature increases ↑ pressure decreases ↓	gases are arranged in layers contains the ozone layer
Mesosphere	temperature decreases ↓ pressure decreases ↓	has the lowest temperature
Thermosphere	temperature increases ↑ pressure decreases ↓	has the highest temperature

TAKE A LOOK

7. Identify At what altitude does the mesosphere end and the thermosphere begin?

8. Identify Sketch clouds in the proper place in the figure.

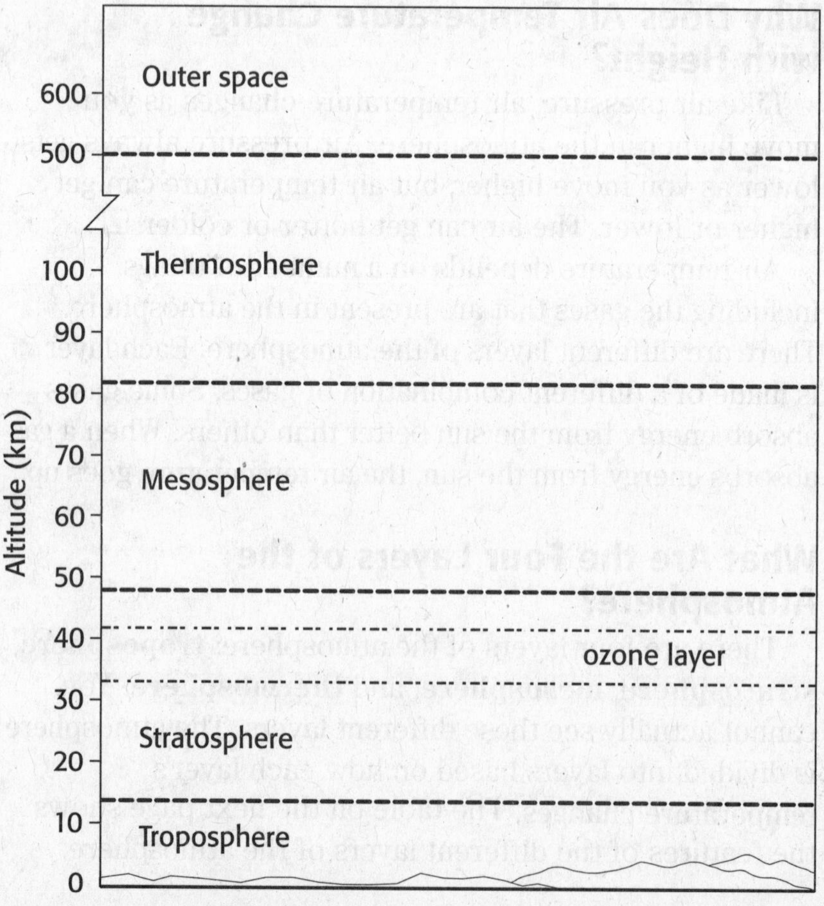
Outer space
600
500
Thermosphere
100
90
80
70
Mesosphere
60
Altitude (km)
50
40
ozone layer
30
Stratosphere
20
10
Troposphere
0

Copyright © by Holt, Rinehart and Winston. All rights reserved.

Name ______________________ Class ______________ Date ____________

Section 1 Review

6.4.b, 6.4.e

SECTION VOCABULARY

air pressure a measure of the force with which air molecules push on a surface

atmosphere a mixture of gases that surrounds a planet or moon

mesosphere the layer of the atmosphere between the stratosphere and the thermosphere and in which temperature decreases as altitude increases

stratosphere the layer of the atmosphere that is above the troposphere and in which temperature increases as altitude increases

thermosphere the uppermost layer of the atmosphere, in which temperature increases as altitude increases

troposphere the lowest layer of the atmosphere, in which temperature decreases as altitude increases

1. Define In your own words, explain what air pressure is.

____

____

2. Explain Why does air temperature change as you move up from the Earth's surface?

____

____

3. Graph Data The graph below shows how the temperature changes as you move up through the atmosphere. On the graph, draw a curve showing how the pressure changes.

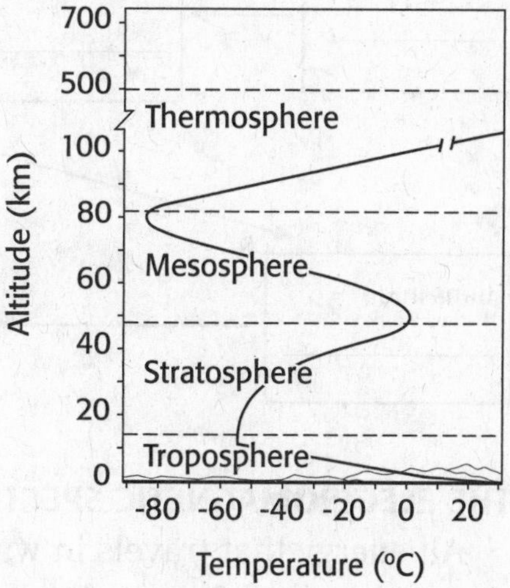

700
500
100
80
60
40
20
0
Thermosphere
Mesosphere
Stratosphere
Troposphere
Altitude (km)
-80 -60 -40 -20 0 20
Temperature (°C)

4. Identify Relationships How does the sun affect air temperatures?

____

____

Copyright © by Holt, Rinehart and Winston. All rights reserved.

Name ______________________ Class ______________ Date ____________

CHAPTER 14 The Atmosphere

SECTION 2

Atmospheric Heating

California Science Standards

6.3.a, 6.3.c, 6.3.d, 6.4.a, 6.4.b, 6.4.d

BEFORE YOU READ

After you read this section, you should be able to answer these questions:

- How does energy travel from the sun to Earth?
- What are the differences between radiation, conduction, and convection?
- Why is Earth's atmosphere as warm as it is?

How Does Energy Travel from the Sun to Earth?

STUDY TIP

Outline In your notebook, write an outline of this chapter. Use the questions in bold to make your outline. As you read, fill in information about each question.

Most of the heat energy on Earth's surface comes from the sun. Energy travels from the sun to Earth by **radiation**, which means that it travels through space as waves. As solar energy (energy from the sun) is absorbed by air, water, and land, it turns into heat energy. This energy from the sun causes winds, the water cycle, ocean currents, and changes in the weather.

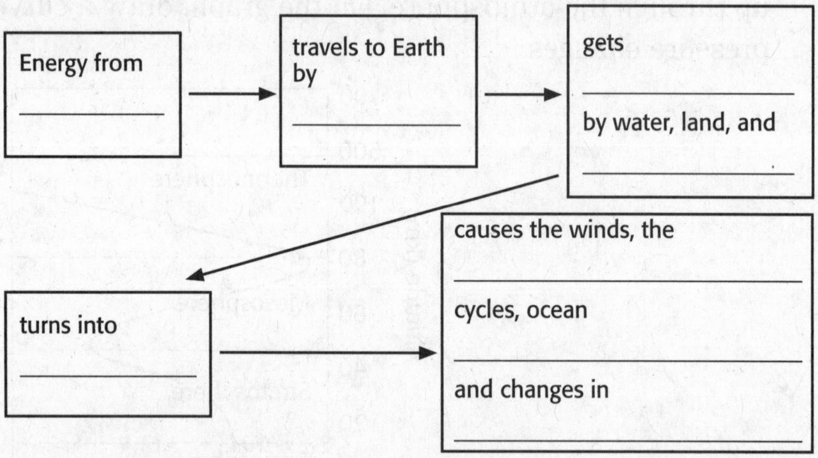

Energy from
travels to Earth by
gets
by water, land, and
turns into
causes the winds, the
cycles, ocean
and changes in

TAKE A LOOK

1. Summarize Use the information in the text to fill in the blanks in the graphic organizer.

THE ELECTROMAGNETIC SPECTRUM

All energy that travels in waves is radiation. The waves that make up all forms of radiation are called *electromagnetic* waves. These waves travel through space at very high speeds—about 300,000 km/s. ☑

READING CHECK

2. Explain What is radiation?

____

____

____

Almost all of the energy that reaches Earth from the sun is in the form of electromagnetic waves. The **electromagnetic spectrum** is made of all the types of electromagnetic waves.

Copyright © by Holt, Rinehart and Winston. All rights reserved.

WAVELENGTHS

Different types of electromagnetic radiation have different wavelengths. A *wavelength* is the distance from any point on a wave to the same point on the next wave.

Visible light is made of light with different wavelengths. People see the different wavelengths as different colors. Ultraviolet rays, X rays, and gamma rays have shorter wavelengths than the wavelengths of visible light. Infrared and radio waves have wavelengths longer than those of visible light.

The Electromagnetic Spectrum

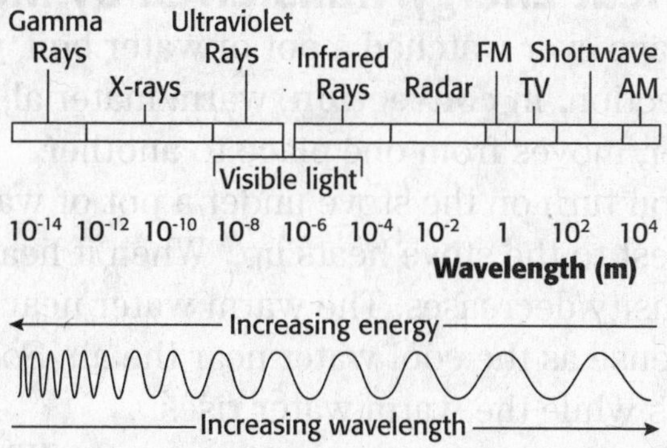
Gamma Rays
X-rays
Ultraviolet Rays
Infrared Rays
Radar
FM TV
Shortwave AM
Visible light
10^{-14} 10^{-12} 10^{-10} 10^{-8} 10^{-6} 10^{-4} 10^{-2} 1 10^{2} 10^{4}
Wavelength (m)
Increasing energy
Increasing wavelength

Math Focus

3. Build Models In the space below, use a ruler to draw a wave that has a wavelength of 1 cm. Label the beginning and the end of one wavelength.

TAKE A LOOK

4. Identify Relationships Which types of radiation have wavelengths shorter than visible light?

____

____

What Happens to Radiation from the Sun?

Not all of the radiation from the sun reaches Earth's surface. Much of it gets absorbed by the atmosphere. Some of it is scattered and reflected by clouds and gases.

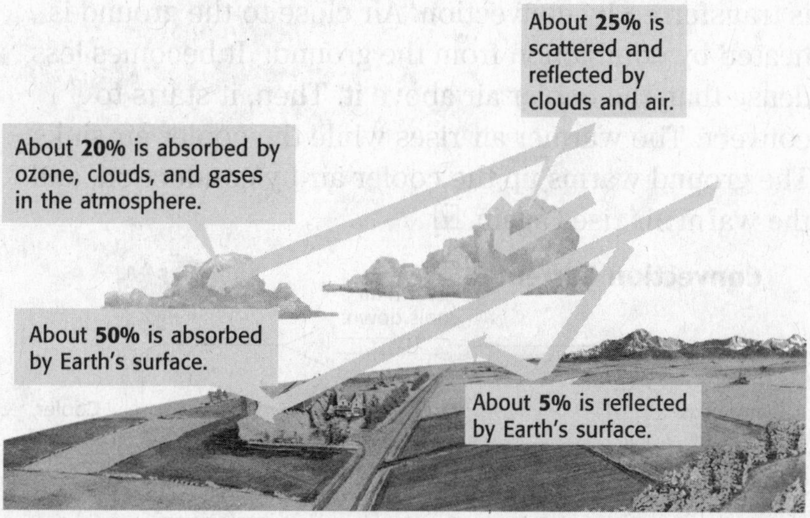
About **25%** is scattered and reflected by clouds and air.
About **20%** is absorbed by ozone, clouds, and gases in the atmosphere.
About **50%** is absorbed by Earth's surface.
About **5%** is reflected by Earth's surface.

TAKE A LOOK

5. Identify How much of the sunlight that gets to Earth is absorbed by Earth's surface?

____

6. Summarize What happens to the sunlight that is not absorbed by Earth's surface?

____

____

____

Copyright © by Holt, Rinehart and Winston. All rights reserved.

How Is Heat Transferred by Contact?

Once sunlight is absorbed by Earth's surface, it is *converted*, or changed, into heat energy. Then, the heat can be transferred to other objects and moved to other places. When a warm object touches a cold object, heat moves from the warm object to the cold one. This movement of heat is called **thermal conduction**.

When you touch the sidewalk on a hot, sunny day, heat energy is conducted from the sidewalk to you. The same thing happens to air molecules in the atmosphere. When they touch the warm ground, the air molecules heat up. ☑

READING CHECK

7. List Name two ways that air gets heated.

____

____

____

How Is Heat Energy Transferred by Motion?

If you have ever watched a pot of water boil, you have seen convection. In **convection**, warm material, such as air or water, moves from one place to another.

When you turn on the stove under a pot of water, the water closest to the stove heats up. When it heats up, the water's density decreases. The warm water near the stove is not as dense as the cool water near the air. So the cool water sinks while the warm water rises.

As it rises, the warm water begins to cool. When it cools, its density increases again. It becomes denser than the layer below, so it sinks back to the stove top. This cycle causes a circular movement called a *convection current*.

Convection currents also move heat through the atmosphere. In fact, most heat energy in the atmosphere is transferred by convection. Air close to the ground is heated by conduction from the ground. It becomes less dense than the cooler air above it. Then, it starts to convect. The warmer air rises while the cooler air sinks. The ground warms up the cooler air by conduction, and the warm air rises again. ☑

READING CHECK

8. Explain How is most heat energy in the atmosphere transferred from one place to another?

____

Convection Current

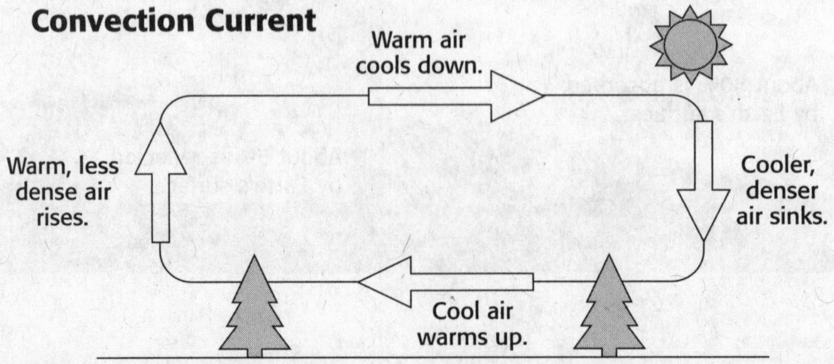
Warm air
cools down.
Cooler,
denser
air sinks.
Cool air
warms up.
Warm, less
dense air
rises.

TAKE A LOOK

9. Describe In a convection current, what happens after cool air sinks?

____

____

Copyright © by Holt, Rinehart and Winston. All rights reserved.

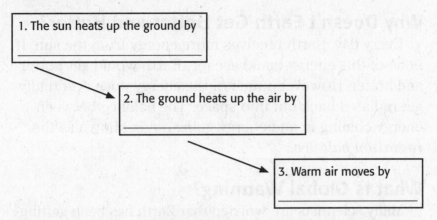

1. The sun heats up the ground by ______________________

2. The ground heats up the air by ______________________

3. Warm air moves by ______________________

TAKE A LOOK
10. Sequence Fill in the blanks in the boxes to show how heat moves in Earth's atmosphere.

Why Doesn't All the Heat in the Atmosphere Escape Back Out into Space?

A gardener who needs to keep plants warm uses a glass building called a greenhouse. Light travels through the glass into the building, and the energy is absorbed by the air and plants inside. The energy is converted to heat, which cannot travel back through the glass as easily as light came in. Much of the heat energy stays trapped within the greenhouse, keeping the air inside warmer than the air outside.

Earth's atmosphere acts like the glass walls of a greenhouse. Sunlight travels through the atmosphere easily, but heat does not. Gases in the atmosphere, such as water vapor and carbon dioxide, absorb heat energy coming from Earth and radiate it back to Earth's surface. This is known as the **greenhouse effect**. ☑

READING CHECK
11. List Name two gases in Earth's atmosphere that absorb heat.

____

____

The Greenhouse Effect

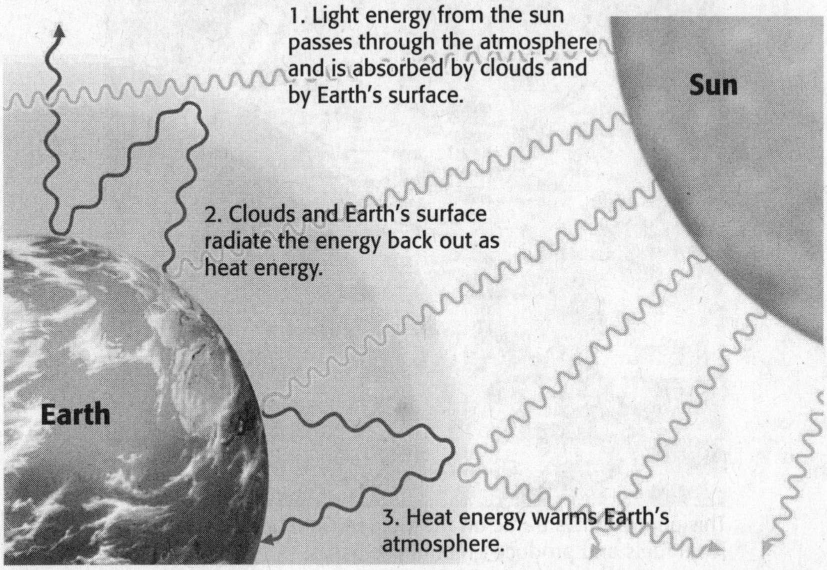
1. Light energy from the sun passes through the atmosphere and is absorbed by clouds and by Earth's surface.
Sun
2. Clouds and Earth's surface radiate the energy back out as heat energy.
Earth
3. Heat energy warms Earth's atmosphere.

TAKE A LOOK
12. Identify On the drawing, label the light coming from the sun with an **L**. Label the heat energy that is trapped by Earth's atmosphere with an **H**.

Copyright © by Holt, Rinehart and Winston. All rights reserved.

Critical Thinking

13. Apply Concepts How will global warming affect the amount of water vapor in the atmosphere? Explain your answer.

____

____

____

____

____

____

Why Doesn't Earth Get Hotter and Hotter?

Every day, Earth receives more energy from the sun. If none of this energy could escape, Earth would get hotter and hotter. However, much of the energy does eventually get radiated back out into space. The balance between energy coming in and energy going out is known as the *radiation balance*.

What Is Global Warming?

Many scientists are worried that Earth has been getting warmer over the past hundred years. This increase in temperatures all over the world is called **global warming**.

Scientists think that human activities may be causing global warming. When we burn fossil fuels, we release greenhouse gases, such as carbon dioxide, into the atmosphere. Because greenhouse gases trap heat in the atmosphere, adding more of them can make Earth even warmer. Global warming can have a strong effect on weather and climate.

Say It

Predict How might global warming affect your community? What can you do to slow global warming? In groups of two or three, discuss how global warming might affect your lives.

This photo shows traffic on the Golden Gate bridge. Cars burn fossil fuels and produce greenhouse gases.

TAKE A LOOK

14. Explain What is one possible reason that Earth is getting warmer?

____

____

Copyright © by Holt, Rinehart and Winston. All rights reserved.

Name ______________________ Class ______________ Date ______________

Section 2 Review

6.3.a, 6.3.c, 6.3.d, 6.4.a, 6.4.b, 6.4.d

SECTION VOCABULARY

convection the transfer of thermal energy by the circulation or movement of a liquid or gas

electromagnetic spectrum all of the frequencies or wavelengths of electromagnetic radiation

global warming a gradual increase in average global temperatures

greenhouse effect the warming of the surface and lower atmosphere of the Earth that occurs when water vapor, carbon dioxide, and other gases absorb and reradiate thermal energy

radiation the transfer of energy as electromagnetic waves

thermal conduction the transfer of energy as heat through a material

1. Explain Explain how energy gets from the sun to Earth. Use at least one of the vocabulary words in the box above.

____

2. Compare Fill in the table below to name and describe the three ways energy is transferred in Earth's atmosphere.

Type of energy transfer	How energy is transferred
	Energy travels as electromagnetic waves.
Conduction	

3. Explain How does most of the heat in Earth's atmosphere move from place to place?

____

____

____

4. Identify Relationships Explain how global warming and the greenhouse effect are related.

____

____

____

Copyright © by Holt, Rinehart and Winston. All rights reserved.

Name ______________________ Class ______________ Date ______________

CHAPTER 14 The Atmosphere

SECTION 3

Air Movement and Wind

California Science Standards

6.4.a, 6.4.d, 6.4.e

BEFORE YOU READ

After you read this section, you should be able to answer these questions:

- What causes wind?
- What is the Coriolis effect?
- What are the major global wind systems on Earth?

What Causes Wind?

STUDY TIP

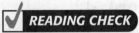

Underline Each head in this section is a question. Underline the answer to each question when you find it in the text.

Wind is moving air. It is caused by differences in air pressure. Air moves from areas of high pressure to areas of low pressure. The greater the pressure difference, the faster the air moves, and the stronger the wind blows. ☑

READING CHECK

1. Define What is wind?

____

____

____

You can see how air moves if you blow up a balloon and then let it go. The air inside the balloon is at a higher pressure than the air around the balloon. If you open the end of the balloon, air will rush out.

TAKE A LOOK

2. Identify On the drawing, label the high-pressure area with an **H** and the low-pressure area with an **L**.

Copyright © by Holt, Rinehart and Winston. All rights reserved.

What Causes Differences in Air Pressure?

Differences in air pressure on Earth are caused mainly by differences in temperature. The air over some parts of Earth gets more energy from the sun than others. For example, the sun shines more directly on the equator than on the poles. As a result, the air is warmer near the equator.

The warm air near the equator is not as dense as the cool air near the poles. Because it is less dense, the air at the equator rises and forms an area of low pressure. The cold air near the poles sinks and forms areas of high pressure. The air moves in large circular patterns called *convection cells*. These convection cells are shown from the side in the drawing below.

Convection Cells

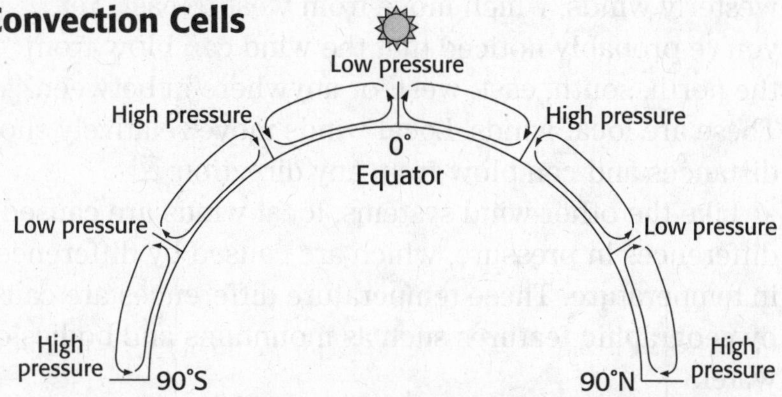
Low pressure
High pressure
High pressure
0°
Equator
Low pressure
Low pressure
High pressure
High pressure
90°S
90°N

CALIFORNIA STANDARDS CHECK

6.4.A Students know the sun is the major source of energy for phenomena on Earth's surface; it powers winds, ocean currents, and the water cycle.

Word Help: major
of great importance or large scale

Word Help: energy
what makes things happen

Word Help: phenomena
any facts or events that can be sensed or described scientifically

3. Explain How does the sun cause the wind?

____

____

____

____

What Are the Major Global Wind Systems?

Global winds are large-scale wind systems. There are three major global wind systems: polar easterlies, prevailing westerlies, and trade winds. The direction the winds blow is controlled by the convection cells and by Earth's rotation.

Global Winds

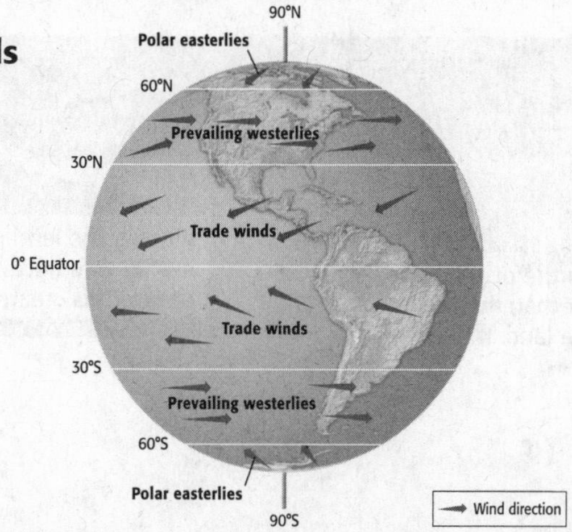
90°N
Polar easterlies
60°N
Prevailing westerlies
30°N
Trade winds
0° Equator
Trade winds
30°S
Prevailing westerlies
60°S
Polar easterlies
90°S
Wind direction

TAKE A LOOK

4. Identify Which wind system do you live in?

____

5. Identify In which direction do all of the wester-lies blow?

____

____

Copyright © by Holt, Rinehart and Winston. All rights reserved.

Why Do Winds Curve?

Pressure differences cause air to move from the poles toward the equator. However, if you look at the global wind map, you will see that the winds don't actually travel straight north or south. The winds have curved paths.

Winds traveling from the equator to the poles curve to the east. Winds traveling from the poles toward the equator curve to the west. This is known as the **Coriolis effect**. The Coriolis effect happens because Earth is rotating, or turning on its axis.

Critical Thinking

6. Predict If air is moving south from California, which way will it tend to curve?

____

What Are Local Winds?

Most of the United States is in the belt of prevailing westerly winds, which move from west to east. Yet, you've probably noticed that the wind can blow from the north, south, east, west, or anywhere in between. These are local winds. Local winds move relatively short distances and can blow from any direction. ☑

Like the other wind systems, local winds are caused by differences in pressure, which are caused by differences in temperature. These temperature differences are caused by geographic features such as mountains and bodies of water.

READING CHECK

7. Compare Describe one difference between global winds and local winds.

____

____

Day
Cool air over the ocean flows toward the land. This creates a sea breeze.
Air over land heats up quickly and rises.

Night
Air over the ocean stays warmer than the air over the land. It rises.
Air over the land cools off quickly. It sinks and flows out toward the ocean. This creates a land breeze.

TAKE A LOOK

8. Identify In the figures, label the high-pressure areas with an **H** and the low-pressure areas with an **L**.

Say It

Share Experiences Have you ever been in a very strong wind? In groups of two or three, discuss the strongest or worst wind.

Copyright © by Holt, Rinehart and Winston. All rights reserved.

Name ______________________ Class ______________ Date ______________

Section 3 Review

6.4.a, 6.4.d, 6.4.e

SECTION VOCABULARY

Coriolis effect the apparent curving of the path of a moving object from an otherwise straight path due to the Earth's rotation	**wind** the movement of air caused by differences in air pressure

1. Identify The drawing below shows a convection cell. Put arrows on the cell to show which way the air is moving. Label high pressure areas with an **H** and low pressure areas with an **L**. Label cold air with a **C** and warm air with a **W**.

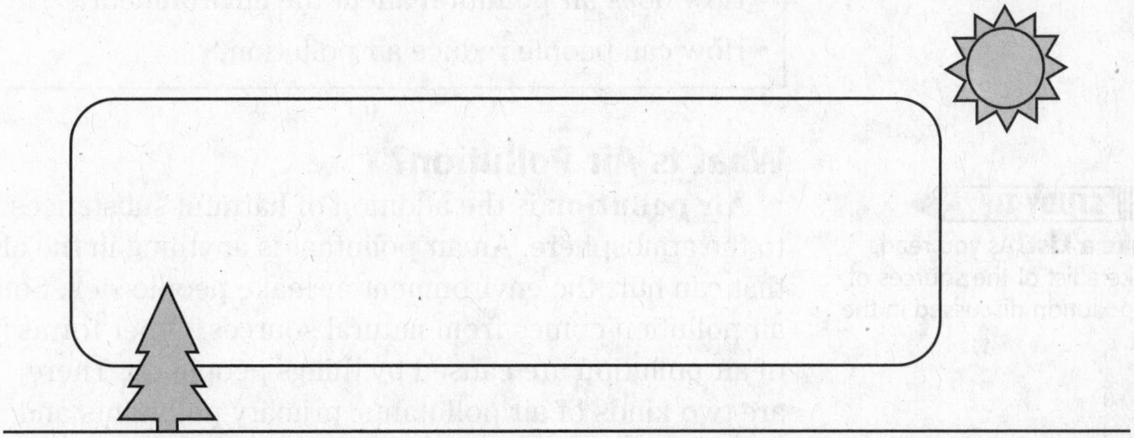

2. Identify Cause and Effect Complete the Cause-and-Effect Table below.

Cause	Effect
	differences in air pressure
Differences in air pressure	
	winds that curve to the east as they flow toward the poles

3. Identify Name the three global wind systems.

____

____

4. Apply Concepts Would there be winds if Earth's surface were the same temperature everywhere? Explain your answer.

____

____

____

Copyright © by Holt, Rinehart and Winston. All rights reserved.

Name ______________________ Class ______________ Date ____________

CHAPTER 14 The Atmosphere

SECTION 4

The Air We Breathe

California Science Standards

6.6.a, 6.6.b

BEFORE YOU READ

After you read this section, you should be able to answer these questions:

- What is air pollution?
- What causes air pollution?
- How does air pollution affect the environment?
- How can people reduce air pollution?

What Is Air Pollution?

STUDY TIP

Make a List As you read, make a list of the sources of air pollution discussed in the text.

Air pollution is the addition of harmful substances to the atmosphere. An air pollutant is anything in the air that can hurt the environment or make people sick. Some air pollution comes from natural sources. Other forms of air pollution are caused by things people do. There are two kinds of air pollutants: primary pollutants and secondary pollutants.

Primary pollutants are pollutants that are put directly into the air. Dust, sea salt, volcanic ash, and pollen are primary pollutants that come from natural sources. Smoke, chemicals from paint, and vehicle exhaust are primary pollutants that come from human activities.

Secondary pollutants form when primary pollutants react with each other or with other substances in the air. Ozone is an example of a secondary pollutant. It forms on sunny days when chemicals from burning gasoline react with each other and with the air. Ozone damages human lungs and can harm other living things, as well. ☑

READING CHECK

1. Explain Why is ozone called a secondary pollutant?

____

____

____

____

TAKE A LOOK

2. Fill In Use the information in the text to fill in the table.

Pollutant	Primary pollutant or secondary pollutant?	Natural or human-caused?
Car exhaust	primary	human-caused
Dust		
Ozone		
Paint chemicals		
Pollen		
Sea salt		
Volcanic ash		

Copyright © by Holt, Rinehart and Winston. All rights reserved.

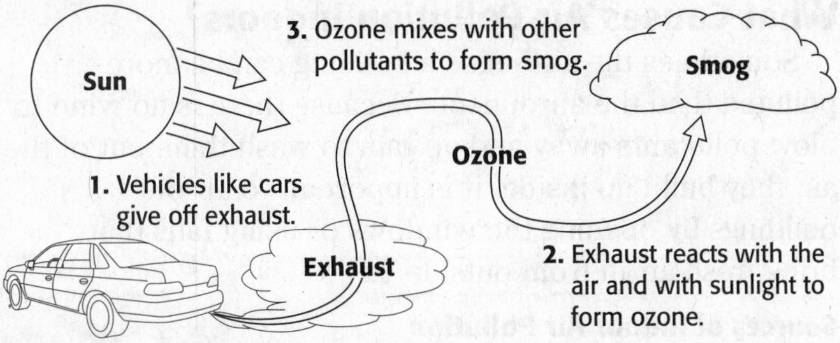
3. Ozone mixes with other pollutants to form smog.
Sun
Smog
Ozone
1. Vehicles like cars give off exhaust.
Exhaust
2. Exhaust reacts with the air and with sunlight to form ozone.

TAKE A LOOK

3. Identify What is the primary pollutant in this figure?

____

What Is Smog?

On a hot, still, sunny day, yellowish brown air can cover a city. This is called *smog*. Smog forms when ozone mixes with other pollutants. During summer in cities such as Los Angeles, a layer of warm air can trap smog near the ground. In the winter, a storm can clear the air.

This is Los Angeles on a clear day.

Here Los Angeles is covered in smog.

Say It

Discuss In a small group, discuss how the pollution shown in this photograph formed.

How Do Humans Cause Air Pollution?

Many of our daily activities cause air pollution. The main source of human-caused air pollution in the United States is motor vehicles. Cars, motorcycles, trucks, buses, trains, and planes all give off exhaust. Exhaust contains pollutants that create ozone and smog.

Factories and power plants that burn coal, oil, and gas also give off pollutants. Businesses that use chemicals, such as dry cleaners and auto body shops, can add to air pollution. ☑

READING CHECK

4. Identify What is the main source of human-caused air pollution in the United States?

____

Copyright © by Holt, Rinehart and Winston. All rights reserved.

What Causes Air Pollution Indoors?

Sometimes the air inside a building can be more polluted than the air outside. Because there is no wind to blow pollutants away and no rain to wash them out of the air, they build up inside. It is important to air out buildings by opening the windows or using fans that bring fresh air in from outside. ☑

READING CHECK

5. Explain Why can air pollution indoors be worse than air pollution outdoors?

____

____

____

Sources of Indoor Air Pollution

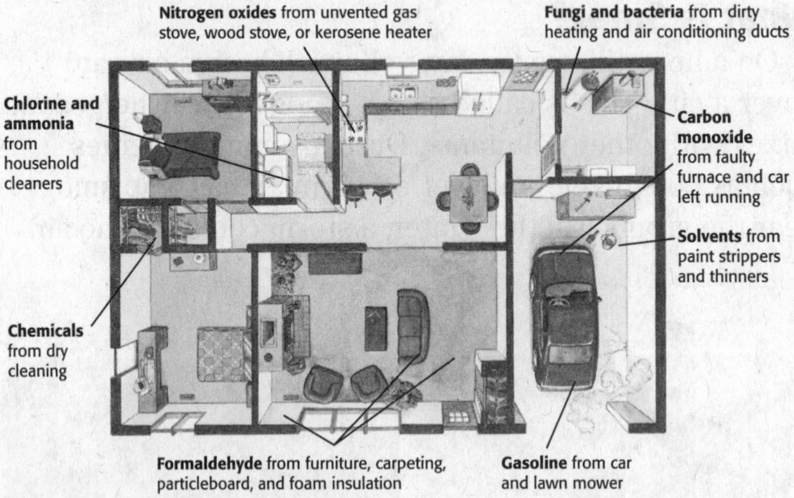
Nitrogen oxides from unvented gas stove, wood stove, or kerosene heater
Fungi and bacteria from dirty heating and air conditioning ducts
Chlorine and ammonia from household cleaners
Carbon monoxide from faulty furnace and car left running
Solvents from paint strippers and thinners
Chemicals from dry cleaning
Formaldehyde from furniture, carpeting, particleboard, and foam insulation
Gasoline from car and lawn mower

TAKE A LOOK

6. Identify Name two sources of indoor air pollution shown here that may be in your own home.

____

____

What Is Acid Precipitation?

Acid precipitation is rain, sleet, or snow that contains acids from air pollution. When we burn fossil fuels, pollutants such as sulfur dioxide are released into the air. These pollutants combine with water in the atmosphere to form acids.

Acid precipitation can kill or damage plants, change what soil is made of, and poison water. When acid rain flows into lakes, it can kill fish and other aquatic life.

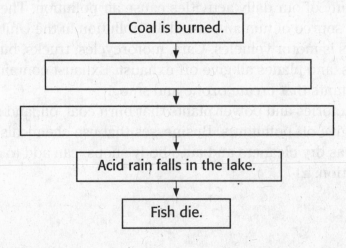

Coal is burned.

↓

____

↓

____

↓

Acid rain falls in the lake.

↓

Fish die.

TAKE A LOOK

7. Sequence Complete the graphic organizer to show how burning coal can cause fish to die.

Copyright © by Holt, Rinehart and Winston. All rights reserved.

What Is the Ozone Hole?

Close to the ground, ozone is a pollutant formed by human activities. However, high in the stratosphere, ozone is an important gas that forms naturally. The ozone layer absorbs harmful ultraviolet radiation from the sun. Ultraviolet (UV) rays harm living things. UV light can cause skin cancer in humans. ☑

In the 1980s, scientists noticed that the ozone layer over the poles was getting thinner. This hole in the ozone layer was being caused by chemicals called CFCs, which destroy ozone. CFCs were being used in air conditioners and chemical sprays. Many CFCs are now banned. The ozone layer may slowly recover, but it will take a long time. CFCs can remain in the atmosphere for 60 to 120 years.

READING CHECK

8. Explain How is ozone helpful to humans?

Ozone in the statosphere	**Ozone near the ground**
Naturally formed	
Not a pollutant	
	harmful to living things

TAKE A LOOK

9. Compare Fill in the chart to show the differences between ozone in the atmosphere and ozone near the ground.

How Does Air Pollution Affect Human Health?

Air pollution can cause many health problems. Some are short-term problems. They happen quickly and go away when the air pollution clears up or the person moves to a cleaner location. Other are long-term health problems. They develop over long periods of time and are not cured easily. The table below lists some of the effects of air pollution on human health. ☑

READING CHECK

10. Compare What is the difference between short-term effects and long-term effects?

Short-term effects	**Long-term effects**
headache	emphysema (a lung disease)
nausea and vomitting	lung cancer
eye, nose, and throat irritation	asthma
coughing	permanent lung damage
difficulty breathing	heart disease
upper respiratory infections	skin cancer
asthma attacks	
worsening of emphysema	

Copyright © by Holt, Rinehart and Winston. All rights reserved.

What Can We Do About Air Pollution?

Believe it or not, air pollution in the United States is not as bad now as it was 30 years ago. People are now much more aware of how they can cause air pollution. Air pollution can be reduced by new laws, by technology, and by people changing their lifestyles.

The United States government and the governments of other countries have passed laws to control air pollution. These laws limit the amount of pollution that sources such as cars and factories are allowed to release. For example, factories and power plants now have scrubbers on smokestacks. A scrubber is a tool that helps remove pollutants from smoke before it leaves the smokestack.

Cars are more efficient now than they used to be, so they run on less fuel. Individuals can do a lot on their own to reduce air pollution, as well. For example, we can walk or bike instead of driving. ☑

READING CHECK

11. Identify Name two things you can do to limit air pollution.

____

____

____

In Copenhagen, Denmark, companies lend bicycles for anyone to use for free. The program helps reduce automobile traffic and reduce air pollution.

Critical Thinking

12. Analyze Processes Electric cars don't give off any exhaust. They don't cause pollution in the cities where they are driven. However, driving them can cause pollution in other places. How?

____

____

____

____

____

Copyright © by Holt, Rinehart and Winston. All rights reserved.

Name ______________________ Class ______________ Date ______________

Section 4 Review

6.6.a, 6.6.b

SECTION VOCABULARY

acid precipitation rain, sleet, or snow that contains a high concentration of acids	**air pollution** the contamination of the atmosphere by the introduction of pollutants from human and natural sources

1. Identify Relationships How are fossil fuels related to air pollution and acid precipitation?

____

____

____

____

2. Compare Complete the table below to compare different pollutants.

Pollutant	Source	Negative effects	Solutions
CFCs			
Ozone			
Sulfur dioxide			

3. List Name three things, other than humans, that could be hurt by air pollution.

____

____

4. Explain Why is the ozone hole dangerous?

____

____

____

____

Copyright © by Holt, Rinehart and Winston. All rights reserved.

SECTION 1
Water in the Air

California Science Standards

6.4.a, 6.4.e

BEFORE YOU READ

After you read this section, you should be able to answer these questions:

- What is weather?
- How does water in the air affect the weather?
- What are the forms of water found in the air?

What Is Weather?

STUDY TIP

Outline Before you read, make an outline of this section using the question in bold. As you read, fill in the main ideas of the chapter in your outline.

Knowing about the weather is important in our daily lives. Your plans to play outside can change if it rains. Being prepared for extreme weather conditions, such as hurricanes and tornadoes, can even save lives.

Weather is the condition of the atmosphere at a certain time and place. The condition of the atmosphere depends a lot on the amount of water in the air. Therefore, to understand weather, you need to understand how water moves through the atmosphere. ☑

READING CHECK

1. Define Write your own definition for *weather*.

____

____

____

____

THE WATER CYCLE

The movement of water between the atmosphere, the land, and the oceans is called the *water cycle*. The sun is the main source of energy for the water cycle. The sun's energy heats Earth's surface. This causes liquid water to *evaporate*, or change into water vapor (a gas). When the water vapor cools, it may change back into a liquid and form clouds. This is called **condensation**. The liquid water may fall as rain on the land.

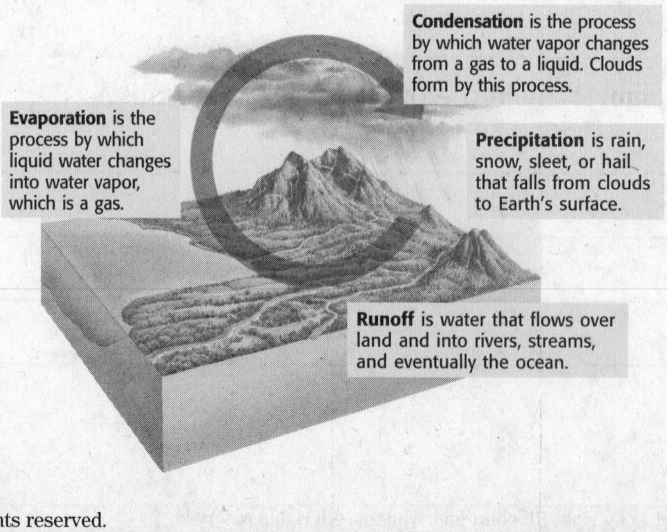
Condensation is the process by which water vapor changes from a gas to a liquid. Clouds form by this process.
Evaporation is the process by which liquid water changes into water vapor, which is a gas.
Precipitation is rain, snow, sleet, or hail that falls from clouds to Earth's surface.
Runoff is water that flows over land and into rivers, streams, and eventually the ocean.

TAKE A LOOK

2. Describe How does water move from oceans into the atmosphere?

____

Copyright © by Holt, Rinehart and Winston. All rights reserved.

What Is Humidity?

Water vapor makes up only a few percent of the mass of the atmosphere. However, this small amount of water vapor has an important effect on weather and climate.

When the sun's energy heats up Earth's surface, some of the water in oceans and water bodies changes to a gas, water vapor. The amount of water vapor in the air is called **humidity**. ☑

Humidity depends on the rates of evaporation and condensation. In general, evaporation happens faster when air temperature is higher. The rate of condensation is affected by vapor pressure. Water vapor, like all gases in the atmosphere, adds to the total atmospheric pressure. The portion of air pressure that is caused by water vapor is called *vapor pressure*. ☑

As vapor pressure increases, the rate of condensation increases. When evaporation and condensation happen at the same rate, the air is said to be *saturated* with water. The temperature at which this happens is called the **dew point**. At temperatures below the dew point, liquid water droplets can form from the water vapor in the air.

READING CHECK

3. Identify What causes water in the oceans to turn into water vapor?

____

READING CHECK

4. Explain What two things does humidity depend on?

____

____

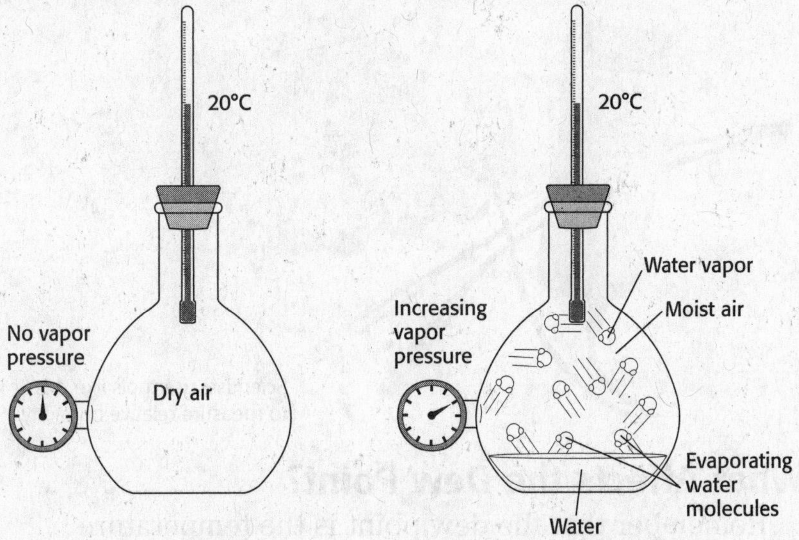
20°C
No vapor pressure
Dry air
20°C
Water vapor
Moist air
Increasing vapor pressure
Evaporating water molecules
Water

The dry air in this flask has no water in it. Therefore, it has no water vapor pressure.

When water comes into contact with dry air, some of the water evaporates. The water molecules in the air exert a pressure called vapor pressure.

TAKE A LOOK

5. Describe How does air pressure change as water vapor pressure increases?

____

____

Copyright © by Holt, Rinehart and Winston. All rights reserved.

RELATIVE HUMIDITY

Scientists often describe the amount of water in the air using the term relative humidity. **Relative humidity** is the ratio of the amount of water vapor in the air to the greatest amount the air can hold. In other words, relative humidity is a measure of how close air is to its dew point.

Math Focus

6. Calculate What is the relative humidity of 25°C air that contains 10 g/m³ of water vapor? Show your work.

____

____

____

To calculate relative humidity, divide the amount of water in the air by the maximum amount of water the air can hold. Then, multiply by 100 to get a percentage. For example, 1 m³ of air at 25°C can hold up to about 23 g of water vapor. If air at 25°C in a certain place contains only 18 g/m³ of water vapor, then the relative humidity is:

$$\frac{18\ g/m^3}{23\ g/m^3} \times 100 = 78\%\ \text{relative humidity}$$

Scientists measure relative humidity using special tools. These tools have sensors that absorb water from the air. When humidity increases, the sensors absorb more water. When humidity decreases, the sensors release water to the air. As the sensors absorb and release water, they send electrical signals to a computer. The computer interprets the signals as a measurement of relative humidity.

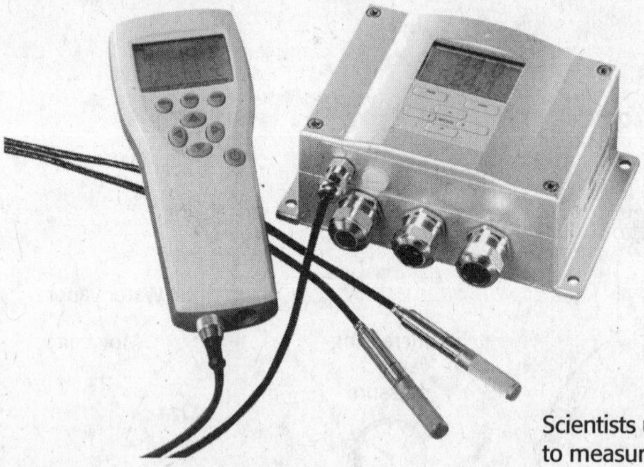

Scientists use tools like this to measure relative humidity.

What Affects the Dew Point?

Remember that the dew point is the temperature at which air is saturated with water. At the dew point, condensation and evaporation happen at the same rate. When the temperature of the air is below the dew point, liquid water droplets can condense from water vapor. ☑

READING CHECK

7. Explain What happens when the temperature of air is below its dew point?

____

____

Copyright © by Holt, Rinehart and Winston. All rights reserved.

REACHING THE DEW POINT

Condensation happens when air is saturated with water vapor. Air can become saturated if water evaporates and enters the air as water vapor. Air can also become saturated when it cools below its dew point.

You have probably seen examples of air becoming saturated because of a temperature decrease. For example, when you add ice cubes to a glass of water, the temperatures of the water and of the glass decrease. The glass absorbs heat from the air, so the temperature of the air near the glass decreases. When the air's temperature drops below its dew point, water vapor condenses on the glass. The condensed water forms droplets on the glass.

The glass absorbs heat from the air. The air cools to below its dew point. Water vapor condenses onto the side of the glass.

When air is nearly saturated, only a small temperature drop is needed for condensation to happen. During the night, objects can lose heat and become cooler. Air may cool below its dew point when the air touches the cool objects. The water droplets that form are called *dew*.

How Do Clouds Form?

A cloud is made of millions of tiny water droplets or ice crystals. Clouds form as air rises and cools. When air cools below the dew point, water droplets or ice crystals form. Water droplets form when water condenses above 0°C. Ice crystals form when water condenses below 0°C.

Clouds are classified by their shapes and by their altitudes. The three main cloud shapes are stratus clouds, cumulus clouds, and cirrus clouds. The three altitude groups are low clouds, middle clouds, and high clouds. The figure on the top of the next page shows these different cloud types. ☑

Critical Thinking

8. Apply Concepts People who wear glasses may notice that their glasses fog up when they go indoors on a cold day. Why does this happen?

____

____

____

____

____

____

TAKE A LOOK

9. Describe Where did the liquid water on the outside of the glass come from?

____

____

☑ READING CHECK

10. Explain How are clouds classified?

____

____

Copyright © by Holt, Rinehart and Winston. All rights reserved.

Name ______ Class ______ Date ______

Say It

Observe and Describe Look at the clouds every day for a week. Each day, write down the weather and what the clouds looked like. At the end of the week, share your observations with a small group. How was the weather related to the kinds of clouds you saw each day?

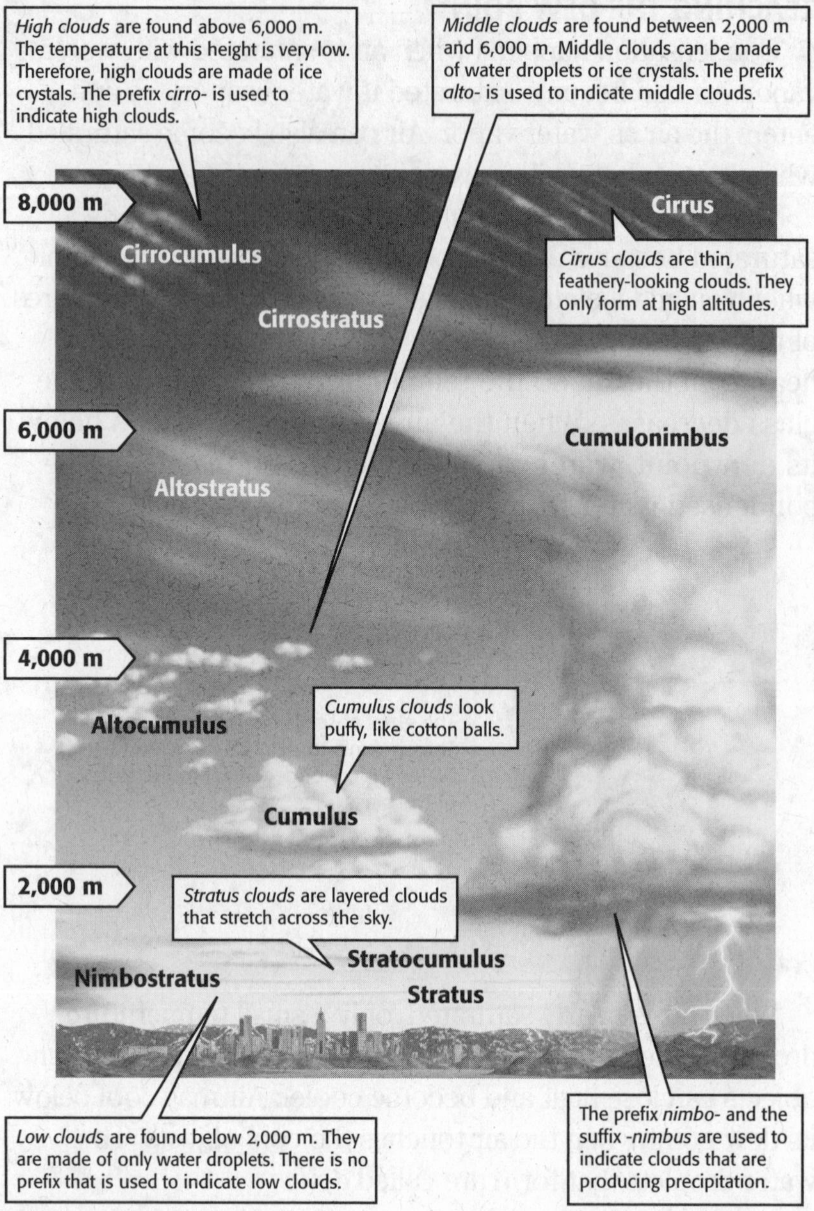
High clouds are found above 6,000 m. The temperature at this height is very low. Therefore, high clouds are made of ice crystals. The prefix *cirro-* is used to indicate high clouds.
Middle clouds are found between 2,000 m and 6,000 m. Middle clouds can be made of water droplets or ice crystals. The prefix *alto-* is used to indicate middle clouds.
8,000 m
Cirrus
Cirrocumulus
Cirrus clouds are thin, feathery-looking clouds. They only form at high altitudes.
Cirrostratus
6,000 m
Cumulonimbus
Altostratus
4,000 m
Altocumulus
Cumulus clouds look puffy, like cotton balls.
Cumulus
2,000 m
Stratus clouds are layered clouds that stretch across the sky.
Stratocumulus
Nimbostratus
Stratus
Low clouds are found below 2,000 m. They are made of only water droplets. There is no prefix that is used to indicate low clouds.
The prefix *nimbo-* and the suffix *-nimbus* are used to indicate clouds that are producing precipitation.

TAKE A LOOK

11. Compare How is a nimbostratus cloud different from a stratus cloud?

______

______

______

PRECIPITATION

Water droplets in clouds are very tiny. As droplets combine, they become larger. When a droplet reaches a certain size, it falls to Earth's surface as precipitation. **Precipitation** is any form of water that falls to Earth's surface from clouds. ☑

There are four main forms of precipitation: rain, snow, sleet, and hail. Rain and snow are the most common kinds of precipitation. Hail and freezing rain are less common.

READING CHECK

12. Define What is precipitation?

______

______

Copyright © by Holt, Rinehart and Winston. All rights reserved.

Name ______________________ Class ______________ Date ______________

Section 1 Review

6.4.a, 6.4.e

SECTION VOCABULARY

condensation the change of state from a gas to a liquid

dew point at constant pressure and water vapor content, the temperature at which the rate of condensation equals the rate of evaporation

humidity the amount of water vapor in the air

precipitation any form of water that falls to Earth's surface from the clouds

relative humidity the ratio of the amount of water vapor in the air to the amount of water vapor needed to reach saturation at a given temperature

weather the short-term state of the atmosphere, including temperature, humidity, precipitation, wind, and visibility

1. Identify Relationships How is dew point related to condensation?

____

____

2. Identify What is the main source of energy for the water cycle?

____

____

3. Describe How does vapor pressure affect condensation?

____

____

4. Explain How do clouds form?

____

____

____

____

5. Apply Concepts Fill in the spaces in the table to describe different kinds of clouds.

Name	Altitude	Shape	Precipitation?
Cirrostratus	high		no
Altocumulus		puffy	
Nimbostratus			
Cumulonimbus	low to middle		

Copyright © by Holt, Rinehart and Winston. All rights reserved.

SECTION 2
Fronts and Weather

California Science Standards

6.2.d, 6.4.a, 6.4.e

BEFORE YOU READ

After you read this section, you should be able to answer these questions:

- What causes weather?
- What is severe weather?

What Are Fronts?

STUDY TIP

Summarize As you read, make a chart comparing the four kinds of fronts. In your chart, describe how each kind of front forms and what kind of weather it can cause.

Have you ever been caught outside when it suddenly started to rain? What causes such an abrupt change in the weather? Changes in weather are caused by the movement of bodies of air called air masses. An **air mass** is a very large volume of air that has the same temperature and moisture content.

When air masses meet, the less-dense air mass rises over the denser air mass. Warm air is less dense than cold air. Therefore, a warm air mass will generally rise above a cold air mass. The place where two or more air masses meet is called a **front**. There are four main kinds of fronts: cold fronts, warm fronts, occluded fronts, and stationary fronts. ☑

READING CHECK

1. Define What is a front?

____

____

COLD FRONTS

A *cold front* forms when a cold air mass moves under a warm air mass. The cold air pushes the warm air mass up. The cold air mass replaces the warm air mass. Cold fronts can move quickly and bring heavy precipitation. When a cold front has passed, the weather is usually cooler. This is because a cold, dry air mass moves in behind the cold front.

TAKE A LOOK

2. Describe What happens to the warm air mass at a cold front?

____

____

____

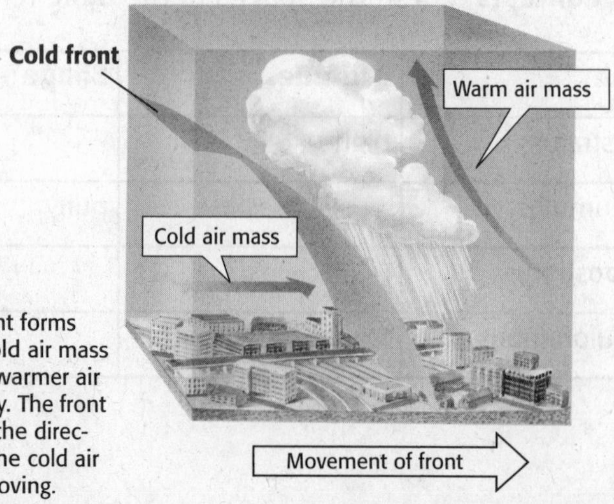
Cold front
Warm air mass
Cold air mass
Movement of front

A cold front forms when a cold air mass pushes a warmer air mass away. The front moves in the direction that the cold air mass is moving.

Copyright © by Holt, Rinehart and Winston. All rights reserved.

WARM FRONTS

A *warm front* forms when a warm air mass moves over a cold air mass that is leaving an area. The warm air replaces the cold air as the cold air moves away. Warm fronts can bring light rain. They are followed by clear, warm weather. ☑

READING CHECK

3. Define What is a warm front?

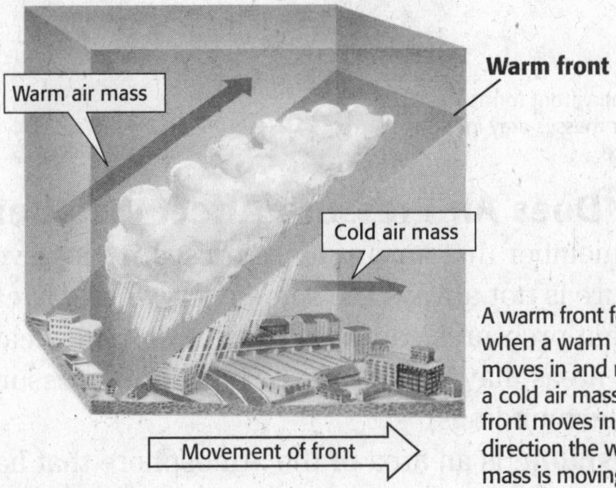
Warm air mass
Warm front
Cold air mass
Movement of front

A warm front forms when a warm air mass moves in and replaces a cold air mass. The front moves in the direction the warm air mass is moving.

OCCLUDED FRONTS

An *occluded front* forms when a warm air mass is caught between two cold air masses. Occluded fronts bring cool temperatures and large amounts of rain and snow.

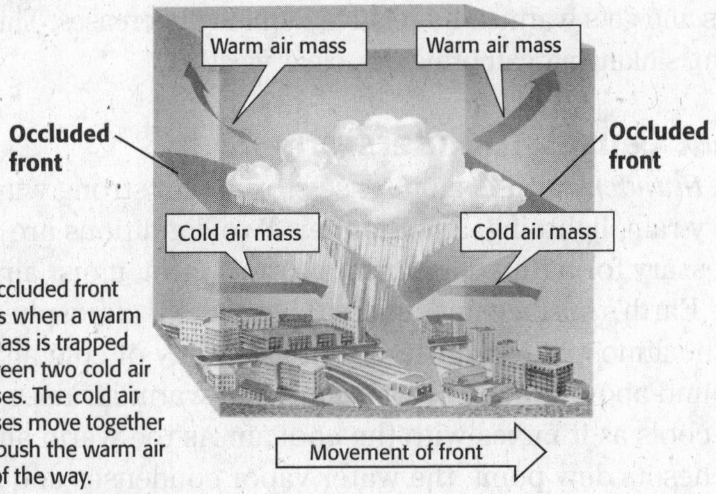
Warm air mass
Warm air mass
Occluded front
Occluded front
Cold air mass
Cold air mass
Movement of front

An occluded front forms when a warm air mass is trapped between two cold air masses. The cold air masses move together and push the warm air out of the way.

TAKE A LOOK

4. Describe What happens to the warm air mass in an occluded front?

STATIONARY FRONTS

A *stationary front* forms when a cold air mass and a warm air mass move toward each other. Neither air mass has enough energy to push the other out of the way. Therefore, the two air masses remain in the same place. Stationary fronts cause many days of cloudy, wet weather.

Copyright © by Holt, Rinehart and Winston. All rights reserved.

TAKE A LOOK

5. Infer What do you think is the reason that stationary fronts bring many days of the same weather?

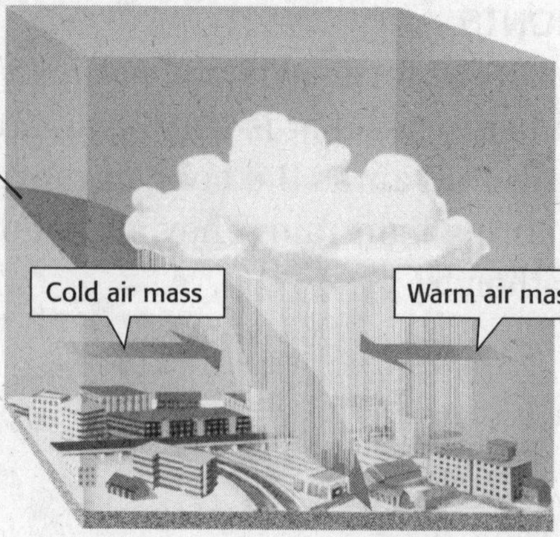
Stationary front
Cold air mass
Warm air mass

A stationary front forms when air masses stay in one place.

How Does Air Pressure Affect Weather?

Remember that air produces pressure. However, air pressure is not always the same everywhere. Areas with different pressures can cause changes in the weather. These areas may have lower or higher air pressure than their surroundings.

A **cyclone** is an area of the atmosphere that has lower pressure than the surrounding air. The air in the cyclone rises. As the air rises, it cools. Clouds can form and may cause rainy or stormy weather.

An *anticyclone* is an area of the atmosphere that has higher pressure than the surrounding air. Air in anticyclones sinks and gets warmer. Its relative humidity decreases. This warm, sinking air can bring dry, clear weather.

CALIFORNIA STANDARDS CHECK

6.4.e Students know differences in pressure, heat, air movement, and humidity result in changes of weather.

6. Describe Why do clouds often form in a cyclone?

What Causes Thunderstorms?

A *thunderstorm* is an intense storm with strong winds, heavy rain, lightning, and thunder. Two conditions are necessary for a thunderstorm to form: warm, moist air near Earth's surface and an unstable area of atmosphere.

The atmosphere is unstable when a body of cold air is found above a body of warm air. The warm air rises and cools as it mixes with the cool air. As the warm air reaches its dew point, the water vapor condenses and forms cumulus clouds. If the warm air keeps rising, the cloud may become a dark cumulonimbus cloud.

Critical Thinking

7. Infer Why does air near the surface have to be moist in order for a thunderstorm to form?

Copyright © by Holt, Rinehart and Winston. All rights reserved.

LIGHTNING AND THUNDER

As the cloud grows bigger, parts of it begin to develop electrical charges. The upper parts of the cloud tend to become positively charged. The lower parts tend to become negatively charged. When the charges get big enough, electricity flows from one area to the other. Electricity may also flow between the clouds and the ground. These electrical currents are **lightning**. ☑.

READING CHECK

8. Describe How does lightning form?

____

____

____

____

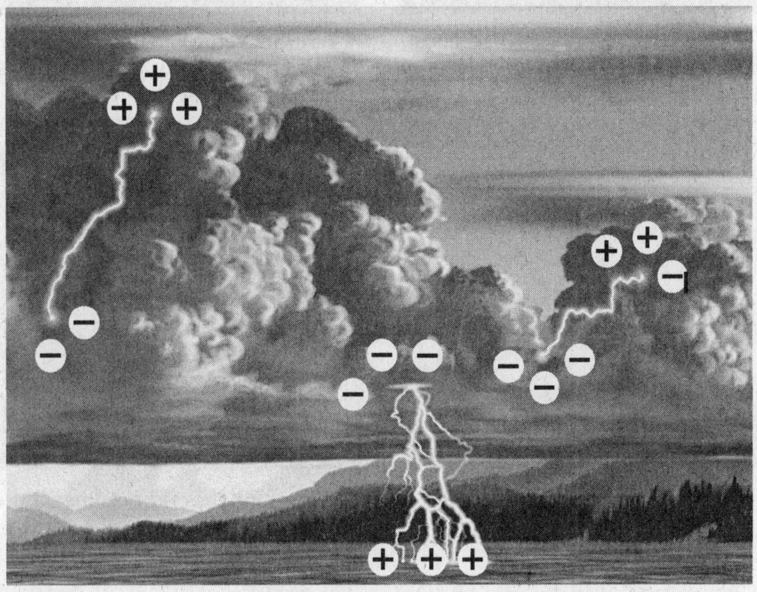
+
+
+
−
−
−
−
−
−
−
−
+
+
−
+
+
+

Different parts of thunderclouds and the ground can have different electrical charges. When electricity flows between these areas, lightning forms.

You have probably seen large lightning bolts that travel between the clouds and the ground. When the lightning moves through the air, the air gets very hot. The hot air expands rapidly. As it expands, it makes the air vibrate. The vibrations release energy in the form of sound waves. The result is **thunder**. ☑

READING CHECK

9. Define What is thunder?

____

____

____

____

How Do Tornadoes Form?

Fewer than 1% of thunderstorms produce tornadoes. A **tornado** can form when a rapidly spinning column of air, called a *funnel cloud*, touches the ground. The air in the center of a tornado has low pressure. When the area of low pressure touches the ground, material from the ground can be sucked up into the tornado.

A tornado begins as a funnel cloud that pokes through the bottom of a cumulonimbus cloud. The funnel cloud becomes a tornado when the funnel cloud touches the ground. The figures on the top of the next page show how a tornado forms.

Copyright © by Holt, Rinehart and Winston. All rights reserved.

❶ Wind moving in opposite directions causes a layer of air in the middle of a cloud to begin to spin.

❷ Strong vertical winds cause the spinning column of air to turn into a vertical position.

❸ The spinning column of air moves to the bottom of the cloud and forms a funnel cloud.

❹ The funnel cloud becomes a tornado when it touches down on the ground.

TAKE A LOOK

10. Describe When does a funnel cloud become a tornado?

____

____

How Do Hurricanes Form?

A *hurricane* is a large, rotating tropical weather system. Hurricanes have wind speeds of over 120 km/h. They can be 160 km to 1,500 km in diameter and can travel for thousands of miles. They are the most powerful storms on Earth.

Most hurricanes form between 5°N and 20°N latitude or between 5°S and 20°S latitude. They form over the warm, tropical oceans found at these latitudes. At higher latitudes, the water is too cold for hurricanes to form. ☑

READING CHECK

11. Explain Why don't hurricanes form at high latitudes?

____

____

Hurricanes are powered by solar energy. The sun's energy causes ocean water to evaporate into water vapor. As the water vapor rises in the air, it cools and condenses. A group of thunderstorms form. They produce a large, spinning storm. A hurricane forms as the storm gets stronger.

The hurricane will continue to grow as long as it is over warm ocean water. When the hurricane moves over colder waters or over land, the storm loses energy. This is why hurricanes are not common in California. Hurricanes that move toward California quickly lose their energy over the cold coastal waters.

Copyright © by Holt, Rinehart and Winston. All rights reserved.

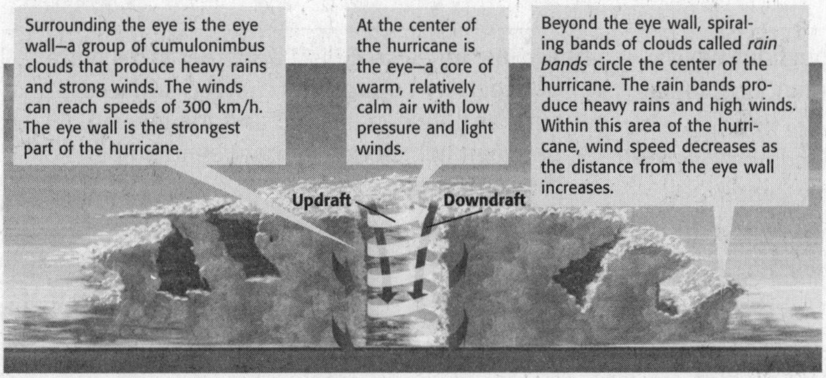
Surrounding the eye is the eye wall—a group of cumulonimbus clouds that produce heavy rains and strong winds. The winds can reach speeds of 300 km/h. The eye wall is the strongest part of the hurricane.
At the center of the hurricane is the eye—a core of warm, relatively calm air with low pressure and light winds.
Beyond the eye wall, spiraling bands of clouds called *rain bands* circle the center of the hurricane. The rain bands produce heavy rains and high winds. Within this area of the hurricane, wind speed decreases as the distance from the eye wall increases.
Updraft
Downdraft

Although hurricanes are very destructive storms, the eye at the center of the tropical storm is relatively calm.

TAKE A LOOK

12. Define What is the eye of a hurricane?

____

____

____

____

What Are the Dangers of Severe Weather?

Severe weather can cause property damage, injury, and sometimes even death. Hail, lightning, high winds, tornadoes, and flash floods are all part of severe weather. Strong winds can knock down power lines and damage homes. Lightning starts thousands of fires every year.

Heavy rains can cause flooding and major property damage. Flash floods are also responsible for many weather-related deaths. Much of the damage caused by hurricanes comes from the heavy rains and storm surge that these storms produce. A *storm surge* is a rise in sea level that happens during a storm. The storm-surge flooding from Hurricane Katrina in 2005 caused more damage than the high-speed winds from the storm.

SEVERE-WEATHER SAFETY

During severe weather, it is important for you to listen to a local TV or radio station. Severe-weather announcements will tell you where a storm is and if it is getting worse. ☑

During most kinds of severe weather, it is safest to stay inside, away from windows. However, there are times when you will need to *evacuate*, or leave your home. For example, if your home is in a low-lying area, you should not stay there during a flash-flood warning. Instead, you should leave and move to higher ground. However, you should never enter floodwaters. Even shallow floodwater can be dangerous.

READING CHECK

13. Explain Why should you listen to weather reports during severe weather?

____

____

Copyright © by Holt, Rinehart and Winston. All rights reserved.

Name _______________ Class _______________ Date _______________

Section 2 Review

6.2.d, 6.4.a, 6.4.e

SECTION VOCABULARY

air mass a large body of air throughout which temperature and moisture content are similar

cyclone an area in the atmosphere that has lower pressure than the surrounding areas and has winds that spiral toward the center

Wordwise: The root *cycl* means "circle" or "wheel." Other examples are *bicycle*, *cyclic*, and *cycle*.

front the boundary between air masses of different densities and usually different temperatures

lightning an electric discharge that takes place between two oppositely charged surfaces, such as between a cloud and the ground, between two clouds, or between two parts of the same cloud

thunder the sound caused by the rapid expansion of air along an electrical strike

tornado a destructive, rotating column of air that has very high wind speeds and that may be visible as a funnel-shaped cloud

1. Identify Relationships How are fronts and air masses related?

____

____

2. Compare Fill in the table to describe cyclones and anticyclones.

Name	The pressure in the middle is . . .	The air inside it . . .	The kind of weather it causes . . .
Cyclone	. . . lower than the surrounding pressure.		
Anticyclone		. . . sinks and warms.	

3. Explain Why do thunder and lightning usually happen together?

____

____

4. Describe How can flooding affect people?

____

____

5. Analyze How does energy from the sun power hurricanes?

____

____

____

____

Copyright © by Holt, Rinehart and Winston. All rights reserved.

Name ______________________ Class ______________ Date ____________

CHAPTER 15 Weather and Climate

SECTION 3

What Is Climate?

BEFORE YOU READ

After you read this section, you should be able to answer these questions:

- What is climate?
- What factors affect climate?
- How does climate differ around the world?

California Science Standards

6.4.b, 6.4.d, 6.4.e

What Is Climate?

How is climate different from weather? *Weather* is the condition of the atmosphere at a certain time. The weather can change from day to day. However, **climate** describes the average weather conditions in a region over a long period of time. The climate of an area includes the area's average temperature and amount of precipitation. Different parts of the world have different climates.

STUDY TIP

Ask Questions As you read this section, write down any questions that you have. When you finish reading, talk about your questions in a small group.

What Factors Affect Climate?

Climate is mainly determined by temperature and precipitation. Temperature and precipitation can be affected by four factors: latitude, wind patterns, landforms, and ocean currents. ☑

READING CHECK

1. List What are the two main things that determine climate?

____

____

SOLAR ENERGY AND LATITUDE

The **latitude** of an area is its distance north or south of the equator. In general, the temperature of an area depends on its latitude: higher latitudes tend to have colder climates. Latitude affects temperature because latitude determines how much direct solar energy an area gets. This is shown in the figure below.

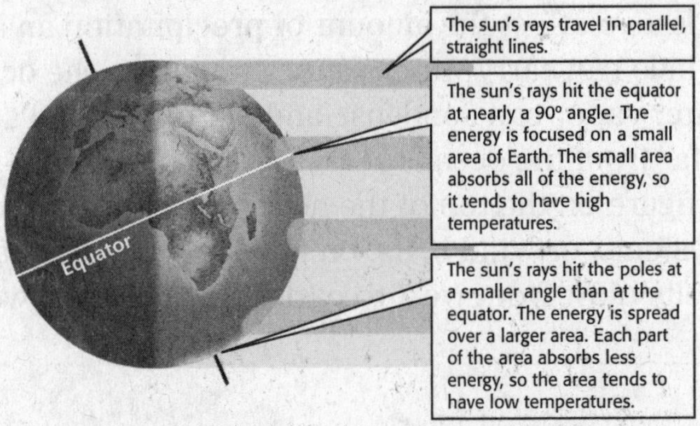

The sun's rays travel in parallel, straight lines.
The sun's rays hit the equator at nearly a 90° angle. The energy is focused on a small area of Earth. The small area absorbs all of the energy, so it tends to have high temperatures.
Equator
The sun's rays hit the poles at a smaller angle than at the equator. The energy is spread over a larger area. Each part of the area absorbs less energy, so the area tends to have low temperatures.

TAKE A LOOK

2. Explain Why do areas near the equator tend to have high temperatures?

____

____

Copyright © by Holt, Rinehart and Winston. All rights reserved.

LATITUDE AND SEASONS

Most places in the United States have four seasons during the year. However, some places in the world do not have such large seasonal changes. For example, places near the equator have about the same temperatures and amount of daylight all year.

Seasons happen because Earth is tilted on its axis by about 23.5°. The tilt affects how much solar energy an area gets as Earth orbits the sun. The figure below shows how Earth's tilt affects the seasons.

TAKE A LOOK

3. Explain Why don't areas near the equator have large seasonal changes in weather?

____

____

____

____

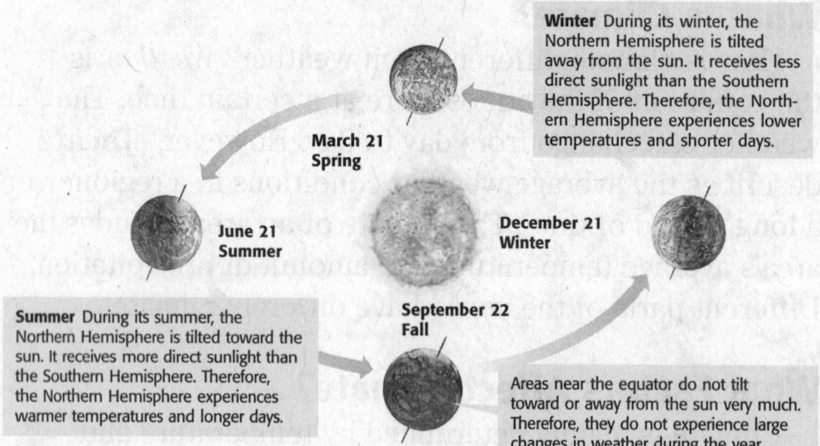
Winter During its winter, the Northern Hemisphere is tilted away from the sun. It receives less direct sunlight than the Southern Hemisphere. Therefore, the Northern Hemisphere experiences lower temperatures and shorter days.
March 21 Spring
June 21 Summer
December 21 Winter
September 22 Fall
Summer During its summer, the Northern Hemisphere is tilted toward the sun. It receives more direct sunlight than the Southern Hemisphere. Therefore, the Northern Hemisphere experiences warmer temperatures and longer days.
Areas near the equator do not tilt toward or away from the sun very much. Therefore, they do not experience large changes in weather during the year.

WINDS

Winds move in predictable patterns over Earth's surface. The wind patterns on Earth are caused by the uneven heating of Earth's surface. This uneven heating forms areas with different air pressures. Wind forms when air moves from areas of high pressure to areas of low pressure. ☑

READING CHECK

4. Identify What causes wind to form?

____

____

Winds affect climate and weather because they move solar energy from one place to another. This can cause the temperature in one place to decrease and the temperature in another place to increase.

Winds also affect the amount of precipitation an area gets. Winds can carry water vapor away from the oceans. The water vapor can condense and fall to the land somewhere far from the ocean.

The figure on the top of the next page shows the major wind systems, or wind belts, on Earth. Notice that most wind belts blow from west to east or from east to west.

Copyright © by Holt, Rinehart and Winston. All rights reserved.

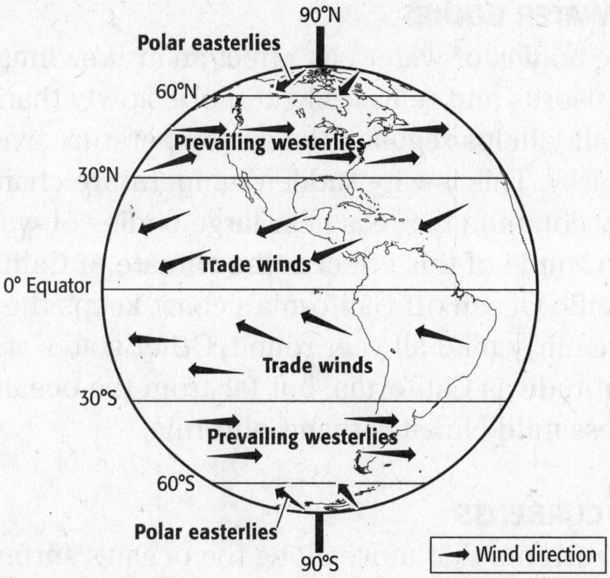
90°N
Polar easterlies
60°N
Prevailing westerlies
30°N
Trade winds
0° Equator
Trade winds
30°S
Prevailing westerlies
60°S
Polar easterlies
90°S
Wind direction

TAKE A LOOK

5. Read a Map In which direction do the Prevailing Westerlies blow?

____

TOPOGRAPHY

The topography of an area affects its climate. *Topography* is the sizes and shapes of the land-surface features of a region. Topography can affect the temperature and precipitation of an area. For example, elevation can have a large impact on temperature. **Elevation** is the height of an area above sea level. As elevation increases, temperature tends to decrease. ☑

READING CHECK

6. Describe In general, how does elevation affect temperature?

____

____

____

Mountains can also affect precipitation. As air rises to move over a mountain, the air cools. The cool air condenses, forming clouds. Precipitation may fall. This process causes the *rain-shadow effect*, which is illustrated in the figure below.

Air rises to flow over mountains. The air cools as it rises, and water vapor can condense to form clouds. The clouds can release the water as precipitation. Therefore, this side of the mountain tends to be wetter, with more vegetation.

The air on this side of the mountain contains much less water vapor. As the air sinks down the side of the mountain, it becomes warmer. The warm air absorbs moisture from the land. Therefore, this side of the mountain tends to be drier and more desert-like.

TAKE A LOOK

7. Explain Why do clouds form as air moves over a mountain?

____

____

____

Copyright © by Holt, Rinehart and Winston. All rights reserved.

LARGE WATER BODIES

Large bodies of water can affect an area's climate. Water absorbs and releases heat more slowly than land. This quality helps regulate the air temperature over the land nearby. This is why sudden temperature changes are not very common in areas near large bodies of water. ☑

An example of this effect is the climate of California. The Pacific Ocean off California's coast keeps the temperature fairly mild all year round. Other states at the same latitude as California, but far from the oceans, have much less mild climates than California.

READING CHECK

8. Explain Why aren't sudden temperature changes common near large bodies of water?

____

____

____

____

____

OCEAN CURRENTS

The currents that move along the ocean's surface can have a big effect on a region's climate. **Surface currents** are paths of flowing water found near the surface of the ocean. As surface currents move, they carry warm or cool water to different places. The temperature of the water affects the temperature of the air above it. For example, warm currents can heat the surrounding air.

California's climate is affected by a cool ocean current. Cold water from the northern Pacific Ocean moves south toward Mexico in the California current. You can see other currents in the figure below.

Critical Thinking

9. Describe Processes Cool surface currents can cause the air above them to become cooler. Explain how this happens.

____

____

____

____

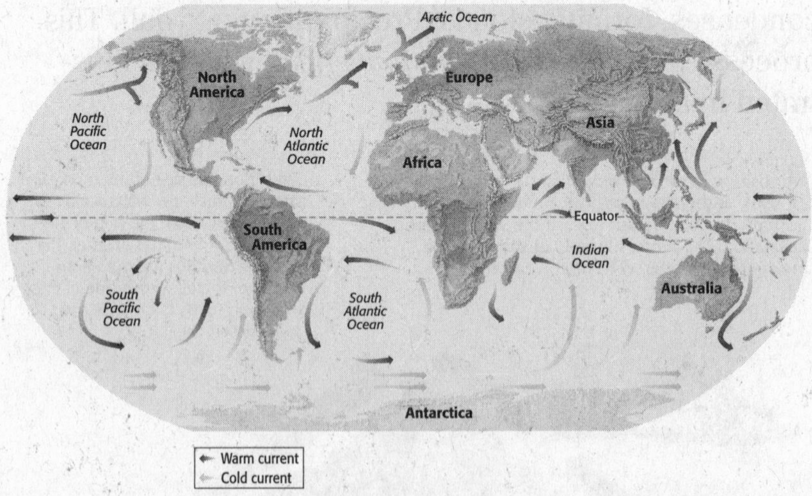
Arctic Ocean
North America
Europe
Asia
North Pacific Ocean
North Atlantic Ocean
Africa
Equator
South America
Indian Ocean
Australia
South Pacific Ocean
South Atlantic Ocean
Antarctica
Warm current
Cold current

TAKE A LOOK

10. Identify What kind of surface current is found off the East Coast of the United States?

____

Copyright © by Holt, Rinehart and Winston. All rights reserved.

What Are the Different Climates Around the World?

Earth has three major climate zones: tropical, temperate, and polar. The figure below shows where these zones are found.

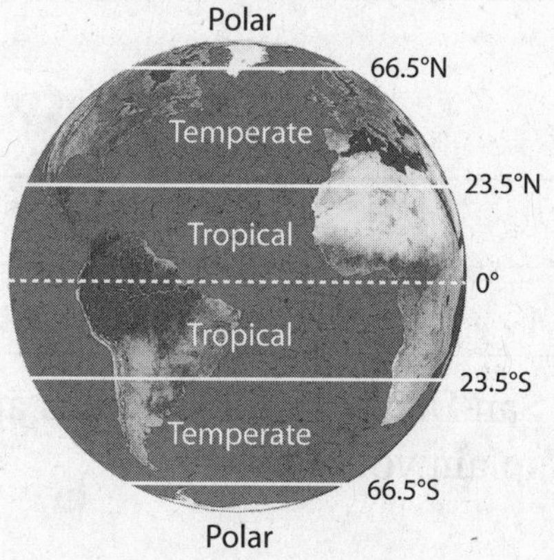
Polar
66.5°N
Temperate
23.5°N
Tropical
0°
Tropical
23.5°S
Temperate
66.5°S
Polar

Earth's three major climate zones are determined by latitude.

TAKE A LOOK

11. Identify What determines Earth's major climate zones?

____

Each climate zone has a certain range of temperatures. The tropical zone, near the equator, has the highest temperatures. The polar zones, located at latitudes above 66.5°, have the lowest temperatures. ☑

Each climate zone contains several different kinds of climates. The different climates are the result of topography, winds, and ocean currents. The figure below shows the different kinds of climates on Earth.

✓ READING CHECK

12. Describe Which climate zone has the highest temperatures?

____

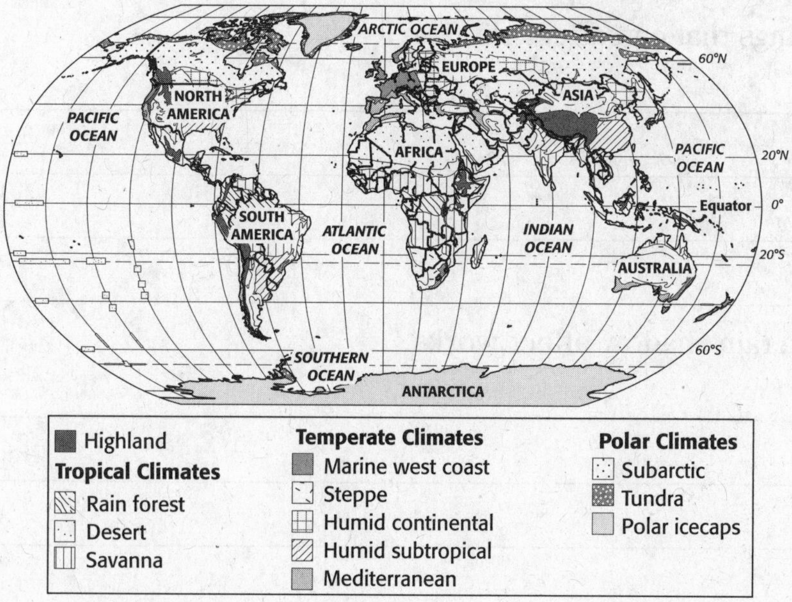
ARCTIC OCEAN
EUROPE
60°N
NORTH AMERICA
ASIA
PACIFIC OCEAN
AFRICA
PACIFIC OCEAN
20°N
Equator
0°
SOUTH AMERICA
ATLANTIC OCEAN
INDIAN OCEAN
20°S
AUSTRALIA
60°S
SOUTHERN OCEAN
ANTARCTICA
Highland
Tropical Climates
Rain forest
Desert
Savanna
Temperate Climates
Marine west coast
Steppe
Humid continental
Humid subtropical
Mediterranean
Polar Climates
Subarctic
Tundra
Polar icecaps

TAKE A LOOK

13. Explain Why are there different climates in each climate zone, even though the areas are found at similar latitudes?

____

____

Copyright © by Holt, Rinehart and Winston. All rights reserved.

Name ______________________ Class ______________ Date ______________

Section 3 Review

6.4.b, 6.4.d, 6.4.e

SECTION VOCABULARY

climate the average weather conditions in an area over a long period of time

elevation the height of an object above sea level

latitude the distance north or south from the equator; expressed in degrees

surface current a horizontal movement of ocean water that is caused by wind and that occurs at or near the ocean's surface

1. Describe How is climate different from weather?

____

____

____

2. Apply Concepts Nome, Alaska, lies at 64°N latitude. San Diego, California, lies at 32°N latitude. Which city receives more sunlight? Explain your answer.

____

____

____

3. Explain What causes some places on Earth to have seasons?

____

____

4. Identify What are four things that can affect climate?

____

____

____

____

5. Explain Describe how the rain-shadow effect works.

____

____

____

____

____

Copyright © by Holt, Rinehart and Winston. All rights reserved.

Name ______________________ Class ______________ Date ______________

CHAPTER 15 Weather and Climate

SECTION 4

Changes in Climate

BEFORE YOU READ

After you read this section, you should be able to answer these questions:

- How has Earth's climate changed over time?
- What factors can cause climates to change?

California Science Standards

6.2.d, 6.4.a, 6.4.b, 6.6.a

How Was Earth's Climate Different in the Past?

The geologic record shows that Earth's climate in the past was different from its climate today. During some periods in the past, Earth was much warmer. During other periods, Earth was much colder. In fact, much of Earth was covered by sheets of ice during some times in the past.

An **ice age** happens when ice at high latitudes expands toward lower latitudes. Scientists have found evidence of many major ice ages in Earth's history. The most recent one began about 2 million years ago. ☑

Many people think of an ice age as a time when the temperature is always very cold. However, during an ice age, there can be periods of colder or warmer weather. A period of colder weather is called a *glacial period*. A period of warmer weather is called an *interglacial period*.

During glacial periods, large sheets of ice grow. These ice sheets form when ocean water freezes. Therefore, sea level drops during glacial periods. The figure below shows the coastlines of the continents during the last glacial period. Notice that the continental coastlines extended further into the ocean than they do today.

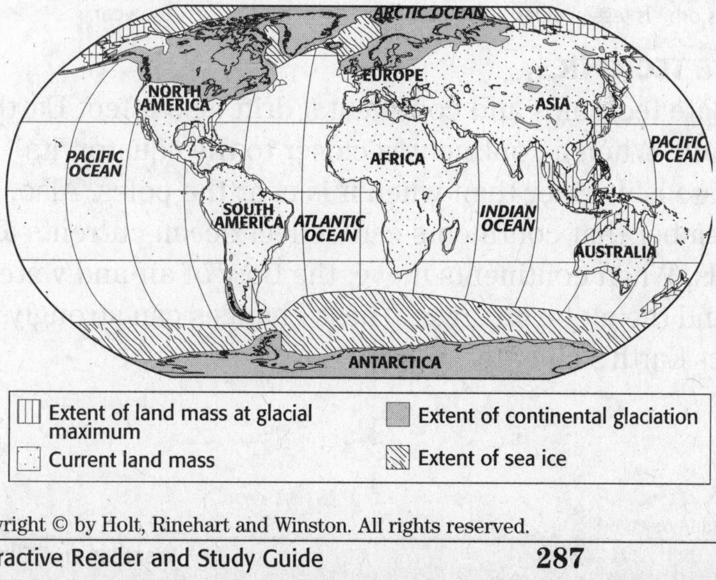
ARCTIC OCEAN
NORTH AMERICA
EUROPE
ASIA
PACIFIC OCEAN
AFRICA
PACIFIC OCEAN
SOUTH AMERICA
ATLANTIC OCEAN
INDIAN OCEAN
AUSTRALIA
ANTARCTICA
Extent of land mass at glacial maximum
Extent of continental glaciation
Current land mass
Extent of sea ice

STUDY TIP

Learn New Words As you read, underline any words that you don't know. When you figure out what they mean, write the words and their definitions in your notebook.

READING CHECK

1. Define Write your own definition for *ice age*.

____

____

____

____

TAKE A LOOK

2. Explain Why is more land exposed during glacial periods than at other times?

____

____

____

____

Copyright © by Holt, Rinehart and Winston. All rights reserved.

What Factors Can Cause Climates to Change?

Scientists have several theories to explain ice ages and other forms of climate change. Factors that can cause climate change are Earth's orbit, plate tectonics, the sun's cycles, asteroid impacts, volcanoes, and human activities.

CHANGES IN EARTH'S ORBIT

A Serbian scientist, Milutin Milankovitch, found that changes in Earth's orbit and tilt can affect Earth's climate. He modeled the way Earth moves in space and found that Earth's movements change in a regular way. These changes happen over tens of thousands of years. For example, Earth's orbit around the sun is more circular at some times than others.

These variations in Earth's orbit and tilt affect how much sunlight Earth gets. Therefore, they can also affect climate. The figure below shows how these factors can change the amount of sunlight Earth gets.

Critical Thinking

3. Infer Could changes in climate over 100 years be caused by changes in Earth's orbit and tilt? Explain your answer.

____

____

____

____

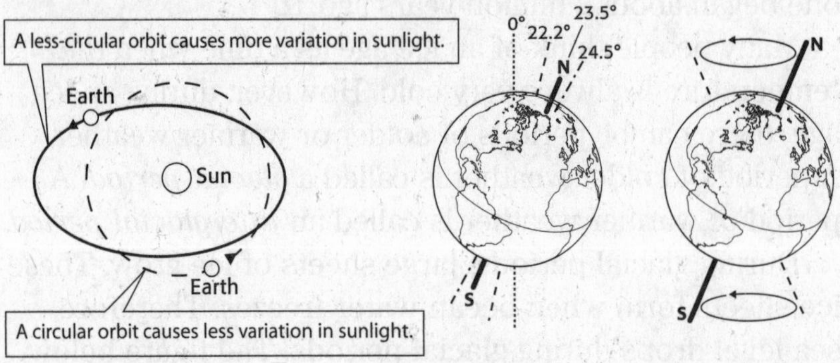
A less-circular orbit causes more variation in sunlight.
Earth
Sun
Earth
A circular orbit causes less variation in sunlight.
0°
22.2°
23.5°
24.5°
N
S
N
S

Earth's orbit is more circular at some times than at other times. The amount of solar energy that Earth gets from the sun varies more when Earth's orbit is less circular.

Earth's tilt on its axis can vary. When the tilt is greater, the poles get more solar energy.

Earth's axis wobbles slightly. This affects how much sunlight Earth's surface gets at different times of the year.

TAKE A LOOK

4. Identify How does the shape of Earth's orbit change?

____

____

PLATE TECTONICS

Plate tectonics and continental drift also affect Earth's climate. When a continent is closer to the equator, its climate is warmer than when it is near the poles. Also, remember that continents can deflect ocean currents and winds. When continents move, the flow of air and water around the globe changes. These changes can strongly affect Earth's climate.

Copyright © by Holt, Rinehart and Winston. All rights reserved.

Name ______________________ Class ______________ Date ____________

THE SUN

Some changes in Earth's climate are caused by changes in the sun. Many people think that the sun is always the same, but this is not true. In fact, the amount of energy that the sun gives off can change over time. The sun follows a regular cycle in how much energy it gives off. Because the sun's energy drives most cycles on Earth, these changes can affect Earth's climate. ☑

READING CHECK

5. Explain Why do changes in the sun's energy affect the climate on Earth?

____

____

____

IMPACTS

Sometimes, objects from outer space, such as asteroids, crash into Earth. An *asteroid* is a small, rocky object that orbits the sun. If a large asteroid crashed into Earth, the climate of the whole planet could change

When a large object hits Earth, particles of dust and rock fly into the atmosphere. This material can block some sunlight from reaching Earth's surface. This can cause temperatures on Earth to go down. In addition, plants may not be able to survive with less sunlight. Without plants, many animals would die off. Many scientists believe that an asteroid impact may have caused the dinosaurs to become extinct.

Critical Thinking

6. Identify Relationships Why may animals die off if there are fewer plants around?

____

____

VOLCANIC ERUPTIONS

Volcanic eruptions can affect Earth's climate for a short time. They send large amounts of dust and ash into the air. As with an asteroid impact, the dust and ash block sunlight from reaching Earth's surface. The figure below shows how volcanic dust can affect sunlight.

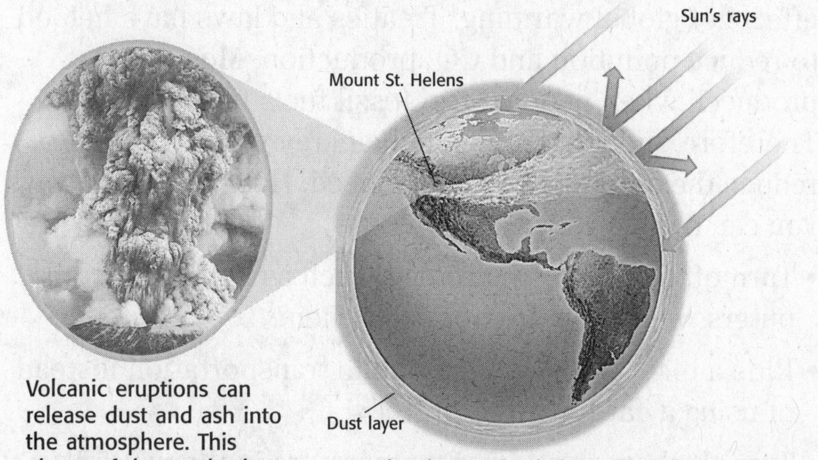
Sun's rays
Mount St. Helens
Dust layer

Volcanic eruptions can release dust and ash into the atmosphere. This plume of dust and ash was produced by the eruption of Mount St. Helens, in Washington, in 1980.

The dust and ash from the volcano can spread throughout the atmosphere. Sunlight reflects off this dust and ash. Less sunlight reaches Earth's surface.

TAKE A LOOK

7. Compare How are the effects on climate of volcanic eruptions and asteroid impacts similar?

____

____

____

____

____

Copyright © by Holt, Rinehart and Winston. All rights reserved.

HUMAN ACTIVITIES

A slow increase in global temperatures is called **global warming**. One thing that can cause global warming is an increase in the greenhouse effect. The **greenhouse effect** is Earth's natural heating process. During this process, gases in the atmosphere absorb energy in sunlight. This energy is released as heat, which helps to keep Earth warm. Without the greenhouse effect, Earth's surface would be covered in ice. ☑

READING CHECK

8. Define What is global warming?

____

____

One of the gases that absorbs sunlight in the atmosphere is carbon dioxide (CO_2). If there is more CO_2 in the atmosphere, the greenhouse effect can increase. This can cause global warming.

Much of the CO_2 in the atmosphere comes from natural processes, such as volcanic eruptions and animals breathing. However, human activities can also increase the amount of CO_2 in the atmosphere.

When people burn fossil fuels for energy, CO_2 is released into the atmosphere. When people burn trees to clear land for farming, CO_2 is released. In addition, plants use CO_2 for food. Therefore, when trees are destroyed, we lose a natural way of removing CO_2 from the atmosphere.

Many scientists think that if global warming continues, the ice at Earth's poles could melt. This could cause sea level to rise. Many low-lying areas could flood. Global warming could also affect areas far from the oceans. ☑

READING CHECK

9. Explain Why may sea level rise if global warming continues?

____

____

____

____

WHAT PEOPLE CAN DO

Many countries are working together to reduce the effects of global warming. Treaties and laws have helped to reduce pollution and CO_2 production. Most CO_2 is produced when people burn fossil fuels for energy. Therefore, reducing how much energy you use can reduce the amount of CO_2 produced. Here are some ways you can reduce your energy use:

- Turn off electrical equipment, such as lights and computers, when you are not using them.
- Ride a bike, walk, or take public transportation instead of using a car to travel.
- Turn the heat to a lower temperature in the winter.
- Turn the air conditioner to a higher temperature in the summer.

Copyright © by Holt, Rinehart and Winston. All rights reserved.

Name ______________________ Class ______________ Date ____________

Section 4 Review

6.2.d, 6.4.a, 6.4.b, 6.6.a

SECTION VOCABULARY

global warming a gradual increase in average global temperature **greenhouse effect** the warming of the surface and lower atmosphere of Earth that occurs when water vapor, carbon dioxide, and other gases absorb and reradiate thermal energy	**ice age** a long period of climatic cooling during which the continents are glaciated repeatedly

1. Identify Relationships How is global warming related to the greenhouse effect?

____

____

____

2. Describe What did Milutin Milankovitch's research show can affect Earth's climate?

____

____

____

3. Identify Give two ways that plate tectonics can affect an area's climate.

____

____

____

____

4. Predict Consequences How could global warming affect cities near the oceans? Explain your answer.

____

____

____

5. List Give three ways that human activities can affect the amount of CO_2 in the atmosphere.

____

____

____

Copyright © by Holt, Rinehart and Winston. All rights reserved.

SECTION 1

Everything Is Connected

California Science Standards

6.5.b, 6.5.e

BEFORE YOU READ

After you read this section you should be able to answer these questions:

- What do organisms in an ecosystem depend on for survival?
- What are biotic and abiotic factors?
- What are the levels of organization in the environment?

STUDY TIP

Underline As you read, underline any new science terms. Find their definitions in the section review or a dictionary. Make sure you learn what each term means before you move to the next section.

What Is the Web of Life?

All organisms, or living things, are linked together in the web of life. In this web, energy and resources pass between organisms and their surroundings. The study of how different organisms interact with one another and their environment is **ecology**.

A hungry bear may go to a river to catch a fish or look for berries to eat. In both examples, the bear is interacting with the environment, which includes other organisms. The bear depends on the other organisms, and they depend on the bear. For example, after the bear eats berries, it spreads the berry plant's seeds to new areas in its waste.

TAKE A LOOK

1. Identify List three things the bear in this figure depends on to survive in its environment.

____

____

____

An environment has both biotic and abiotic factors.

READING CHECK

2. Compare What is the difference between biotic and abiotic factors?

____

____

____

What Are the Two Parts of an Environment?

An organism's environment is made up of biotic and abiotic factors. **Biotic** factors are the living parts of the environment, such as fish. **Abiotic** factors are the nonliving parts of the environment, such as rivers. Organisms need both biotic and abiotic factors to live. ☑

Copyright © by Holt, Rinehart and Winston. All rights reserved.

How Is the Environment Organized?

The environment can be organized into six levels. Individual organisms are at the first level. The higher levels include more and more parts of the environment. The highest level is the largest. It is the biosphere.

1. An *individual* is a single organism.

2. A **population** is a group of individuals of the same species in the same area. For example, all the black bears in the same forest make a population. The whole population uses the same area for food and shelter.

3. A **community** is made up of all the different populations that live and interact in the same area. The different populations in a community depend on each other. For example, bears eat fish and berries. Berry plants depend on bears to eat their berries, which moves their seeds to other areas. River plants are food for some water animals, and they shelter fish that are hiding from bears.

4. An **ecosystem** is made up of a community and its abiotic environment. The abiotic factors provide resources for all the organisms and energy for some. A river, for example, can provide water for river plants and many animals, and shelter for water insects. It can provide nutrients for plants, as well as food for fish and bears.

Critical Thinking

3. Making Inferences Could a community be made up of only one population of organisms? Explain.

____

____

____

____

TAKE A LOOK

4. Identify Use colored pencils to make circles on the picture.
Circle an individual in red.
Circle a population in blue.
Circle a community in brown.
Circle an ecosystem in green.

Copyright © by Holt, Rinehart and Winston. All rights reserved.

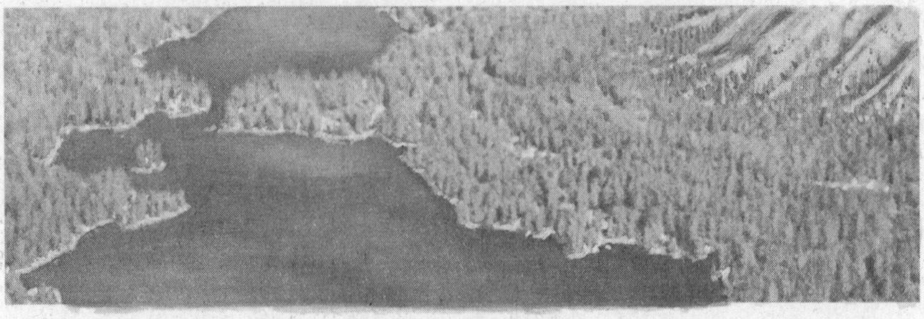

5. A **biome** is a large area made of many similar ecosystems with the same climate. The climate determines the kinds of organisms that live there. The organisms in a biome must be adapted to live in the conditions in that biome. For example, a temperate deciduous forest biome is warm and moist in summer but dry and cold in winter. Only animals and plants that can live through the winter will survive there. Bears and many other animals there sleep for most of the winter, while food is harder to find. Deciduous trees shed their leaves so that they won't dry out.

Say It

Interpret Look at the map and read the text above it. With a partner, discuss what the map shows, and what the different patterns on the map mean.

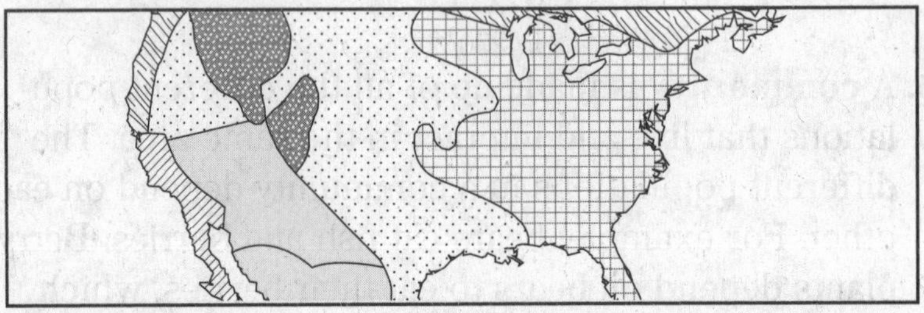

6. The **biosphere** is the part of Earth where life exists. The biosphere is the largest environmental level. It reaches from the bottom of the ocean and the Earth's crust to high in the sky. Scientists study the biosphere and its abiotic factors, such as atmosphere, water, soil, and rock, to learn how life survives.

Copyright © by Holt, Rinehart and Winston. All rights reserved.

Name ______________________ Class ______________ Date ____________

Section 1 Review

6.5.b, 6.5.e

SECTION VOCABULARY

abiotic describes the nonliving part of the environment, including water, rocks, light, and temperature

biome a large region characterized by a specific type of climate and certain types of plant and animal communities

biosphere the part of Earth where life exists

biotic describes living factors in the environment

community all of the populations of species that live in the same habitat and interact with each other

ecology the study of the interactions of living organisms with one another and with their environment

ecosystem a community of organisms and their abiotic, or nonliving, environment

population a group of organisms of the same species that live in a specific geographical area

1. Compare What is the difference between a community and an ecosystem?

____

____

____

2. Organize Complete the chart below to describe the six levels of the environment, from smallest to largest.

Level	Description
	a single organism
Population	
	all of the populations of species that live in the same habitat and interact with one another
Ecosystem	
	a large region with the same climate and same types of organisms
Biosphere	

3. Identify What two kinds of factors does an organism depend on for survival?

____

____

Copyright © by Holt, Rinehart and Winston. All rights reserved.

Name _______________ Class _______________ Date _______________

CHAPTER 16 Interactions of Living Things

SECTION 2

Living Things Need Energy

California Science Standards

6.5.a, 6.5.b, 6.5.c, 6.5.e

BEFORE YOU READ

After you read this section, you should be able to answer these questions:

- How do producers, consumers, and decomposers get energy?
- What is a food web?
- How does the amount of resources in an area affect the organisms living there?

How Do Organisms Get Energy?

Eating gives organisms two things they cannot live without—energy and nutrients. Prairie dogs, for example, eat grasses and seeds to get their energy and nutrients. Like all organisms, prairie dogs need energy to live.

Organisms in any community can be separated into three groups based on how they get energy. These groups are producers, consumers, and decomposers.

STUDY TIP

Circle Choose different colored pencils for producers, primary consumers, secondary consumers, and decomposers. As you read, circle these terms in the text with the colors you chose. Use the same colors to circle animals in any figures that are examples of each group.

PRODUCERS

Producers are organisms that use the energy from sunlight to make their own food. This process is called *photosynthesis*. Most producers are green plants, such as grasses on the prairie and trees in a forest. Some bacteria and algae also photosynthesize to make food. ☑

READING CHECK

1. Explain Why is sunlight important to producers?

____

____

____

CONSUMERS

Consumers cannot make their own food as producers can. Consumers need to eat other organisms to obtain energy and nutrients.

Consumers can be put into four groups. A **herbivore** is a consumer that eats only plants. Prairie dogs and bison are herbivores. A **carnivore** is a consumer that eats other animals. Eagles and cougars are carnivores. An **omnivore** is a consumer that eats both plants and animals. Bears and raccoons are omnivores. A *scavenger* is a consumer that eats dead plants and animals. Turkey vultures are scavengers. They will eat animals and plants that have been dead for days. They will also eat what is left over after a carnivore has had a meal.

Critical Thinking

2. Apply Concepts What types of consumers are the following organisms?

tigers _______________

deer _______________

humans _______________

Copyright © by Holt, Rinehart and Winston. All rights reserved.

DECOMPOSERS

Decomposers recycle nature's resources. Decomposers get energy by breaking down dead organisms into simple materials. These materials, such as carbon dioxide and water, can then be used by other organisms. Many bacteria and fungi are decomposers.

CALIFORNIA STANDARDS CHECK

6.5.c Students know populations of organisms can be categorized by the functions they serve in an ecosystem.

Word Help: categorize
to put into groups or classes

Word Help: function
use or purpose

3. Define What is the role of decomposers in an ecosystem?

____

____

____

____

What Is a Food Chain?

A **food chain** is the sequence of organisms that pass energy and nutrients from one to the next. Producers form the beginning of the food chain. Energy passes through the rest of the chain as one organism eats another.

- Producers are eaten by primary consumers.
- Primary consumers are eaten by secondary consumers.
- Secondary consumers are eaten by tertiary consumers.

In the food chain below, the grasses are the producers. The grasses are eaten by prairie dogs, which are the primary consumers. The prairie dogs are eaten by coyotes, which are the secondary consumers. When coyotes die, they are eaten by turkey vultures, which are the tertiary consumers. The tertiary consumer is usually the end of the food chain.

A Prairie Ecosystem Food Chain

TAKE A LOOK

4. Identify Label the food chain diagram with the following terms: energy, producer, primary consumer, secondary consumer, tertiary consumer, decomposer.

Copyright © by Holt, Rinehart and Winston. All rights reserved.

What Is a Food Web?

Simple food chain diagrams do not show how energy really moves through an ecosystem. A **food web** is a system of many connected food chains in an ecosystem. Organisms in the different food chains feed upon one another. A food web is more accurate than a food chain because most animals eat more than one kind of food. ☑

In a food web, energy moves from one organism to the next in one direction. The energy in an organism that is eaten goes into the body of the organism that eats it.

READING CHECK

5. Explain Why does a food web show feeding relationships better than a food chain?

____

____

____

Simple Food Web

CALIFORNIA STANDARDS CHECK

6.5.a Students know energy entering ecosystems as sunlight is transferred by producers into chemical energy through photosynthesis and then from organism to organism through food webs.

Word Help: transfer
to carry or cause to pass from one thing to another

6. Predict Consequences What do you think would happen if all of the plants were taken out of this food web?

____

____

____

____

____

MANY AT THE BASE

Only a small part of the energy stored in an organism's body is passed on to the next consumer in a food chain or web. Because of this, many more organisms have to be at the base, or bottom, of the food chain than at the end, or top. For example, in a prairie community, there is more grass than prairie dogs and there are more prairie dogs than coyotes.

Copyright © by Holt, Rinehart and Winston. All rights reserved.

What Is an Energy Pyramid?

Energy is lost as it passes through a food chain. An **energy pyramid** is a diagram that shows energy loss in an ecosystem. The bottom of the pyramid is larger than the top. There is less energy for use at the top of the pyramid than at the bottom. This is because most of the energy is used up at the lower levels. Only about 10% of the energy at each level of the energy pyramid passes on to the next level.

Math Focus

7. Calculate How much energy is lost at each level of the energy pyramid?

____

Energy Pyramid

TAKE A LOOK

8. Explain In which level of this energy pyramid do you think deer would belong? Explain.

____

____

____

EFFECT OF ONE SPECIES

A single species can change the flow of energy in an ecosystem. For example, gray wolves are at the top of their food chains. They eat a lot of different organisms but are usually not eaten by any other animal. By eating other organisms, wolves help control the size of those populations.

At one point, wolves were found across the United States. As settlers moved west, most wolves moved away or were killed. With few wolves left to feed on the primary consumers, the population of elk began to grow. The elk ate all the grass, and there was none left for the smaller herbivores, such as hares. As these small herbivores died, there was less food for the secondary consumers. When wolves were removed from the food web, the whole ecosystem was affected. ☑

READING CHECK

9. Summarize Why did a change in the wolf population affect the other organisms in the community?

____

____

____

____

____

Copyright © by Holt, Rinehart and Winston. All rights reserved.

Name ______________________ Class ______________ Date ______________

Section 2 Review

6.5.a, 6.5.b, 6.5.c, 6.5.e

SECTION VOCABULARY

carnivore an organism that eats animals

energy pyramid a triangular diagram that shows an ecosystem's loss of energy, which results as energy passes through the ecosystem's food chain

food chain the pathway of energy transfer through various stages as a result of the feeding patterns of a series of organisms

food web a diagram that shows the feeding relationships between organisms in an ecosystem

herbivore an organism that eats only plants

omnivore an organism that eats both plants and animals

1. **Explain** Why are producers important in an ecosystem?

____

____

____

2. **Connect** Make a food chain using the following organisms: mouse, snake, grass, hawk. Draw arrows showing how energy flows through the chain. Identify each organism as a producer, primary consumer, secondary consumer, or tertiary consumer.

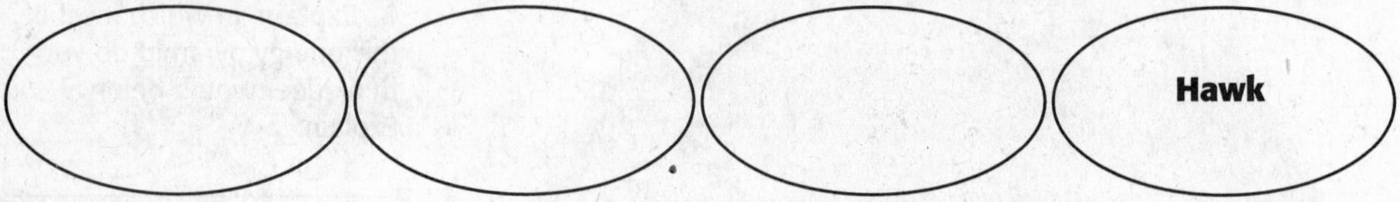
Hawk

__________ __________ Secondary consumer __________

3. **Make Inferences** Do you think you could find a food chain that had 10 organisms? Explain.

____

____

____

4. **Decide** A food chain off the California coast includes kelp, which is eaten by sea urchins. The urchins are eaten by sea otters, which are eaten by sharks. Can you combine this food chain with the one in question 2 to make a food web? Explain your answer.

Copyright © by Holt, Rinehart and Winston. All rights reserved.

Name ______________________ Class ______________ Date __________

SECTION 3

Types of Interactions

BEFORE YOU READ

After you read this section, you should be able to answer these questions:

- What determines an area's carrying capacity?
- Why does competition occur?
- How do organisms avoid being eaten?
- What are three kinds of symbiotic relationships?

California Science Standards

6.5.a, 6.5.b, 6.5.c, 6.5.e

How Does the Environment Control Population Sizes?

Most living things have more offspring than will survive. A female frog, for example, may lay hundreds of eggs in a small pond. If all of the eggs became frogs, the pond would soon become very crowded. There would not be enough food for the frogs or other organisms in the pond. But in nature, this usually does not happen. The biotic and abiotic factors in the pond control the frog population so that it does not get too large.

Populations cannot grow without stopping, because the environment has only a certain amount of food, water, space, and other resources. A resource that keeps a population from growing forever is called a *limiting factor*. Food is often a limiting factor in an ecosystem.

STUDY TIP

Make a List As you read this section, write down any questions you may have. Work with a partner to find the answers to your questions.

What Is Carrying Capacity?

The largest number of organisms that can live in an environment is called the **carrying capacity**. When a population grows beyond the carrying capacity, limiting factors will cause some individuals to leave the area or to die. As individuals die or leave, the population decreases.

The carrying capacity of an area can change if the amount of the limiting factor changes. For example, the carrying capacity of an area will be higher in seasons when more food is available.

CALIFORNIA STANDARDS CHECK

6.5.e Students know the number and types of organisms an ecosystem can support depends on the resources available and on abiotic factors, such as quantities of light and water, a range of temperatures, and soil composition.

Word Help: resource
anything that can be used to take care of a need

Word Help: available
that can be used

Word Help: range
a scale or series between limits

1. Define What is a limiting factor?

____

____

____

Copyright © by Holt, Rinehart and Winston. All rights reserved.

How Do Organisms Interact in an Ecosystem?

Populations are made of individuals of the same species. Communities are made of different populations that interact. There are three main ways that individuals and populations affect one another in an ecosystem: in competition, as predator and prey, and through symbiosis. ☑

READING CHECK

2. List What are three ways that organisms in an ecosystem interact?

____

____

____

Why Do Organisms Compete?

Competition happens when more than one individual or population tries to use the same resource. There may not be enough resources, such as food, water, shelter, or sunlight, for all the organisms in an environment. When one individual or population uses a resource, there is less for others to use.

Competition can happen between organisms in the same population. For example, in Yellowstone National Park, elk compete with one another for the same plants. In the winter, when there are not many plants, competition is much higher. Some elk will die because there is not enough food. In spring, when many plants grow, there is more food for the elk, and competition is lower.

Competition can also happen between populations. In a forest, different types of trees compete to grow in the same area. All of the plant populations must compete for the same resources: sunlight, space, water, and nutrients.

Critical Thinking

3. Prediction In a prairie ecosystem, which two of the following organisms most likely compete for the same food source: elk, coyotes, prairie dogs, vultures?

____

____

How Do Predators and Prey Interact?

Another way organisms interact is when one organism eats another to get energy. The organism that is eaten is called the **prey**. The organism that eats the prey is called the **predator**. When a bird eats a worm, for example, the bird is the predator, and the worm is the prey.

PREDATORS

Predators have traits or skills that help them catch and kill their prey. Different types of predators have different skills and traits. For example, a cheetah uses its speed to catch prey. On the other hand, lions have colors that let them blend with the environment so that prey cannot see them easily.

Copyright © by Holt, Rinehart and Winston. All rights reserved.

PREY

Prey generally have some way to protect themselves from being eaten. Different types of organisms protect themselves in different ways:

1. **Run Away** When a prairie dog is chased, it runs underground.

2. **Travel in Groups** Some animals, such as bison, travel in herds, or groups. Many fishes, such as anchovies, travel in schools. All the animals in these groups can help one another by watching for predators.

3. **Show Warning Colors** Some organisms have bright colors that act as a warning. The colors warn predators that the prey might be poisonous. A brightly colored fire salamander, for example, sprays a poison that burns.

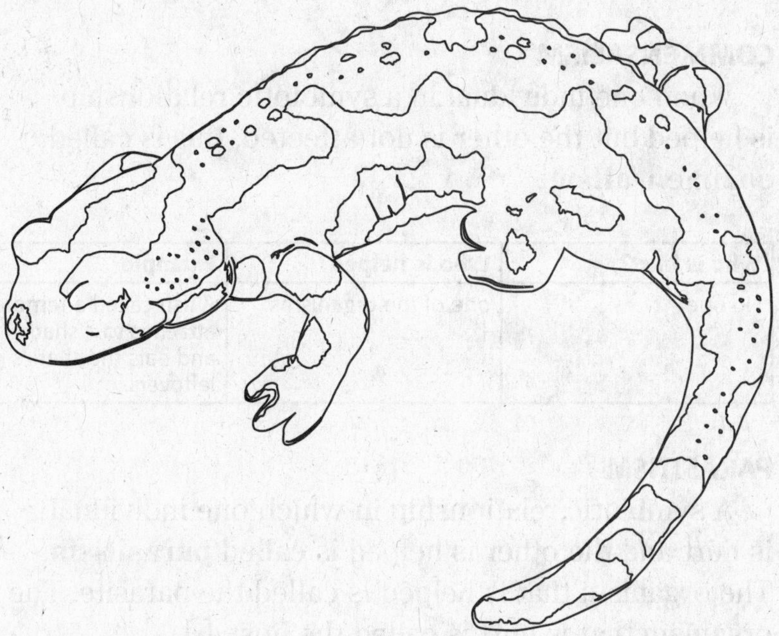

4. **Use Camouflage** Some organisms can hide from predators by blending in with the background. This is called *camouflage*. A rabbit's natural colors, for example, may help it blend in with dead leaves or shrubs so that it cannot be seen. Some animals may look like twigs, stone, or bark.

Say It

Give examples In small groups, talk about other animals that escape predators in the four ways described in the text.

TAKE A LOOK

4. Color A fire salamander has a black body with bright orange or yellow spots. Use colored pencils to give this salamander its warning colors.

Copyright © by Holt, Rinehart and Winston. All rights reserved.

What Is Symbiosis?

Some species have very close interactions with other species. A close association between two or more species is called **symbiosis**. Each individual in a symbiotic relationship may be helped, hurt, or not affected by another individual. Often, one species lives on or in another species. Most symbiotic relationships can be divided into three types: mutualism, commensalism, and parasitism. ☑

READING CHECK

5. List List the three types of symbiotic relationships.

____

____

____

MUTUALISM

When both individuals in a symbiotic relationship are helped, it is called **mutualism**. You can see mutualism in the relationship between a bee and a flower.

Who is hurt?	Who is helped?	Example
No one	both organisms	A bee transfers pollen for a flower; a flower provides nectar to a bee.

COMMENSALISM

When one individual in a symbiotic relationship is helped but the other is not affected, this is called **commensalism**.

Who is hurt?	Who is helped?	Example
No one	one of the organisms	A fish called a remora attaches to a shark and eats the shark's leftovers.

PARASITISM

A symbiotic relationship in which one individual is hurt and the other is helped is called **parasitism**. The organism that is helped is called the parasite. The organism that is hurt is called the host. ☑

READING CHECK

6. Define In parasitism, is the host helped or hurt?

____

____

Who is hurt?	Who is helped?	Example
Host	parasite	A flea is a parasite on a dog.

Copyright © by Holt, Rinehart and Winston. All rights reserved.

Name ______________________ Class ______________ Date ______________

Section 3 Review

6.5.b, 6.5.e

SECTION VOCABULARY

carrying capacity the largest population that an environment can support at any given time

commensalism a relationship between two organisms in which one organism benefits and the other is unaffected

mutualism a relationship between two species in which both species benefit

parasitism a relationship between two species in which one species, the parasite, benefits from the other species, the host, which is harmed

predator an organism that kills and eats all or part of another organism

prey an organism that is killed and eaten by another organism

symbiosis a relationship in which two different organisms live in close association with each other

1. Identify What are two resources for which organisms are likely to compete?

____

____

____

2. Identify Relationships Complete the chart below to describe the different kinds of symbiotic relationships.

Example organisms	Type of symbiosis	Organism(s) helped	Organism(s) hurt
Flea and dog			host (dog)
Bee and flower	mutualism		
Remora and shark			none

3. Describe Give an example of a predator-prey relationship. Be sure to identify which is the predator and which is the prey. Choose a example different from the one in the text.

____

____

____

____

____

Copyright © by Holt, Rinehart and Winston. All rights reserved.

Name ______________________ Class ______________ Date ____________

CHAPTER 17 Biomes and Ecosystems

SECTION 1

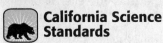

Studying the Environment

California Science Standards

6.5.a, 6.5.b, 6.5.c, 6.5.d, 6.5.e

BEFORE YOU READ

After you read this section, you should be able to answer these questions:

- What are abiotic factors?
- What are the two roles organisms play in an environment?
- Why do similar biomes occur in similar climates?

STUDY TIP

Summarize After reading each of the subsections, write a short answer to the question asked by the subsection title.

How Are Environments Different?

How is a desert different from a forest? Is it that a forest gets more rainfall than a desert? Is it that the kinds of organisms in a desert and a forest are different? Actually, it is both of these, along with other characteristics, that help separate these two types of environment from each other.

You may know that the *biotic factors* of the environment include organisms, or living things. The nonliving parts, such as rainfall, are called *abiotic factors*. Organisms interact with the abiotic parts of their surroundings in different ways. The table below lists some abiotic factors and tells why they are important.

Abiotic factor	Importance to organisms
Sunlight	Plants use sunlight to make food. If sunlight is lacking, plants cannot live. For example, no plants live in the deep ocean, because no sunlight reaches there.
Temperature	Most living things survive better in warm environments. Colder areas have fewer living things than warmer areas. For example, tropical rain forests have more kinds of organisms than colder forests in the north or south.
Rainfall	Areas that get a lot of rain generally have more plants than areas that are dry. Areas with many plants have more animals than areas with few plants.
Soil	Plants rely on chemical nutrients in the soil to grow. If the soil is lacking nutrients, plants do not grow as well.

Critical Thinking

1. Apply Concepts Explain why adding chemical fertilizers to soil can improve crop growth.

____

____

____

____

There are two main kinds of abiotic factors. Some abiotic factors in an area are resources that organisms can use there. For example, organisms can use nutrients in soil as a resource to help them grow. Other abiotic factors are conditions that organisms must adapt to in order to live there. Organisms must adapt to the temperature in an area to live there.

Copyright © by Holt, Rinehart and Winston. All rights reserved.

How Are Biomes Different from Ecosystems?

Ecosystems are made up of the organisms and abiotic factors in an area. *Biomes* are made up of many connected land and water ecosystems. Therefore, most biomes are much larger than ecosystems. Biomes have certain types of plants that support certain types of animals.

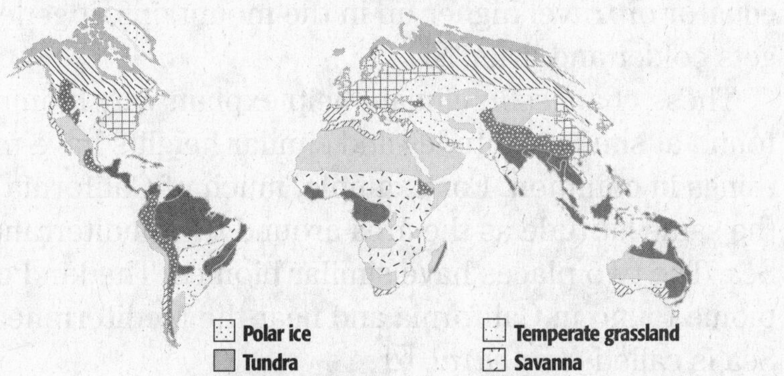
Polar ice
Tundra
Coniferous forest
Tropical rain forest
Temperate deciduous forest
Temperate grassland
Savanna
Desert
Chaparral
Mountains

This map shows some of the major land biomes on Earth.

TAKE A LOOK

2. Infer Find the tundra biomes on the map. What kind of climate do you think tundra biomes have? Explain your answer.

____

____

____

The climate of a biome affects what kinds of organisms can live there. The *climate* of an area is the overall weather conditions in a region over a long time. Climate is caused mainly by a few abiotic factors, such as the amount of rainfall and the temperatures at different times.

What Are the Roles of Organisms in an Ecosystem?

The biomes and ecosystems on Earth contain many kinds of organisms. However, each organism has one of two main functions, or ecological roles, in an ecosystem. Those roles are producer and consumer.

Producers are organisms that make their own food. Plants and many kinds of bacteria are producers. Most producers make their food from sunlight. Other producers, like some bacteria, make their food from chemicals that other organisms cannot use for food-making.

Consumers are organisms that cannot make their own food. All animals are consumers. One important kind of consumer is a *decomposer*. Decomposers get their food by breaking down dead organisms. As they eat the dead organisms, decomposers also break them down into simple chemicals that other organisms can use.

CALIFORNIA STANDARDS CHECK

6.5.c Students know that populations of organisms can be categorized by the functions they serve in an ecosystem.

Word Help: categorize
to put into groups or classes

Word Help: function
use or purpose

3. Explain Predators are animals that kill and eat other organisms. Why are both predators and decomposers categorized as consumers?

____

____

____

____

Copyright © by Holt, Rinehart and Winston. All rights reserved.

Why Do Similar Biomes Occur in Similar Climates?

In places with similar climates, the biomes have similar plant and animal communities. The latitude and height above sea level affect the temperature and rainfall of a location. As you move away from the equator or travel higher up in the mountains, it generally gets colder and drier. ☑

These changes in climate help explain why biomes found at similar latitudes and similar heights have many things in common. For example, much of California is at the same latitude as the area around the Mediterranean Sea. The two places have similar biomes. The kind of biome found in California and near the Mediterranean Sea is called *chapparal.* ☑

READING CHECK

4. Explain In general, what happens to temperature as you move away from the equator?

______

READING CHECK

5. Identify Why is the climate near the Mediterranean Sea similar to the climate in much of California?

______

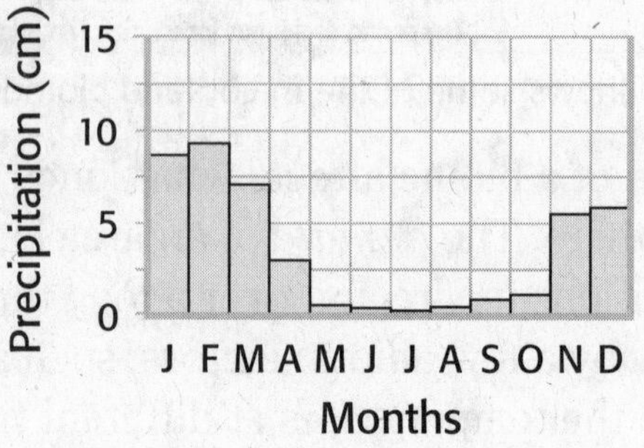
Precipitation (cm)
15
10
5
0
J F M A M J J A S O N D
Months

This graph shows the average monthly precipitation in the chaparral biome in Santa Barbara, California.

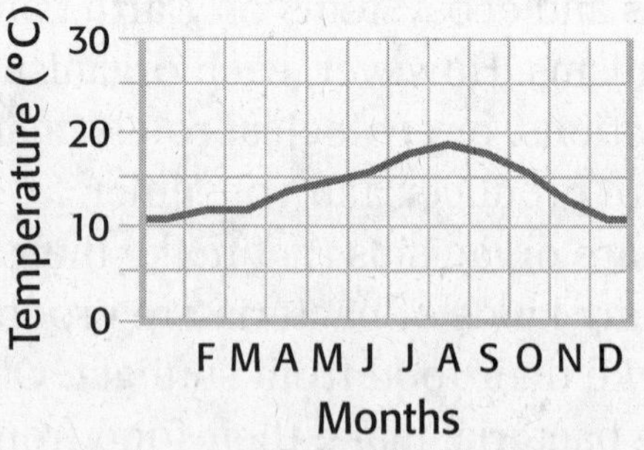
Temperature (°C)
30
20
10
0
F M A M J J A S O N D
Months

This graph shows the average monthly temperature in the chaparral biome in Santa Barbara, California.

Math Focus

6. Read a Graph In which month does Santa Barbara have the highest average temperature?

______

7. Read a Graph About how much precipitation does Santa Barbara get in the month of February?

______

Copyright © by Holt, Rinehart and Winston. All rights reserved.

Name ______________________ Class ______________ Date ______________

Section 1 Review

6.5.a, 6.5.b, 6.5.c, 6.5.d, 6.5.e

1. **Compare** What can producers do that consumers cannot do?

2. **Identify** How are biomes and ecosystems different?

3. **Describe** How do abiotic factors affect living things?

4. **Explain** Why do fewer kinds of organisms live in desert biomes than in forest biomes?

5. **Apply Concepts** Give two examples of producers. The examples that you give should be from different biomes.

6. **Infer** What would happen to an ecosystem if all of the decomposers disappeared? Explain your answer.

Copyright © by Holt, Rinehart and Winston. All rights reserved.

SECTION 2 Land Biomes

California Science Standards

6.5.c, 6.5.d, 6.5.e

BEFORE YOU READ

After you read this section, you should be able to answer these questions:

- What are the eight kinds of land biomes?
- How are organisms adapted to survive in each land biome?
- What organisms fill the ecological roles of producer and consumer in each land biome?

What Is a Desert?

STUDY TIP

Summarize As you read about each type of land biome, write down its important features. When you finish reading this section, write a summary of the features of each type of land biome into your notebook.

Why are some kinds of organisms common in some areas but not other areas? The reason is that each of Earth's biomes has different abiotic factors. The organisms that live in a biome are adapted to the biome's abiotic factors.

Deserts are very dry biomes, and most are very hot. The Mojave Desert in California is an example of a desert. The organisms that can live in a desert have special features that let them survive in the desert.

Many desert plants have roots that spread near the surface. This allows them to take up water quickly after a rain, before it evaporates.

Desert animals also have ways to survive the hot, dry desert conditions. Some live underground, where it is cooler. They come only out at night, when air temperatures are lower. Others, such as the fringe-toed lizard, bury themselves in the loose sand to escape the heat and avoid predators. ☑

READING CHECK

1. List What are two ways that burrowing under the ground helps the fringe-toed lizard?

____

____

Math Focus

2. Calculate About how many inches of rain does a desert get every year? Show your work.

____

____

Desert

- **Average Yearly Rainfall less than 25 cm**
- **Average Temperatures Summer: 38°C (100°F) Winter: 7°C (45°F)**

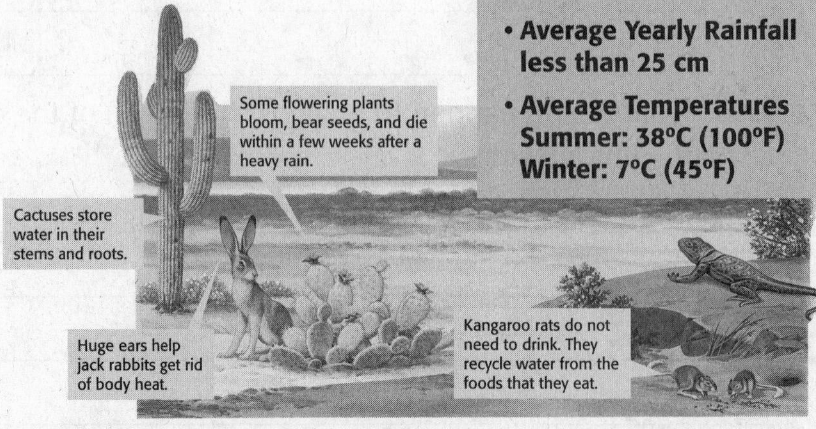

Some flowering plants bloom, bear seeds, and die within a few weeks after a heavy rain.
Cactuses store water in their stems and roots.
Huge ears help jack rabbits get rid of body heat.
Kangaroo rats do not need to drink. They recycle water from the foods that they eat.

Copyright © by Holt, Rinehart and Winston. All rights reserved.

What Is a Chaparral?

A biome that has a fairly dry climate but receives more rainfall than a desert is called a **chaparral**. In chaparral biomes, summers are warm and dry, and winters are mild and wet. Chaparral biomes can be found in California and around the Mediterranean Sea. ☑

The plants in a chaparral grow close to the ground. Many are small evergreen trees and shrubs that have wide, flat leaves. *Evergreen* plants are plants that keep their leaves all year round. Many chaparral plants have leaves that help them store water.

During natural fires, chaparral shrubs and trees are destroyed. After a fire, the shrubs grow back more quickly than the trees. Natural fires prevent the trees from competing with the shrubs for resources. Therefore, natural fires are an abiotic factor that helps maintain the chaparral.

The animals of the chaparral also have features to help them survive. Lizards, chipmunks, and mule deer are usually gray and brown, which helps them blend into their surroundings and hide from predators.

Mule deer are herbivores that eat grasses and shrubs. Lizards and chipmunks are omnivores that eat insects and parts of plants. Bobcats are carnivores that eat other animals.

Chaparral

- **Average Yearly Rainfall 25 to 43 cm (10 to 17 in.)**
- **Average Temperatures Summer: 22°C (71.6°F) Winter: 17.8°C (64°F)**

Plants of the chaparral are adapted to recover quickly after a natural fire.

READING CHECK

3. Identify Where are chaparral biomes found?

____

____

Critical Thinking

4. Predict How would putting out all natural fires affect the types of plants found in a chaparral?

____

____

____

____

____

TAKE A LOOK

5. Identify What kind of plant is most common in the chaparral?

____

Copyright © by Holt, Rinehart and Winston. All rights reserved.

What Are Two Kinds of Grassland?

A **grassland** is a biome made up mainly of grasses, small flowering plants, and a few trees. The two main kinds of grassland are temperate grasslands and savannas.

TEMPERATE GRASSLANDS

In temperate grasslands, the summers are warm and the winters are cold. The soils of temperate grasslands are very rich in nutrients. Fires, droughts, and grazing prevent trees and shrubs from growing.

Many seed-eating animals, such as prairie dogs and mice, live in this kind of grassland. They use camouflage and burrows to hide from predators, such as coyotes. ☑

READING CHECK

6. Identify What are two ways small animals hide from predators?

____

____

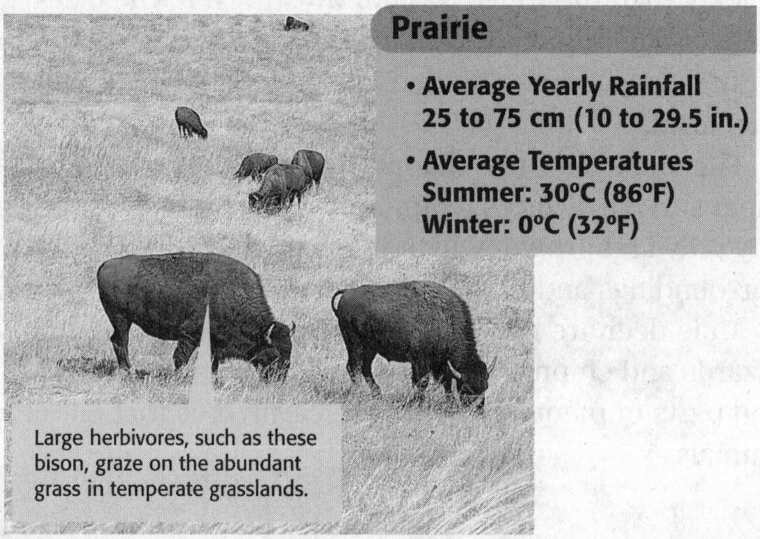
Prairie
• Average Yearly Rainfall 25 to 75 cm (10 to 29.5 in.)
• Average Temperatures Summer: 30°C (86°F) Winter: 0°C (32°F)

Large herbivores, such as these bison, graze on the abundant grass in temperate grasslands.

SAVANNAS

The savanna is a grassland that has a lot of rainfall during some seasons and very little rainfall in other seasons. During the dry season, savanna grasses dry out and turn yellow. However, the roots can live for many months without water.

TAKE A LOOK

7. Compare How does the amount of rainfall in a temperate grassland compare with the amount of rainfall in a savanna?

____

____

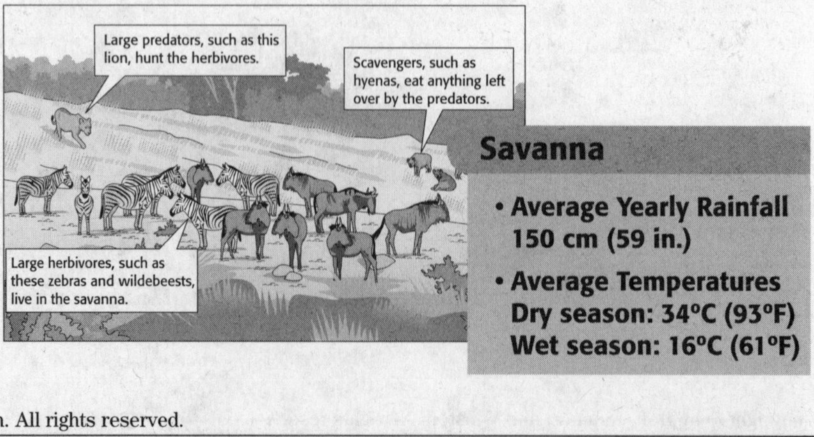
Large predators, such as this lion, hunt the herbivores.
Scavengers, such as hyenas, eat anything left over by the predators.
Large herbivores, such as these zebras and wildebeests, live in the savanna.
Savanna
• Average Yearly Rainfall 150 cm (59 in.)
• Average Temperatures Dry season: 34°C (93°F) Wet season: 16°C (61°F)

Copyright © by Holt, Rinehart and Winston. All rights reserved.

What Is a Tundra?

Imagine a place on Earth that is too cold for trees to grow. A **tundra** is a biome that has very cold temperatures and little rainfall. Tundras can be found near the North and South Poles. In a tundra, the layer of soil below the surface stays frozen all year long. This layer is called *permafrost*.

During the short, cool summers, only the water in the soil at the surface melts. This surface soil is too shallow for most plants. Only plants with shallow roots, such as grasses and small shrubs, are common. Mosses and lichens grow beneath these plants. Growing close to the ground helps the plants resist the wind and the cold. ☑

Animals of the tundra also have ways to live in this biome. In the winter, food is hard to find and the weather is very harsh. Some animals, such as bears, sleep through much of the winter. Other animals, like the caribou, travel long distances to find food. Many animals have extra layers of fat to keep them warm.

During the summer, the soil above the permafrost becomes muddy from melting ice and snow. Insects, such as mosquitoes, lay eggs in the mud. Birds that prey on these insects are carnivores. Other carnivores, such as wolves, prey on herbivores, such as caribou and musk oxen.

Tundra

- **Average Yearly Rainfall**
  30 to 50 cm (12 to 20 in.)
- **Average Temperatures**
  Summer: 12°C (54°F)
  Winter: –26°C (–15°F)

Caribou are large herbivores that live in the tundra.

Say It

Share Experiences Have you ever been to a very cold place? In a group, discuss what it was like.

READING CHECK

8. Explain How does growing close to the ground help tundra plants?

____

____

____

TAKE A LOOK

9. Describe How are the ecological roles of a caribou in a tundra and a zebra in a savanna similar?

____

____

____

Copyright © by Holt, Rinehart and Winston. All rights reserved.

What Are Three Forest Biomes?

Three forest biomes are coniferous forests, temperate deciduous forests, and tropical rain forests. All of these forests receive plenty of rain, and the temperatures are not extreme. The kind of forest that forms in an area depends on the area's climate.

CONIFEROUS FORESTS

The coniferous forest gets its name from *conifers*, the main type of tree that grows there. Conifers produce seeds in cones. They have special needle-shaped leaves covered in a thick waxy coating. These features help the tree conserve water during the dry winters. The waxy coating also protects the needles from being damaged by cold weather. Most conifers are evergreens. ☑

READING CHECK

10. Identify Where do conifers produce their seeds?

____

In a coniferous forest, decomposition is slow. The ground is often covered by a thick layer of needles. The conifers stop most of the sunlight from reaching the ground. Because there is little light there, not many plants live under the conifer trees.

The animals in a coniferous forest are adapted for cold winters. Like animals in the tundra, they may hibernate or travel to warmer climates in the winter.

TAKE A LOOK

11. Explain Why is it important for trees in a coniferous forest to conserve water?

____

____

____

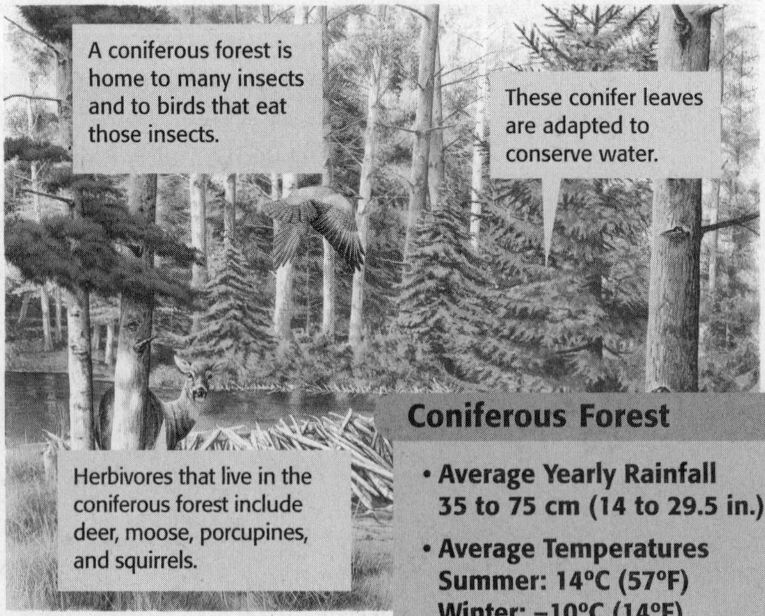
A coniferous forest is home to many insects and to birds that eat those insects.
These conifer leaves are adapted to conserve water.
Herbivores that live in the coniferous forest include deer, moose, porcupines, and squirrels.

Coniferous Forest

- **Average Yearly Rainfall**
  35 to 75 cm (14 to 29.5 in.)
- **Average Temperatures**
  Summer: 14°C (57°F)
  Winter: –10°C (14°F)

Copyright © by Holt, Rinehart and Winston. All rights reserved.

TEMPERATE DECIDUOUS FORESTS

Have you seen trees with leaves that change color and then fall off the trees? If so, you have seen deciduous trees. The word *deciduous* comes from a Latin word that means "to fall off." Deciduous trees shed their leaves to save water during the winter. Temperate deciduous forests contain many deciduous trees.

Since deciduous trees lose their leaves, the ground in deciduous forests gets a lot of sunlight during part of the year. Therefore, small trees and shrubs can grow on the forest floor. ☑

Animals live in different parts of the deciduous forest. Black bears and rabbits live on the forest floor. Bears are omnivores that eat nuts, berries, and small animals. Rabbits are herbivores.

READING CHECK

12. Compare Why does the floor of a temperate deciduous forest have more small plants than the floor of a coniferous forest?

____

____

____

____

____

In forests, plant growth happens in layers. The leafy tops of the trees reach high above the forest floor, where the leaves can get sunlight.
Woodpeckers are omnivores. They eat insects, nuts, and tree sap.
Woody shrubs catch the light that filters through the trees.
Temperate Deciduous Forest
• Average Yearly Rainfall 75 to 125 cm (29.5 to 49 in.)
• Average Temperatures Summer: 28°C (82°F) Winter: 6°C (43°F)
Ferns and mosses are scattered across the forest floor. Flowering plants often bloom in early spring, before the trees grow new leaves.
Squirrels are omnivores that eat nuts, seeds, and insects.

TAKE A LOOK

13. Identify What are two producers in a temperate deciduous forest?

____

____

Copyright © by Holt, Rinehart and Winston. All rights reserved.

TROPICAL RAIN FORESTS

Tropical rain forest biomes contain the greatest variety of plants and animals on Earth. The warm temperatures and high rainfall allow a lot of plants to grow. Trees grow very tall and dense. Their leaves prevent much light from reaching the forest floor, so not many small plants live on the forest floor. The plants support a large diversity of animals.

Most of the animals in a rain forest live in the trees. Birds such as toucans are omnivores that eat fruits, reptiles, and other birds. Carnivores, such as harpy eagles, eat other animals, such as howler monkeys. Howler monkeys are herbivores that eat fruits, nuts, and leaves. ☑

You may think that the soil in a rain forest is very rich in nutrients because of all the plants that live there. However, most of the nutrients in the tropical rain forest are found in plants, not in soil. The soil is so thin that many trees grow roots above-ground for support.

READING CHECK

14. Describe Where in a rain forest are most animals found?

______

______

Trees form a continuous green roof, or canopy, that may extend 60 m above the forest floor.
Woody vines climb the tree trunks to reach sunlight.
Little light reaches the ground. Low-growing plants in the rain forest don't need a lot of light.

Tropical Rain Forest

- **Average Yearly Rainfall up to 400 cm (157.5 in.)**
- **Average Temperatures Daytime: 34°C (93°F) Nighttime: 20°C (68°F)**

TAKE A LOOK

15. Infer Why do many vines grow on tree branches instead of on the forest floor?

______

______

______

______

Copyright © by Holt, Rinehart and Winston. All rights reserved.

Name ______________________ Class ______________ Date ____________

Section 2 Review

6.5.c, 6.5.d, 6.5.e

SECTION VOCABULARY

chaparral a type of vegetation that includes broad-leaved evergreen shrubs and that is located in areas with hot, dry summers and mild, wet winters

desert a region that has little or no plant life, long periods without rain, and extreme temperatures: usually found in hot climates

grassland a region that is dominated by grasses, that has few woody shrubs and trees, that has fertile soils, and that receives moderate amounts of seasonal rainfall

tundra a treeless plain found in the Arctic, in the Antarctic, or on the tops of mountains that is characterized by very low winter temperatures and short, cool summers

1. Explain The tundra has been called a "frozen desert." Explain why this is a good name for the tundra.

____

____

____

2. Compare Compare the temperate grassland and the savanna by filling in the blank spaces in the table below.

	Temperate grassland	Savanna
Abiotic factors		constant warmth with seasonal rains
Types of producer	grass with a few trees	
Types of consumer		

3. List What are some of the adaptations that allow desert plants to live in such a hot, dry environment?

____

____

____

Copyright © by Holt, Rinehart and Winston. All rights reserved.

SECTION 3

Marine Ecosystems

California Science Standards

6.5.a, 6.5.c, 6.5.e

BEFORE YOU READ

After you read this section, you should be able to answer these questions:

- What abiotic factors affect marine ecosystems?
- What are the major zones found in the ocean?
- What kinds of organisms are found in marine ecosystems?

STUDY TIP

Compare Create a table comparing the locations, abiotic factors, and organisms for each marine ecosystem.

What Abiotic Factors Affect Marine Ecosystems?

Oceans cover almost three-fourths of Earth's surface! Scientists call the ecosystems in the ocean *marine ecosystems*. Marine ecosystems, like all ecosystems, are affected by abiotic factors.

WATER DEPTH AND SUNLIGHT

Two abiotic factors that affect marine ecosystems are water depth and sunlight. The average depth of the oceans is 4,000 m, but sunlight does not reach deeper than 200 m. Producers that carry out photosynthesis, such as algae, can live only above 200 m.

Phytoplankton are tiny producers that float near the surface of the ocean. Algae and phytoplankton are the producers at the base of most ocean food chains. Large consumers, such as whales, feed on these tiny producers.

Math Focus

1. Calculate What percentage of the water in the oceans receives sunlight? Show how you got your answer.

____

____

Marine ecosystems support many different organisms. Both large humpback whales and tiny phytoplankton live near the surface of the ocean.

TAKE A LOOK

2. Identify In the picture, which organism is the producer and which is the consumer?

____

____

____

Copyright © by Holt, Rinehart and Winston. All rights reserved.

TEMPERATURE

A third abiotic factor in marine ecosystems is the temperature of the water. The water near the surface is much warmer that the rest of the ocean because it is heated by the sun. Ocean water becomes colder as it gets deeper.

Water temperatures at the surface are also affected by latitude. Water near the equator is generally warmer than water closer to the poles. The water at the surface is also warmer in summer than winter. ☑

Temperature affects the animals that live in marine ecosystems. For example, fish that live near the poles have a special chemical in their blood that keeps them from freezing. Most animals that live in coral reefs need warm water to live.

Most marine organisms live in the warm waters near the surface. Some animals, like whales, migrate from cold areas near the poles to warm areas near the equator to reproduce.

Water temperature also affects whether some animals, such as barnacles, can eat. If the water is too hot or too cold, these animals may not be able to eat. If the temperature changes suddenly, they may die. ☑

Ocean Temperature and Depth

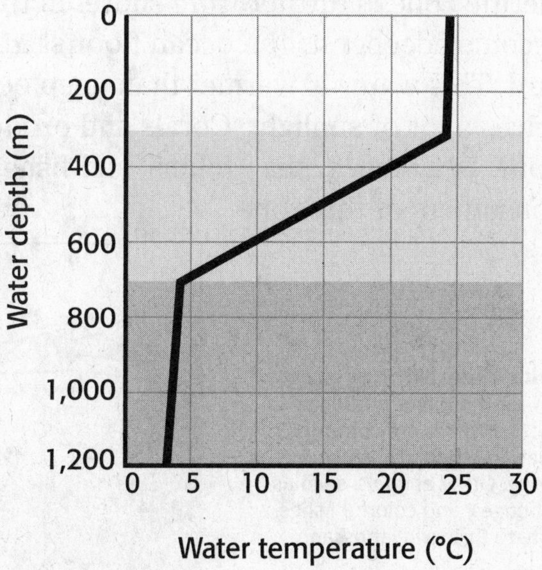
Water depth (m)
0
200
400
600
800
1,000
1,200
0
5
10
15
20
25
30
Water temperature (°C)

READING CHECK

3. List What are the three main abiotic factors that affect marine ecosystems?

____

____

____

READING CHECK

4. Identify Why do most barnacles live in water with a constant temperature?

____

____

____

TAKE A LOOK

5. Analyze About how much colder is ocean water at 600 m depth than at 400 m depth?

____

Copyright © by Holt, Rinehart and Winston. All rights reserved.

What Are the Major Zones in the Ocean?

The ocean can be divided into zones based on things such as distance from the shore, water depth, the amount of sunlight, and water temperature.

Say It

Share Experiences In a group, discuss the abiotic factors and the living organisms you have seen or might see at the beach.

THE INTERTIDAL ZONE

The intertidal zone is where the ocean meets the shore. The organisms of the intertidal zone are covered with water at high tide and exposed to air at low tide.

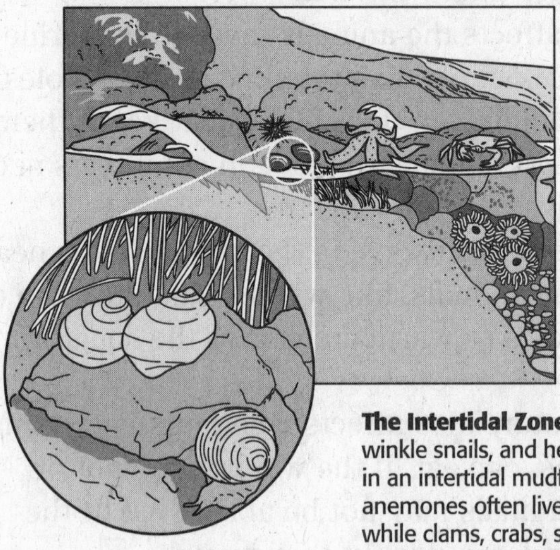

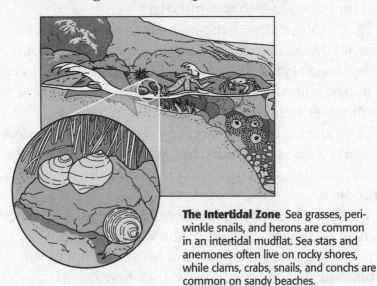

The Intertidal Zone Sea grasses, periwinkle snails, and herons are common in an intertidal mudflat. Sea stars and anemones often live on rocky shores, while clams, crabs, snails, and conchs are common on sandy beaches.

TAKE A LOOK

6. Explain Why is it difficult for many sea creatures to live in the intertidal zone?

____

____

THE NERITIC ZONE

The neritic zone is further from shore. In this zone, the water becomes deeper as the ocean floor starts to slope downward. This water is warmer than deep ocean water and receives a lot of sunlight. Corals and producers thrive in this zone. Sea turtles, sea urchins, and fishes are some of the consumers of this zone.

The Neritic Zone Although phytoplankton are the major producers in this zone, seaweeds are common, too. Sea turtles and dolphins live in the neritic zone. Other animals, such as corals, sponges, and colorful fishes, contribute to this vivid landscape.

TAKE A LOOK

7. Identify What are the two main kinds of producer in the neritic zone?

____

____

Copyright © by Holt, Rinehart and Winston. All rights reserved.

THE OCEANIC ZONE

In the oceanic zone, the sea floor drops off quickly. The oceanic zone extends from the surface to the deep water of the open ocean. Phytoplankton live near the surface, where there is sunlight. ☑

Consumers such as fishes, whales, and sharks live in the oceanic zone. Some of the animals live in deep waters, where there is no sunlight. These animals feed on each other and on material that sinks from the surface waters.

READING CHECK

8. Explain Why do phytoplankton need to live near the surface?

____

____

The Oceanic Zone Many unusual animals are adapted for the deep ocean. Whales and squids can be found in this zone. Also, fishes that glow can be found in very deep, dark water.

TAKE A LOOK

9. Explain How can the consumers that live in deep waters survive if there are no producers present?

____

____

____

THE ABYSSAL ZONE

The abyssal zone is the ocean floor. It does not get any sunlight and is very cold. Fishes, worms, and crabs have special features to live in this zone. Many of them feed on material that sinks from above.

Some organisms, such as angler fish, eat smaller fish. Other organisms, such as bacteria, are decomposers and help break down dead organisms.

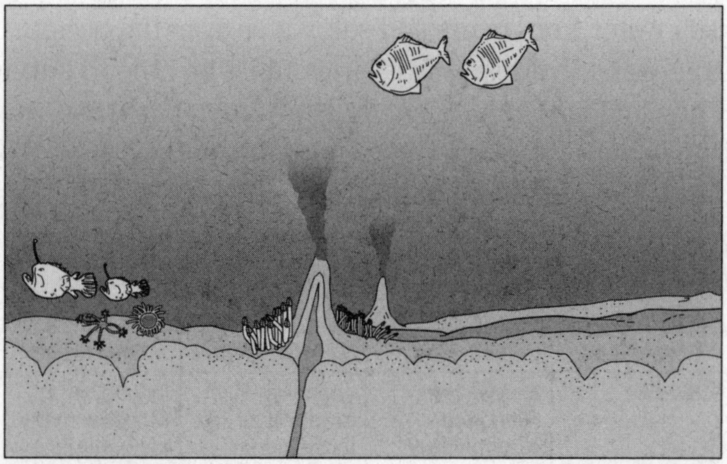

The Abyssal Zone Organisms such as bacteria, worms, and sea urchins thrive on the sea floor.

TAKE A LOOK

10. Describe What abiotic factors do organisms that live in the abyssal zone need to adapt to?

____

____

Copyright © by Holt, Rinehart and Winston. All rights reserved.

Critical Thinking

11. Predict Consequences How would humans be affected if there were no oceans?

____

____

____

____

____

What Kinds of Ecosystems Are in the Marine Biome?

Life on Earth depends on the ocean. The water that evaporates from the ocean creates most of the rain and snow that falls on land. The ocean affects world climates and wind patterns. Many people depend on the ocean for food.

Many different kinds of organisms live in the ocean. They live in the many ecosystems in the different zones of the ocean.

INTERTIDAL ECOSYSTEMS

Organisms in intertidal ecosystems must be able to live both underwater and in the air. Those that live in mudflats and sandy beaches may dig into the ground during low tide.

On rocky shores, organisms have adaptations to keep from being swept away by crashing waves. For example, seaweeds use structures called *holdfasts* to attach themselves to rocks. Other organisms, such as barnacles, attach themselves to rocks with a special cement. Sea stars feed on these organisms. ☑

READING CHECK

12. Describe How do organisms of intertidal ecosystems protect themselves from being washed away by waves?

____

____

____

Ecosystem	Location	Abiotic factors	Organisms
Intertidal	intertidal zone: mudflats, sandy beaches, rocky shores	times under water, times above water	worms, clams, crabs, birds, seaweed

ESTUARIES

An **estuary** is an area where fresh water from streams and rivers flows into the ocean. The water in an estuary is a mix of fresh water from rivers and salt water from the ocean.

Organisms that live in estuaries must be able to survive the changing amounts of salt in the water. The fresh water that flows into an estuary is rich in nutrients washed from the land. The nutrients in the water support large numbers of producers, such as algae. The algae are able to support many consumers, such as fish, oysters, and crabs.

TAKE A LOOK

13. Compare What unusual abiotic factor do organisms in an estuary need to adapt to?

____

____

Ecosystem	Location	Abiotic factors	Organisms
Estuaries	areas where rivers and streams flow into oceans	changing amounts of salt in the water; nutrient-rich water	large amount of algae, many consumers

Copyright © by Holt, Rinehart and Winston. All rights reserved.

CORAL REEFS

Coral reefs are named for the small animals called *corals* that form the reefs. Many of these tiny animals live together in a colony, or group

Most coral reefs are built on the skeletons of a kind of coral called *stony coral*. The skeleton of a stony coral is made up of a hard chemical called calcium carbonate. When the corals die, their hard skeletons remain. New corals grow on these hard remains.

Over time, layers of skeletons build up and form a rock-like structure called a *reef*. This reef provides a home for many marine animals. These organisms include fishes, sponges, sea stars, and sea urchins. Because so many kinds of organisms live there, coral reefs are some of the most diverse ecosystems on Earth. ☑

Ecosystem	Location	Abiotic factors	Organisms
Coral reefs	shallow areas of the neritic zone	warm water, sunlight	corals, algae, fishes, sponges, sea stars, sea urchins

READING CHECK

14. Explain How is a coral reef both a living and a nonliving structure?

____

____

____

____

____

KELP FORESTS

Kelp forests are found in cold, shallow areas of the neritic zone. Kelp are large brown algae that grow well in nutrient-rich waters, such as areas along the coast of California. Some kelp can be as much as 33 m (108 ft) long!

Many consumers, such as sea otters, fishes, crabs, and sea urchins, are found in kelp forests. Sea urchins eat the algae growing on rocks. Sea otters eat many of the organisms in the kelp forest, including sea urchins. ☑

Ecosystem	Location	Abiotic factors	Organisms
Kelp forests	shallow areas of the neritic zone	cold water, sunlight, nutrient-rich water	kelp, sea otters, fishes, crabs, sea urchins

READING CHECK

15. Identify What are the ecological roles of the sea urchin and sea otter in the kelp forest?

____

____

____

Copyright © by Holt, Rinehart and Winston. All rights reserved.

DEEP SEA HYDROTHERMAL VENTS

In some places on the ocean floor, heat from within the Earth heats the ocean water. The water can get as hot as 350°C! The hot water moves out of cracks in the ocean floor called *hydrothermal vents*.

These vents release chemicals from beneath the surface into the seawater. The chemicals can be made into food by special kinds of bacteria. These bacteria are considered producers. Many unusual consumers, such as tube worms, rely on the bacteria for energy. ☑

READING CHECK

16. Explain Why are bacteria at hydrothermal vents producers if they do not use sunlight to make food?

____

____

____

____

____

Ecosystem	Location	Abiotic factors	Organisms
Deep sea hydrothermal vents	abyssal zone	hot water with chemicals, high pressure	bacteria, vent worms, clams

THE SARGASSO SEA

Floating mats of algae in the middle of the Atlantic Ocean make up the Sargasso Sea ecosystem. Algae called *sargassums* are at the base of this ecosystem. Sargassums float at the surface because parts of the algae are filled with air.

Many animals live in this ecosystem. Most of these are omnivores that can eat many different organisms throughout the year.

TAKE A LOOK

17. Explain Why is it helpful for the organisms living in sargassums to be omnivores?

____

____

Ecosystem	Location	Abiotic factors	Organisms
Sargasso Sea	Atlantic Ocean near the surface	much sunlight	floating sargassum algae, fishes, mostly omnivores

POLAR ICE

The icy waters near the poles are rich in nutrients that support large numbers of phytoplankton. These producers can support many types of consumers. One of these is a small shrimplike organism called krill. Larger consumers eat krill and, in turn, serve as food for other consumers.

Ecosystem	Location	Abiotic factors	Organisms
Polar ice	Arctic Ocean, Antarctica	cold, nutrient-rich water	phytoplankton, krill, fishes, seals, whales, polar bears

Copyright © by Holt, Rinehart and Winston. All rights reserved.

Name ______________ Class ______________ Date ______________

Section 3 Review

6.5.a, 6.5.c, 6.5.e

SECTION VOCABULARY

estuary an area where fresh water mixes with salt water from the ocean	**phytoplankton** the microscopic, photosynthetic organisms that float near the surface of marine or fresh water **Wordwise** The prefix *phyto-* means "plant."

1. Describe What are some of the different kinds of producers found in the marine biome?

2. Apply Concepts Draw a food chain that shows the flow of energy through a polar ice ecosystem.

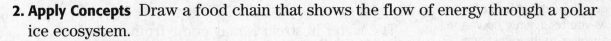

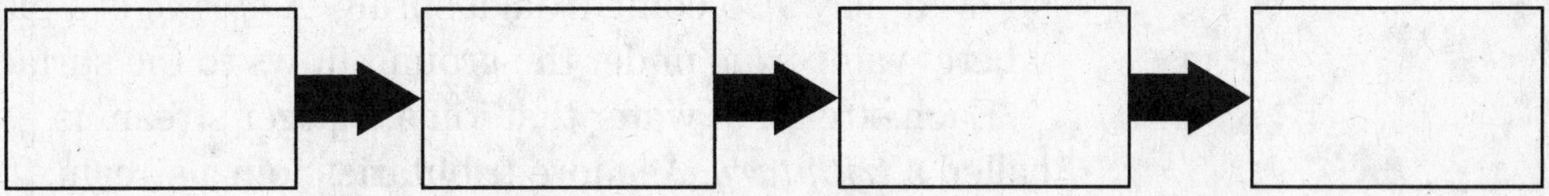

3. Explain Why are there few producers below 200 m in the ocean?

4. Identify What abiotic factors make the neritic zone a good home for many different organisms?

5. List Give two ecosystems that are found in the neritic zone.

Copyright © by Holt, Rinehart and Winston. All rights reserved.

Name ______________________________ Class ____________________ Date ________________

CHAPTER 17 Biomes and Ecosystems

SECTION 4

Freshwater Ecosystems

California Science Standards

6.5.c, 6.5.e

BEFORE YOU READ

After you read this section, you should be able to answer these questions:

- What organisms live in stream and river ecosystems?
- What are the three zones in a pond or lake?
- What are the two kinds of wetlands?

STUDY TIP

Answer Questions Before reading this section, write the three Before You Read questions on a piece of paper. As you read, write down the answers to the questions.

What Are Stream and River Ecosystems?

One important abiotic factor that affects freshwater ecosystems is how quickly the water is moving. In rivers and streams, the water is moving faster than in other freshwater ecosystems.

The water in streams may come from melted ice or snow. It may also come from a spring. A *spring* is a place where water from under the ground flows to the surface.

Each stream of water that joins a larger stream is called a *tributary*. As more tributaries join a stream, it becomes stronger and wider. A very strong, wide stream is called a *river*.

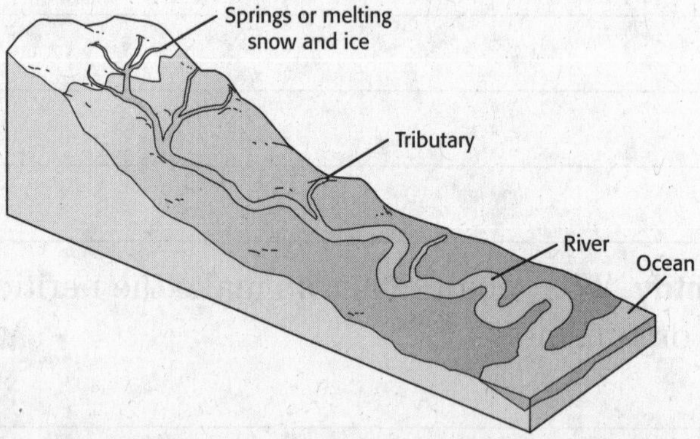

Springs or melting snow and ice
Tributary
River
Ocean

TAKE A LOOK

1. Describe What happens to the size of a stream when a tributary flows into it?

____

Stream and river ecosystems are full of life. Plants live along the edges of streams and rivers. Fish live in the open waters. Clams and snails live in the mud at the bottom.

Organisms that live in fast-moving water have to keep themselves from being washed away. Some producers, such as algae and moss, are attached to rocks. Consumers, such as tadpoles, use suction to hold themselves to rocks. Other consumers, such as crayfish, hide under rocks.

Copyright © by Holt, Rinehart and Winston. All rights reserved.

What Are Pond and Lake Ecosystems?

The water in ponds and lakes is not moving very much compared with rivers and streams. As a result, they have different types of ecosystems. Like marine ecosystems, pond and lake ecosystems are affected by water depth, sunlight, and temperature.

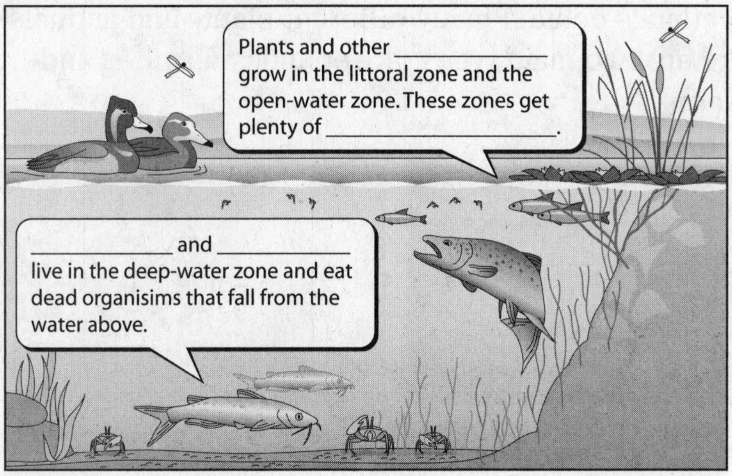
Plants and other ______________ grow in the littoral zone and the open-water zone. These zones get plenty of ______________.
______________ and ______________ live in the deep-water zone and eat dead organisims that fall from the water above.

TAKE A LOOK

2. Identify Fill in the blank spaces in the figure with the correct words.

LIFE NEAR THE SHORE

The area of water near the edge of a pond or lake is called the **littoral zone**. Sunlight reaches the bottom, which allows producers such as algae to grow in this zone. Plants, such as cattails and rushes, grow here too, farther from shore.

Many consumers, such as tadpoles and some insects, eat the algae and plants. Some consumers, such as snails and insects, make their homes in plants. Consumers that live in the mud include clams and worms. Other consumers, such as fishes, also live in this zone. ☑

READING CHECK

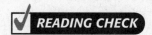

3. Identify Which consumers of the littoral zone are herbivores?

____

LIFE AWAY FROM THE SHORE

The area of a lake or pond away from the littoral zone near the surface is called the **open-water zone**. This zone is as deep as sunlight can reach. Producers such as phytoplankton grow well here. This zone is home to bass, lake trout, and other consumers.

Beneath the open-water zone is the **deep-water zone**, where no sunlight reaches. Photosynthetic organisms cannot live in this zone. Scavengers, such as catfish and crabs, live here and feed on dead organisms that sink from above. Decomposers, such as fungi and bacteria, also help to break down dead organisms. ☑

READING CHECK

4. Explain Why can't producers live in the deep-water zone?

____

____

Copyright © by Holt, Rinehart and Winston. All rights reserved.

What Is a Wetland?

A **wetland** is an area of land that is sometimes underwater or whose soil contains a lot of water. Wetlands help control floods. During heavy rains, wetlands soak up large amounts of water. This water sinks into the ground and helps refill underground water supplies. ☑

Wetlands contain many different plants and animals. There are two main types of wetlands: **marshes** and **swamps**.

READING CHECK

5. Define What is a wetland?

____

____

____

TAKE A LOOK

6. Compare What is a quick way to tell the difference between a marsh and a swamp?

____

____

____

How Can an Ecosystem Change?

Did you know that a pond or lake can disappear? The water flowing into the lake carries sediment. The sediment, along with dead leaves and other materials, sinks to the bottom of the lake.

Bacteria decompose the material at the bottom of the lake. The decay process uses up some of the oxygen in the water. As the amount of oxygen in the water goes down, fewer fish and other organisms can live in it.

Over time, the pond or lake is filled with sediment. New kinds of plants grow in the new soil. Shallow places fill in first, so plants grow closer and closer to the center of the pond or lake. What is left of the pond or lake becomes a wetland. As the soils dry out and the oxygen levels increase, forest plants can grow. In this way, a pond or lake can become a forest.

Critical Thinking

7. Apply Concepts Why is the amount of oxygen in pond water an abiotic factor?

____

____

Copyright © by Holt, Rinehart and Winston. All rights reserved.

Name ______________________ Class ______________ Date ______________

Section 4 Review

6.5.c, 6.5.e

SECTION VOCABULARY

deep-water zone the zone of a lake or pond below the open-water zone, where no light reaches

littoral zone the shallow zone of a lake or pond where light reaches the bottom and nurtures plants

marsh a treeless wetland ecosystem where plants such as grasses grow

open-water zone the zone of a pond or lake that extends from the littoral zone and that is only as deep as light can reach

swamp a wetland ecosystem in which shrubs and trees grow

wetland an area of land that is periodically underwater or whose soil contains a great deal of moisture

1. Compare Why are the kinds of producers in the littoral zone of a lake different from the producers in the open-water zone?

____

____

____

2. Compare How are the producers in a swamp different from those in a marsh?

____

____

3. Explain What ecological role do the fishes in the open-water zone and the fishes in swamps fill?

____

____

4. Describe What abiotic factors do organisms living in rivers and streams have to adapt to?

____

____

____

5. Describe How can a pond become a forest?

____

____

____

____

Copyright © by Holt, Rinehart and Winston. All rights reserved.